CW01221350

Human Factors, Security and Safety

hfes europe chapter

Edited by

D. De Waard
J. Godthelp
F.L. Kooi
K.A. Brookhuis

2009, Shaker Publishing

Europe Chapter of the Human Factors and Ergonomics Society

europechapter@hfes-europe.org

http://hfes-europe.org

© Copyright Shaker 2009

All rights reserved. No part of this publication may be reproduced, stored in a retrieval system, or transmitted, in any form or by any means, electronic, mechanical, photocopying, recording or otherwise, without the prior permission of the publishers.

Printed in The Netherlands.

D. de Waard, J. Godthelp, F.L. Kooi, and K.A. Brookhuis (Eds.).
Human Factors, Security and Safety
ISBN 978-90-423-0373-7
Chapter and cover illustrations by Eva Fabriek (http://www.lelijkepoppetjes.nl)
Cover design by Klaas Woudstra

Shaker Publishing BV
St. Maartenslaan 26
6221 AX Maastricht
Tel.: +31 43 3500424
Fax: +31 43 3255090
http://www.shaker.nl

Contents

Preface 9
    Dick de Waard, Hans Godthelp, Frank Kooi, & Karel Brookhuis

CRISIS MANAGEMENT 11

Multi-agency operations: cooperation during flooding.. 13
    Richard McMaster & Chris Baber
Performance-based measurement of national crisis management 29
    Marcel van Berlo, Miguette Jadoul, Richelle van Rijk,
    Sjoerd Wartna, & Mart Kösters
On the relationship between physical distance, communication style, and team
situation awareness Preliminary results and implications for collaborative
environments 41
    Marco Camilli, Sara Coletta, Alessandra Talamo, & Francesco Di Nocera

TRAINING 63

Transitional information in simplified mine detection tasks 65
    Chris Baber & Robert J. Houghton
Cognitive and personality variables of operators: individual characteristics and
process control performance 77
    Dina Burkolter, Annette Kluge, & Matthias Brand
The use of simulator training for train drivers in Europe – an overview and new
approaches in 2TRAIN 89
    Christian Maag & Marcus Schmitz
Evaluating the effectiveness of a human factors training –Focus on situation
awareness 101
    Nicki Marquardt, Swantje Robelski, Gwen Jenkins, & Rainer Höger

AVIATION 111

Sound design for auditory guidance in aircraft cockpits 113
    Bernd-Burkhard Borys
Barriers and accidents: the flight of information 123
    Thomas G.C. Griffin, Mark S. Young, & Neville A. Stanton
PAMELA, a portable solution for workflow support and human factors feedback
in the aircraft maintenance environment 131
    Martine Hakkeling-Mesland & Koen van de Merwe
Human Factors in the display and use of aeronautic information from different
sources and of different status 143
    Yvonne Barnard

Transportation - ADAS (Advanced Driver Assistance Systems) 159

Is drivers' situation awareness influenced by a highly automated driving scenario? 161
    Natasha Merat & A. Hamish Jamson

Evaluation of a generic warning for multiple intersection assistance systems 173
    Stephan Thoma, Thomas Lindberg, & Gudrun Klinker

Enhanced information design for high speed train displays: determining goal set operation under a supervisory automated braking system 189
    Anjum Naweed, Bob Hockey, & Sam Clarke

Warning drivers of approaching hazards: the importance of location clues and multi-sensory cues 203
    Michael G. Lenné & Thomas J. Triggs

Different alarm timings for a forward collision warning system and their influence on braking behaviour and drivers' trust in the system 213
    Genya Abe, Makoto Itoh, & Tomohiro Yamamura

Human machine interface for a dual mode vehicle 225
    Amon Rambaldini, Antonella Toffetti, & Claudio Arduino

Transportation - Evaluation 237

Developing a virtual driving environment to test dose related effects of alcohol and drugs on simulated driving performance 239
    Janet L. Veldstra, Karel A. Brookhuis, & Dick de Waard

CarUSE – Developing a framework for IVIS evaluation 251
    Thomas Vöhringer-Kuhnt

Evaluating the distractive power of secondary tasks while driving 261
    Claire Petit, Antoine Clarion, Carolina Ramon, & Christian Collet

Menu interaction in Head-Up Displays 271
    Natasa Milicic & Thomas Lindberg

Representation of driver's mental workload in EEG data 285
    Shengguang Lei, Sebastian Welke, & Matthias Roetting

Effects of a driving assistance system on the driver's functional near-infrared spectroscopy 295
    Kazuki Yanagisawa, Hitoshi Tsunashima, Yoshitaka Marumo Makoto Itoh, & Toshiyuki Inagaki

Go Ahead – I will follow you! Social pull-factors in driving manoeuvres 305
    Jessica Seidenstücker, Ernst Roidl, & Rainer Höger

Attitudes of public bus drivers towards technology integration in the dashboard design 317
    Emine Nazli Özer, Gulsen Töre, & Çigdem Erbuğ

Public Surveillance 331

Searching for threat: factors determining performance during CCTV monitoring . 333
    Christina J. Howard, Tomasz Troscianko, Iain D. Gilchrist, Ardhendu Behera, & David C. Hogg

Expertise and strategies in the detection of firearms via CCTV 341
    Iain Darker, Alastair Gale, Anastasssia Blechko, & Marc Whittle
Improving Situational Awareness in camera surveillance by combining top-view maps with camera images 355
    Frank L. Kooi & René Zeeders
'Sounds like trouble' 369
    Peter W.J. van Hengel, M. Huisman, & J.E. Appel
Detecting deviants within flocks.. 377
    Robert J. Houghton & Chris Baber

THEORIES AND MISCELLANEOUS 389

Using social network analysis to identify sub-groups in the operating room.......... 391
    Tita A. Listyowardojo, Christian E.G. Steglich,
    Stephen Peuchen, & Addie Johnson
Behavioural adaptation: friend or foe? 401
    Oliver Carsten
Spanning attention: what really goes on between perception and action? 411
    Mark S. Young
Performance in manual process control– mediation of time pressure and practice effects by the structure of control behaviour 425
    Stefan Röttger & Dietrich Manzey

Acknowledgement to reviewers.. 437

Preface

Dick de Waard[1], Hans Godthelp[2], Frank Kooi[2], & Karel Brookhuis[1,3]
[1] University of Groningen
[2] TNO Human Factors
[3] Delft University of Technology
The Netherlands

In 2008 the Europe Chapter of the Human Factors and Ergonomics Society celebrated her 25th birthday. For this occasion the Annual Meeting returned to the place of birth: TNO in Soesterberg, the Netherlands. Three days of presentations, posters, and demonstrations with the usual mix of experienced and first time speakers. In this book you will find the contributions of most of the oral presenters categorised into seven chapters. Eva Fabriek, as she did in previous years, kindly provided an illustration for each chapter, and for the cover of the book. I hope you appreciate these enlivenments as much as we do.

We are very grateful for the hospitality of the host of the meeting: TNO Human Factors. We also thank all external reviewers, who helped us to evaluate all the manuscripts. The conference was also supported by the European Office of Aerospace Research and Development of the USAF, under Award No. FA8655-08-1-5084. We are very grateful for their support.

In D. de Waard, J. Godthelp, F.L. Kooi, and K.A. Brookhuis (Eds.) (2009). *Human Factors, Security and Safety* (p. 9). Maastricht, the Netherlands: Shaker Publishing.

Crisis Management

CO-OPERATION DURING FLOODING

In D. de Waard, J. Godthelp, F.L. Kooi, and K.A. Brookhuis (Eds.) (2009). *Human Factors, Security and Safety* (p. 11). Maastricht, the Netherlands: Shaker Publishing.

Multi-agency operations: cooperation during flooding

Richard McMaster & Chris Baber
University of Birmingham
UK

Abstract

This paper presents an investigation of command and control during Multi-Agency Operations and describes a study of the successful combined military and civilian defence of Walham electricity substation from rising floodwater in July 2007; widespread flooding of the English county of Gloucestershire meant that the emergency services were unable to cope and a formal request for military aid was made. The purpose of this study was to explore:
- How the context of the situation forces responding agencies away from formal structures and procedures;
- How command intent is managed across ad hoc and fragmented systems;
- How technology could be used to support multi-agency operations.

Our findings are presented in relation to the concepts of Command Intent and Situation Awareness and we suggest that the development of a deeper, shared understanding of the relevant factors involved in an incident is required, but that this is effortful to achieve and likely to be harder to support with contemporary networking technologies. We also argue that effective coordination requires a high level of trust between responding organisations, as the 'problem space' is likely to extend beyond the expertise of any one group and they are reliant on each other's expertise.

Introduction

Multi-agency incident response: the need for cooperation

Large-scale civil emergencies (also known in the UK as Major Incidents) tend to require a multi-agency response, as no single agency has the expertise or the equipment to implement a fast and appropriate response single-handedly. Whilst personnel from each emergency service are trained to deal with a variety of situations, each service specialises on a particular sub-set of the response. For that reason the specialist skills and training of several different agencies may be required at a Major Incident in order to rapidly resolve the problem. In spite of the difficulties associated with such large scale disasters and often considerable individual acts of courage in the face of danger, an overall lack of coordination between responding agencies is an enduring feature of multi-agency operations (Smith and Dowell, 2000). The multi-agency responses to two recent crises - the South Asian Tsunami and Hurricane Katrina have both received criticisms that responding agencies failed to cooperate or share information and displayed a lack of trust in one another (Bennett, et al., 2006; Chua et al., 2007).

Features of Major Incidents

Crichton, Flin, and Rattray (2000) point out that multi-agency emergency operations share a number of challenging features, including:

- Ad hoc teams that work together only when responding to an emergency incident;
- Multiple objectives which have to be achieved in parallel for the incident to be successfully contained;
- High psychological demands, with people working under time pressure and stressful conditions;
- Role specialisation, with the need to pool different types of expertise (Crichton, Flin, & Rattray, 2000; p208).

Features of the emergency situation itself also make the job of responding agencies more difficult; major emergencies are characterised by high levels of complexity and uncertainty, requiring knowledge outside of traditional emergency response domains and presenting uncertainties across a range of dimensions, including legal and moral issues, as well as situational concerns (de Marchi, 1995; Becerra-Fernandez et al., 2008; von Lubitz, Beakley, & Patricelli 2008). During the initial stages of a large and complex incident, it is unlikely that any single individual or organisation will be in possession of all of the information to fully understand the problem and formulate a solution – various organisations will be in possession of 'pieces of the puzzle' (McMaster, Baber, & Houghton, 2007). Close cooperation between responding agencies is therefore required, in order to enable a coherent response to the emergency.

Recognizing that large-scale emergencies occur with little or no notice, involve temporary organisations of agencies who rarely (if ever) work together and improvised organisational structures (Smith & Dowell, 2000), raises the question of how responding agencies currently negotiate these issues during Major Incidents. This paper uses a contemporary case study of a successful incident in order to further investigate the problems encountered during multi-agency flood responses.

Military Aid to the Civil Authorities

Military Aid to Civil Authorities (MACA) is the name for emergency military support within the UK and forms part of the wider '*Defence Contribution to Resilience*' (MoD, 2002). Use of the armed forces is strictly limited to only the most exceptional circumstances, where civil capability or capacity has been exceeded (MoD, 2002). Though use of the military is restricted, they may still be called upon to deal with a wide range of situations including natural disasters, terrorism or outbreaks of disease; these very different scenarios require the military to be able to effectively cooperate with a variety of other organisations, in order to augment civil authorities and close the capability gap.

Gloucestershire floods July 2007

The summer of 2007 saw widespread flooding in several regions of UK; one of the worst affected areas was Gloucestershire, with widespread flooding across the county. In addition to the extensive water damage caused to residential properties, travel became difficult and as roads and towns flooded, many people became trapped in their homes. The electricity supply to large parts of the county was put at risk, as both the Walham and Castlemeads substations came under threat from rising floodwater (Elliott & Brown, 2007). Walham substation is a site of critical national importance, supplying electricity to over 500,000 homes (approx. 2,000,000 people) in England and Wales (Snow & Manning, 2007); if the site had flooded, then this may have forced the mass evacuation of the county, as critical infrastructure failed (Griffin, 2007).

The emergency services were unable to cope with the scale of the county-wide emergency, so a formal request for military assistance was made. A multi-agency operation was launched to prevent rising flood water from overwhelming the substation at high tide that night (the section of the River Severn near to Walham is tidal). The response involved hundreds of personnel from a number of organisations, including the Fire and Rescue Services, Environment Agency and initially personnel from several Royal Air Force (RAF) bases. The plan of action was to construct a series of flood defences around the critical substation switching room; this included the use of sandbags, 1 kilometre of the Environment Agency's modular flood barrier and the Fire and Rescue Service's specialist high volume pumps.

Despite the short notice and difficult working conditions, the various agencies were able to coordinate an effective response and prevent the flood water from forcing the shutdown of the substation, buying time for a semi-permanent flood defence to be constructed around the site. In comparison, the nearby Castlemead substation had to be shut down, cutting power to around 50,000 homes, before flood defences could be established and power restored (Environment Agency, 2007).

Aims of the research

The purpose of this report is to present our findings relating to coordination and Command Intent during multi-agency operations; we specifically examined three issues:

- How does the context of the situation force responding agencies away from formal structures and procedures?
- How is Command Intent managed across ad hoc and fragmented systems?
- How could technology be used to support multi-agency operations?

In order to explore these issues, four aspects of multi-agency command and control were examined:

- The formulation and communication of Command Intent;
- The Organisation Structures used to share information and coordinate activity within and across agencies;
- Inter-agency Cooperation during the incident response;
- The development and maintenance of Shared Awareness within and across agencies.

Method

This retrospective case study was mainly concerned with the coordination of the response at the scene of the incident, which is known as the Bronze level of command. Whilst a large number of organisations were actively involved in the response to the flooding of Walham substation, we have concentrated our analysis on the main agencies involved in the construction of flood defences and the extraction of water from the site. The critical initial period of activity lasted approximately 12 hours on the night of Sunday 22nd July, 2007; during this time, the agencies involved were notified of the problem, resources were mobilised, the emergency defences were constructed and the substation was successfully protected during the first high tide that threatened to inundate it – providing time for these emergency defences to be consolidated and more permanent protection to be organised.

Table 1. Roles of the six interviewees and the Agencies that they represented

Agency	Interviewee / Role
Avon Fire and Rescue Service	-Incident Commander – 'Bronze' (during the Consolidation Phase)
Gloucestershire Fire and Rescue Service	-Deputy Chief Fire Officer – 'Gold Liaison'
43 (Wessex) Brigade	-Joint Regional Liaison Officer
	-Walham Site Liaison Officer (Brigade Reinforcement Team)
Environment Agency	-Team Leader (Operations Delivery)
	-Specialist Team Member (Operations Delivery)

Interviews were conducted with six individuals that were directly involved in the planning and execution of the incident response for the three main agencies involved (Fire and Rescue, Environment Agency and Military); their roles and organisational affiliations are listed in Table 1. These semi-structured interviews began by asking participants to provide a description of the events that occurred that day. This description was then used to structure the subsequent interview, when the participants were asked for detailed descriptions of particular features of the emergency response, including: organisational structures, inter-agency communication and cooperation, plan formation, significant decision points and difficulties encountered during the response. The revised Critical Decision Method (CDM) probes developed by O'Hare et al. (2000) were then used to explore the decision making processes which were applied during the incident. The interviews were analysed in parallel, to identify points of commonality and divergence; participants were then contacted again (by telephone or email) in order to reconcile any differences in their accounts and to check for accuracy in the analysis. The interview process took place over a 3 month period, which began 3 months after the incident took place.

This approach was deemed appropriate, as the authors were investigating social and cultural processes at the inter-organisation (i.e. macroergonomic) level and it was felt that more standard quantitative ergonomics techniques would not have been adequate for this purpose.

Results

Vignettes

Three vignettes on key themes have been drawn from the combined accounts of the incident given by the interviewees and are presented below. These provide illustrations of the issues which are discussed in the following sections of the paper.

Vignette 1: Requests for Diesel

Due to the protracted nature of the county-wide emergency and the high numbers of resources involved, the re-fuelling of Fire and Rescue appliances became a priority concern. In response, the Gloucestershire Fire and Rescue command structure was modified, by creating the role of 'Pseudo Silver' - a command function entirely dedicated to the specific problem of coordinating the refuelling operation.

The Bronze Commander on the scene at Walham reported that when he made requests for fuel to be sent to the site (in order to protect 'critical national infrastructure'), he was told that other incidents took priority and was not given an estimated time of arrival for the fuel. As a lack of diesel for the pump generators had the potential to lead to the substation flooding (with the potential consequence of a full-scale evacuation of the county), the Bronze Commander was forced to request that the Deputy Chief Fire Officer (acting as 'Gold Liaison') contact Pseudo Silver and use his authority within Gloucestershire Fire and Rescue service in order to ensure that fuel would be delivered in time.

Vignette 2: Perceptions of the incident

The Environment Agency team had a specialist role in the response, namely the deployment of their flood barrier equipment - a task which they were familiar with, having already used the barrier several times that year. From their perspective, the incident was straightforward and they knew what had to be done, but they felt that the Fire and Rescue service were slow to adapt to the pace and nature of the incident; during the early stages, the Environment Agency considered the Fire and Rescue Service to be 'in the way', as they were having difficulty in getting vehicles onto the site. The Environment Agency personnel spoken to asserted that it was their equipment, personnel and knowledge that had been crucial in the defence of Walham ("it was us that did it") and that this was not recognised either by the Fire and Rescue service or in media reports of the incident. They were of the opinion that the Fire and Rescue service had treated the incident as more of a public relations opportunity and "an opportunity to dust off their gear", as they brought in equipment "from all over the country" that was then in the Environment Agency's way.

The Fire and Rescue Bronze Commander was the overall Incident Commander and therefore had responsibility for the coordination of the whole multi-agency response, as well as the safety of all personnel working on the site. Therefore, from the Fire and Rescue Commander's perspective the incident was much more complex, with many factors to consider, including numerous hazards and a number of equally critical aspects to the flood defences, of which the Environment Agency barrier was one part. 8 specialist high volume pumps were brought in from Fire and Rescue services around the country to keep the flood water level within the substation down; this was one of the reasons for the large number of Fire appliances at the scene. The Bronze Commander felt that all of the agencies involved in the response were focussed on the same goal, rather than thinking that their own agenda was more important.

Vignette 3: Multi-agency Coordination

National Grid safety personnel advised responding agencies at Walham on safe working practices; these restricted the use of lifting equipment in parts of the site, meaning that sections of the Environment Agency flood barrier would have to be moved into place by hand. It became clear to the Environment Agency team that they did not have enough personnel to complete the construction of their barrier in the time available. The military teams sent to the site were tasked with moving barrier components into place and assisting the Environment Agency with construction work.

The Bronze Commander kept Liaison Officers from the military and RNLI close by, as he needed to maintain constant contact with these organisations. The Fire Commander initially thought that the Environment Agency were happy to be left to get on with their tasks, leaving him to focus on other aspects of the response. However, the large articulated vehicles bringing Environment Agency equipment had been held up in the queue of traffic outside the incident cordon. Once the Environment Agency Team Leader realised what had happened, he tried to get the lorries into the queue of traffic entering the site, but their size meant that all other vehicles would have to be stopped to let them in and out. This delay put the completion of the barrier before high tide at risk. The Environment Agency Team Leader approached the Fire Commander and they discussed the problem, agreeing that, given the need to complete the barrier on time, these lorries had to be given priority access. All other work and site traffic was stopped to allow the lorries into the site to be unloaded.

Decision-making

Given the hazards posed by the live electricity substation and the presence of flood water, one of the key decisions during the incident was the risk assessment for personnel working on the site. This decision was further complicated by a lack of PPE (Personal Protective Equipment - i.e., life jackets, gloves and high visibility waterproof clothing) for the military teams, who had only ordinary military fatigues. As the flood water level rose, the concern became that water might suddenly break through the temporary defences and inundate the site. The decision to have staff remain on-site was therefore regularly reassessed, before all non-essential staff were finally evacuated prior to high tide. Table 2 summarises the responses to CDM probes from the 3 organisations interviewed in relation to the decision to keep personnel working on the site.

Table 2. Responses to CDM questions from the various organisations in relation to the risk assessment of having staff working inside the electricity substation.

CDM question	Fire and Rescue	Environment Agency	Military
Goal specification What was your overall goal?	Prevent the substation from flooding. They were told: *"We have got to save this if we possibly can."* Usually have everyone thinking their agenda is more important – everyone was focussed on one objective.	Construction of the Environment barrier before high tide.	Told to report to the incident site and provide maximum support to the Bronze Commander - within the confines of the situation and military abilities and taking into consideration the safety of the troops.
Cue identification What features were you looking at when you formulated your decision?	Hazard conditions (advice from RNLI, National Grid, reports from fire fighters) Predicted time and height of flood water at high tide. Lack of PPE for the military. Control measures. Evacuation signals.	Dynamic risk assessment of Environment Agency staff – safe to work on site, safe to walk on and off site. National Grid guidelines on safe working practices. Evacuation signal from Fire and Rescue.	The risk assessment of the Fire and Rescue Service; responsibility for the welfare of those under your command. State of floodwater across approach road – determined this necessitated vehicular transport on and off site.
Conceptual model Are there any situations in which your decision would have turned out differently? Describe the nature of these situations.	Would not have deployed military without PPE had the conditions been different (e.g. if site was nearer river). Evacuated all non essential personnel near high tide, as risk of water overwhelming the defences rose.	Fire Brigade were worried they would not be able to control the water level; they evacuated everyone before high tide.	Continuous review of decision by all parties, under the chairmanship of Bronze Commander.
Influence of uncertainty At any stage, were you uncertain about the appropriateness of the decision?	Constant review of decision; risk to personnel set against priority of goal; measures taken to manage risks.	Staff familiar with the task and experienced in working in water hazard, had constructed the barrier several times that year. Trusted the National Grid as they are experts.	Could see Bronze Commander was hesitant about military commitment to an unpleasant task without PPE (filthy, cold water, dark); was told by one of the military Commanding Officers: *"Don't worry, they'll do what they're told."*
Situation awareness What information did you have available to you at the time of the decision?	Hazard assessment from National Grid: maximum safe flood water level. Water depth and hazard assessment from RNLI and Fire and Rescue personnel. Compliance with PPE. Time of high tide.	Safe working practices from National Grid, experience of EA personnel.	The risk assessment of the Fire and Rescue Service.

Findings

Organisational structures

During the operation to save Walham substation, a number of alterations were made to the standard command structures in order to cope with unique features of the situation; for example, the Fire and Rescue Service command structure for the incident was substantially more complex than the basic Gold, Silver, Bronze structure adopted during Major Incidents in the UK (Figure 1). Whilst some of these changes were pre arranged, many were made on an ad hoc basis – as is demonstrated in Vignette 1 – illustrating how constraints of the emergency situation may force commanders to adapt their organisations to better suit the environment in which they have to work.

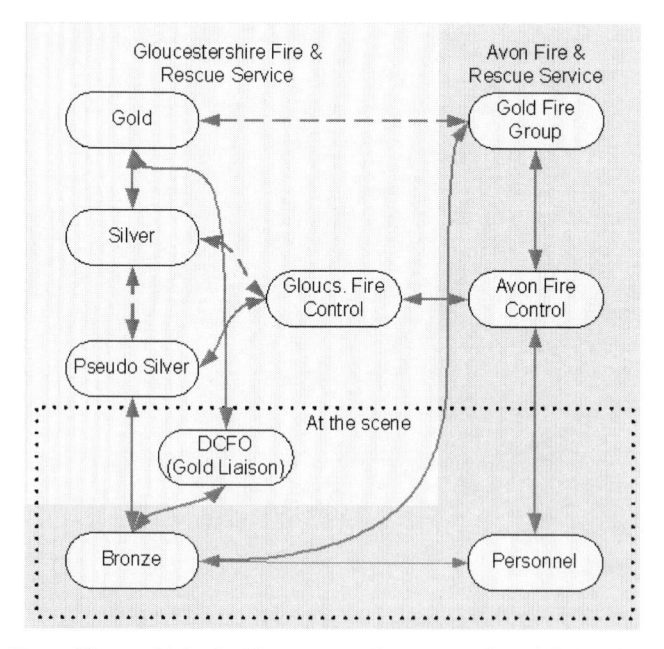

Figure 1. Fire and Rescue Major Incident command structure adopted during the response to the flooding of Walham electricity substation, from the perspective of the Bronze Commander (further probable lines of communication not observed by the Bronze Commander are indicated by dashed lines).

Due to the scale of the emergency, Gloucestershire Fire and Rescue drew support from neighbouring services under the established Mutual Aid scheme; at Walham, many of the personnel on site – including the Bronze Commander – were from Avon Fire and Rescue Service, meaning that the Bronze Commander was not directly part of the Gloucestershire Major Incident command structure. The Deputy Chief Fire Officer (DCFO) from Gloucestershire Fire and Rescue Service was on-site in a non-standard role, acting as 'Gold Liaison'. The DCFO was 'hands off', i.e. he was not directly involved in the command of the incident response, leaving that to the Bronze

Commander; however, he provided direct input to Gold (strategic) Command on the progress of the response to the emergency situation and provided advice and support when the Bronze Commander experienced problems due to working in an unfamiliar county, as can be seen in Vignette 1. The overall Fire and Rescue Command structure in relation to the response to the Walham substation flooding is shown in Figure 1, with lines of communication shown by the arrows.

Shared awareness

Major Incidents have been characterised by their "un-ness", i.e. they are unexpected, unplanned for and unprecedented (Crichton, 2003). As a result, one of the defining features of Major Incidents is the high level of uncertainty surrounding aspects of the situation, including the nature of the incident and the appropriate response. The different training and experience of personnel involved in multi-agency operations can also result in different interpretations of the incident, both across organisations and between levels of command. At Walham, this was most notable in the contrasting perceptions of two of the agencies involved in the emergency response, as described in Vignette 2.

Whilst the Environment Agency and Fire and Rescue Service were concerned with broadly similar elements of the incident, the two organisations went on to form very different perceptions of the problem and the appropriate multi-agency response. Table 2 provides some indication as to why these two agencies might have come to different conclusions regarding the incident; it can be seen that, in deciding whether to allow personnel to work on the site, the various agencies were attending to different information and more significantly had different overall goals, which were likely to have affected how they viewed the incident. In this instance, these contrasting perceptions did not result in catastrophe; however, they serve to illustrate how different agencies working alongside one another on the same incident can have different understandings of the nature of the situation. In other words, developing shared situation awareness requires more than just the possession of relevant incident information, it also calls for a level of shared understanding. Ideally, this would involve a discussion between commanders over how or why they have these perceptions, as they are based on experience and expertise that is not shared across agencies. However, time limitations and limited contact between agencies may preclude this labour intensive process, therefore instead requiring a high level of trust between personnel from different organisations in order to accept and accommodate each others' viewpoints and to agree to cooperate in achieving them.

Inter-agency cooperation

The separate incident command structures adopted during multi-agency operations appear not to foster the coordination of activity. The different agencies involved in this incident communicated on and off site through their own equipment; the incompatibility of these systems meant that high level inter-agency cooperation required physical proximity, which was not easy to achieve, given that the various agencies were engaged in different tasks around the site and movement was restricted by flood water and construction activity.

Coordination of site access between the responding agencies appears to have been problematical, with apparent failures to request information from and present information to other agencies, as is described in Vignette 3. This failure to 'push' and 'pull' information likely stems in part from a lack of awareness between the responding organisations (namely the Environment Agency and Fire and Rescue Service) regarding each other's roles, methods and processes; it would have been difficult to know whether another agency was following a particular course of action because that is "just how they do things", or because they do not have the same understanding of the incident. This is likely to be exacerbated during a Major Incident, where the organisations are faced with a unique problem and it may not be immediately apparent what factors should be concentrated on. In addition, during this incident unfamiliarity between agencies led one agency to interpret another's actions as having selfish or malign intentions, which is unlikely to have motivated individuals from different organisations to cooperate. Thus, there are mental barriers to cooperation between responding agencies.

It would appear that Liaison Officers work as an interface between different organisations, bridging their different languages, practices and perspectives on an incident and helping to build trust between organisations. Indeed, the Military Liaison Officer working on-site commented that his role was to understand and communicate the needs of each agency on site. The value of this role was demonstrated by their widespread and effective use by the military and the number of ad hoc liaison roles that were created within other organisations, in order to address the particular needs of this incident and to ensure continuity of purpose across organisations and levels of command.

Command intent

The large numbers of responding agencies, combined with communications problems and a lack of on-site command support meant that events began to overwhelm the 'control' aspect of the command and control capability. The Fire and Rescue Service therefore adapted their working practices to the situation: command roles, reporting lines and evacuation procedures were altered to take account of the situation they were faced with. This demonstrates how the constraints of an emergency situation may force decision-makers to adapt virtually everything else and the consequent requirement for command structures to be able to adjust to extraordinary situations.

The principle reason for the requirement for multiple agencies to attend an incident is that no single agency has the expertise or the equipment to implement a fast and appropriate response single-handedly. Effective decision-making during multi-agency operations is therefore dependent on effective cooperation and the sharing of information and analyses of the situation. During this incident, one of the key decisions was the ongoing assessment of whether it was safe for personnel to work on site. In order to do this, information on a number of factors was collected and combined to produce an overall risk assessment for the site, as is summarised in Table 2. The National Grid established safe working practices for personnel operating in live electrical areas and defined a maximum depth for floodwater to reach before it would become too dangerous to remain on site. RNLI crews monitored water depths around the site and assessed floodwater risk to personnel, as well as reporting on compliance with the use of PPE by personnel on site. The Fire and Rescue Service took information from all sources, and the Incident Commander assessed the overall risk to personnel working on the site.

On the face of it, all agencies appeared to share the same overall goal - to prevent the substation from flooding - and were clear on what needed to be done; this was certainly the perception of the Fire and Rescue Bronze Commander. However, Table 2 suggests that the responding organisations were actually working to slightly different priorities and making decisions based on different environmental cues, as well as on their own experience and expertise; whilst these priorities were broadly the same, this may explain the mismatch in perspectives of the incident and why the Environment Agency may have felt that they were being obstructed by the Fire and Rescue Service, who were trying to balance the competing requirements of different aspects of the response. In addition, there may have been a more fundamental reason for the mismatch between some of these agencies, to do with the recognition and acceptance of command authority; both the Fire and Rescue Service and Environment Agency are Category 1 responders (under the Civil Contingencies Act, 2004) and are used to being in command of their own operations.

Discussion

It is clear that during Major Incidents, all agencies are concentrating their efforts on resolving the emergency, but there are some problems that are inherent within the nature of the situation and of multi-agency operations. Returning to the aims of this work, our findings can be summarised as follows:

How does the context of the situation force responding agencies away from formal structures and procedures?
As a result of the severe limitations imposed by the restricted location, numerous hazards, short timeframe and environmental conditions, the responding agencies were forced to adapt almost every aspect of their response, from ad hoc alterations of organisational structures, to adaptation of procedures, roles and the use of equipment. In order to meet the demands of an exceptional incident, conventional rules that are enshrined in Standard Operating Procedures (SOPs) had to be circumvented or 'broken'. This can result in problems, such as misunderstandings with other agencies and therefore necessitates close coordination and negotiation in order to achieve a position of shared understanding of the situation and the appropriate response.

How is Command Intent managed across ad hoc and fragmented systems?
The high levels of uncertainty and multi-domain nature of major emergencies requires effective communication and sharing of information between agencies; though awareness of even a 'simple' incident can vary widely and achieving shared understanding across different organisations is labour intensive. This helps explain why common goals at the strategic level can still lead to different (conflicting) tactical and operational responses. Separate command structures may act as a barrier

to cooperation, which is something that organisations attempt to overcome through the use of liaison roles and shared command facilities.

Different perspectives on the nature of the problem and the role of each agency in the response can lead to misinterpretation of intentions and a loss of trust across organisations. This may be due to a lack of experience of working together, which means an unfamiliarity with different agencies' working practices, knowledge and requirements, though there may be potentially more fundamental questions over who is 'in charge' of the incident response. This lack of trust may well adversely impact on inter-agency cooperation during future multi-agency operations.

An idealised multi-agency emergency command group could be seen as a community of practice. Communities of practice are described in terms of three elements: a common domain of interest; a community of members engaged in joint discussions and activities and who develop a shared practice, or common repertoire of resources (Wenger, McDermott, & Snyder, 2002).

The notion of communities of practice has been used in management circles to describe natural associations of individuals that transcend organisational structures and teams and which are argued to bring a number of benefits to collaborative work, including knowledge transfer and retention, skill sharing and the development of trust. A community of practice can be measured according to their level of commitment on each of the three elements of domain, community and practice; the 'community' of responding organisations in emergency response could be said to have a high common commitment to the domain, but their commitment to the community might be lower (with some mistrust amongst agencies), as is the commitment to the practice (lack of common culture, experience and approaches to viewing the problem). If it were possible to nurture the development of communities of practice, then it would seem that this would be a sensible course of action to take with regards to multi-agency operations. However, these communities are thought to develop gradually over a period of time and so the question is how to develop the commitment to the domain, community and practice over a short period of time and with groups of individuals who may not know each other? This may be one area where networking technology could be used to try to augment social processes and speed up the development of the community of practice.

How could technology be used to support multi-agency operations?
A Common Operational Picture (COP) is a single representation of relevant incident information that could be shared across service command centres during a multi-agency response (USJFCOM Glossary). The recent adoption of a TETRA (Terrestrial Trunked Radio) standard secure digital communications network (known as Airwave) by all of the UK emergency services could enable the development of a COP, as it allows for the creation of multi-user talk groups and the sharing of data; this may then lead to faster and more appropriate joint service responses by reducing the level of uncertainty surrounding factors of the incident.

A COP might be useful during emergency response incidents, in order to share statements of intent amongst the responding agencies and to improve information exchange. However, examination of the Walham incident response has illustrated the difficulty of developing a shared understanding of an incident, particularly between organisations that are not familiar with each other's domains of expertise and work practices. This has implications for the development of a COP, as the presentation of information alone would appear insufficient to enable the development of a common understanding. Thus, one might instead argue for a Common *Relevant* Operational Picture (CROP), which would present information in the format familiar to specific agencies. This might require additional functions to 'translate' terms, concepts and procedures. The question therefore becomes one of whether a COP or other networking technology can be used to facilitate the development of a common understanding.

Conclusions

Whilst the response to this emergency was a resounding success, a number of adaptations to standard organisational structures, processes and procedures were necessary; some issues in terms of Shared Awareness and Inter-agency Cooperation were identified, which related to inexperience of personnel and organisations in working together, though which may also stem from a more fundamental question over the recognition of command authority.

This study has shown that effective cooperation across agencies requires more than merely the exchange of information and that developing shared understanding is a crucial - but labour intensive - process. It also highlights the necessity for command and control structures and technologies to be flexible, in order to accommodate the changing demands of unique situations and multi-agency associations. Historically, implementations of communications technology have failed to account for social processes and accommodate them within the solution, with the result that the 'improved' process is actually less effective, for example in the case of the London Ambulance Service's computer-aided despatch system (LASCAD) project failure (Beynon-Davies, 1999). It is therefore suggested that a more sociotechnical systems view, i.e. the optimisation of both social and technical systems (Cherns, 1976; Clegg, 2000; Trist & Bamforth, 1951), should be taken during the design and implementation of future networking technologies, to ensure that the relevance of complex social issues, such as trust, to multi-agency operations is identified and that the unanticipated consequences of even small changes to command and control networks is recognized.

Acknowledgements

This work is supported by a grant from the Human Factors Integration Defence Technology Centre, part-funded by the Human Capability Domain of the UK Ministry of Defence Scientific Research Programme. The authors would like to thank personnel from Avon Fire and Rescue Service, Gloucestershire Fire and Rescue Service, Environment Agency Operations Delivery and 43 (Wessex) Brigade for their assistance with this research.

References

Becerra-Fernandez, I., Xia, W., Gudi, A., & Rocha, J. (2008). Task characteristics, knowledge sharing and integration, and emergency management performance: research agenda and challenges, *Proceedings of the 5th International ISCRAM Conference – Washington, DC, USA, May 2008*. (pp. 88-92) Retrieved 14/3/2008 from: http://www.iscram.org/dmdocuments/ISCRAM2008/papers/ISCRAM2008_Becerra-Fernandez_etal.pdf

Bennett, J., Bertrand, W., Harkin, C., Samarasinghe, S., & Wickramatillake, H. (2006). *Coordination of international humanitarian assistance in tsunami-affected countries*. London: Tsunami Evaluation Coalition.

Beynon-Davies, P. (1999). Human error and information systems failure: The case of the London Ambulance Service Computer-Aided Despatch system project. *Interacting with Computers*, 11, 699-720.

Cherns, A. (1976). The principles of sociotechnical design. *Human Relations*, 29, 783-792.

Chua, A.Y.K., Kaynak, S., & Foo, S.S.B. (2007). An analysis of the delayed response to Hurricane Katrina through the lens of knowledge management. *Journal of the American Society for Information Science and Technology*, 58, 391-403.

Clegg, C.W. (2000). Sociotechnical principles for system design. *Applied Ergonomics*, 31, 463-477.

Crichton, M. (2003). Decision making in emergencies, *NATO / RUSSIA / ARW Forecasting and Preventing Catastrophes Conference*, June 2003. Aberdeen, UK: University of Aberdeen.

Crichton, M.T., Flin, R., & Rattray, W.A.R. (2000). Training decision makers – tactical decision games. *Journal of Contingencies and Crisis Management*, 8, 208-217.

De Marchi, B. (1995). Uncertainty in environmental emergencies: a diagnostic tool. *Journal of Contingencies and Crisis Management*, 3, 103-112.

Elliott, V. & Brown, D. (2007) 10,000 homes flooded, 50,000 without power and 150,000 have no water, *The Times*, July 24th. Retrieved 14/3/2008 from http://timesonline.co.uk/tol/news/weather/article2127616.ece

Environment Agency (2007). Case study: 2007 Summer floods. Retrieved 14/3/2008: from http://www.environment-agency.gov.uk/commondata/acrobat/infrastructurestudy_1917458.pdf

Griffin, C (2007). The Battle for Walham, Glosfire, Winter, 12-13. Liverpool, Benham publishing. Retrieved 14/3/2008 from http://www.glosfire.gov.uk/sections/about_us/downloads/glosfire02.pdf.

McMaster, R., Baber, C., & Houghton, R. (2007). *Analysis of multi-agency intent: An example from the emergency services*, HFI DTC report, HFIDTC/2/WP3.1.4/1. Retrieved 14/3/2008 from: http://www.hfidtc.com/pdf/reports/New%20Reports/ HFIDTC-2-3.1.4-1.pdf

MoD (2002). Operations in the UK: The Defence Contribution to Resilience. Retrieved 14/3/2008 from http://www.mod.uk/NR/rdonlyres/80095393-A7B0-4B3A-69AD13742C29AAF/0/20071218_JDP02_2nd_Edition_U_DCDCIMAPPS.pdf

O'Hare, D., Wiggins, M., Williams, A., & Wong, W. (2000). Cognitive task analyses for decision centred design and training. In J. Annett and N.A. Stanton (Eds.) *Task Analysis* (pp. 170-190). London: Taylor & Francis.

Snow, J. & Manning, L. (2007). Saving Walham. Retrieved 14/3/2008 from http://www.channel4.com/news/articles/society/environment/saving+walham/624157#fold

Smith, W. & Dowell, J. (2000). A case study of co-ordinative decision-making in disaster management, *Ergonomics, 43*, 1153-1166.

Trist, E. & Bamforth, K. (1951). Some social and psychological consequences of the longwall method of coal getting, *Human Relations, 4,* 3-38.

USJFCOM Glossary. Retrieved 14/3/2008 from www.jfcom.mil/about/glossary.htm.

Von Lubitz, D.K.J.E., Beakley, J.E., & Patricelli, F. (2008). 'All hazards approach' to disaster management: the role of information and knowledge management, Boyd's OODA loop, and network-centrality, *Disasters, 32*, 561-585.

Wenger, E., McDermott, R., & Snyder, W. M. (2002). *Cultivating communities of practice: a guide to managing knowledge.* Boston, USA: Harvard Business School Press.

Performance-based measurement of national crisis management

Marcel van Berlo[1], Miguette Jadoul[1], Richelle van Rijk[1],
Sjoerd Wartna[2], & Mart Kösters[2]
[1] Netherlands Organisation for Applied Scientific Research TNO
[2] Netherlands Institute for Physical Security – NIFV
The Netherlands

Abstract

Voyager was the biggest crisis management exercise ever held in The Netherlands. About 2000 professionals, ranging from fire-fighters to ministers and from local teams to the national crisis centre, participated in a complex multi-threat scenario. In order to learn from this exercise, an evaluation was carried out. The individual actors as such were not evaluated, but rather the overarching system of national crisis management. At the system level, three major crisis management functions were described: information exchange, decision-making, and communication with the media. Next, an evaluation format was developed. For each function, goals, criteria and performance indicators were identified. The performance indicators were specified in consultancy with representatives of all major actors during the Voyager exercise. This provided relevant information for the evaluators and made the results of the evaluation study more acceptable to the actors. Based on the evaluation format, observation schemes were designed to support the observers. Overall, 34 trained observers monitored the performance of 28 teams. The evaluation team combined all these observations and additional material into one coherent system evaluation. This paper describes the system approach to the evaluation, the evaluation format, (briefly) the main results of the Voyager exercise, and recommendations for further research.

Introduction

Technological developments have resulted in more sophisticated and complex systems in which humans have to operate. These systems are characterized by a highly dynamic and sometimes hostile environment, the variation of (often conflicting) goals, the incompleteness, uncertainty and ambiguity of information, and the involvement of teams of officers with members having different roles and responsibilities (Rouse, Cannon-Bowers, & Salas, 1992). In these situations, many tasks are conducted by multi-disciplinary teams. Teams are social entities composed of members with high task interdependency and shared and valued common goals (Dyer, 1984). They are usually organized hierarchically and sometimes dispersed geographically; they must integrate, synthesize, and share information; and they need to coordinate and cooperate as task demands shift throughout a performance episode to accomplish their mission (Salas, Cooke, & Rosen, 2008, p. 541).

Team performance is conceptualized as a multilevel process arising as team members engage in managing their individual- and team-level task-work and teamwork processes (Kozlowski & Klein, 2000). A distinction can be made between 'performance' and 'effectiveness'. Performance pertains to how well the task-work and teamwork are carried out. Effectiveness pertains specifically to the accomplishment of the goals, milestones, and objectives as defined by requirements of the context or the stakeholders (Essens, Vogelaar, Mylle, Blendell, Paris, Halpin, & Baranski, 2004, p.16). In short, a team is effective if it meets (or even exceeds) the standards for the products and processes of interested stakeholders (Hackman, 1987).

Team performance is affected by many different variables, both within the team itself and in the organizational and operational context. One of these factors is training. Training of multi-disciplinary emergency management teams is becoming a more common practice. Nevertheless, the value of these trainings and exercises is questionable. Scenarios are quite often realistic and challenging to the trainees: the team members are heavily engaged in doing their jobs in a multi-disciplinary context. But the degree to which they can really learn from these experiences depends on not only the realism: the training situation may be too complex or hectic to get a good understanding of the team's performance. A solid evaluation afterwards is therefore of utmost importance (Van Berlo, Van Dommele, Schneider, Van de Veerdonk, Braakhekke, Hendriks van de Weem, Van Dijkman, & Wartna, 2007). In order to train the crisis management system at a national level, the Ministry of the Interior and Kingdom Affairs organized the Voyager exercise on October 3, 2007. Voyager was the biggest crisis management exercise ever held in The Netherlands at the time.

In the next section the scenario of the Voyager exercise is briefly outlined. Next, the evaluation approach and format are described. This is followed by a brief presentation of the evaluation results. Finally, the lessons learned with respect to this evaluation methodology are discussed.

Outline of the scenario

The main objective of the Voyager exercise was to train the crisis management decision-making process and the interaction between actors in the overall chain of command from the national-political level to the operational level. Voyager was an exercise from fire-fighter to prime minister, from operational to strategic-political level. The complete decision-making column participated, from the local and regional to the national level. About 2000 professionals participated in a complex multi-threat scenario.

The scenario contained several elements: a disaster with chemical pollution, large-scale employment of first responders, many victims, the threat of a terrorist attack and environmental activists blocking roads and railways. The Voyager scenario had three major storylines. The first is the collision between a cargo ship and a passenger ship in the harbour of Rotterdam. Several containers of the cargo ship fall on the deck of the passenger ship. The chemical and contagious contents of the containers enter the water and passengers get contaminated. There are several drowning victims, wounded and dead people. Environmental activists interpret the collision of the two ships as a proof of the devastating results the current politics lead to. They call upon activists all over the country to block bridges and train tracks in the Netherlands.

The second story-line concerns the transport of alleged terrorists from the United Kingdom to the harbour of Vlissingen, and the police entering a house in Utrecht where a terrorist unit is possibly being given shelter.

The third story-line is the threat of a terrorist attack at the site of a (private) petrochemical factory in the industrial area of Rotterdam. Terrorists are threatening to blow up the factory using a truck with explosives. They stress their threat by taking two employees hostage. An explosion of this kind would result in destroying a large part of the factory and the dispersion of dangerous chemicals in the wider environment. For several actors participating in this exercise, these story-lines came together one way or the other, affecting the decisions that had to be made to manage the various crises.

System approach to evaluation

For the evaluation of the Voyager exercise, a system approach to evaluation was followed (see Figure 1). The system evaluation was conducted in order to evaluate the performance of three critical system functions (processes) as identified in national crisis management: Information Management, Decision Making and Communication with the Media and Public.

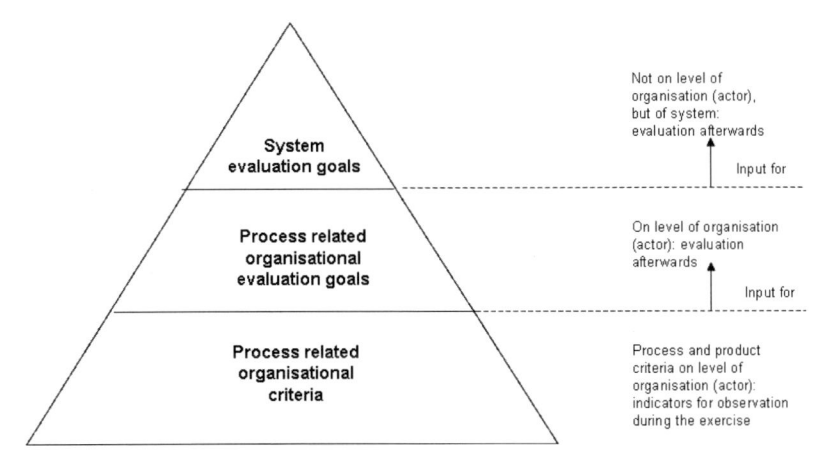

Figure 1. A system approach to evaluation

The focus of the evaluation was not so much on the individual organisations (actors), but on the whole system of national crisis management in the Netherlands. The system of national crisis management has as a goal to effectively and efficiently manage crises on a national level. This system goal is achieved by several system functions (i.e. information management) each enabling the achievement of a part of that system goal. The organisational parts that actually achieve these system goals are the participating organisations (actors), like for instance the various departmental crisis centres. On the level of each organisation (actor) relevant process and product criteria are identified in terms of observable aspects: these guide the observations of the observers during the exercise. The observations are input for the evaluation afterwards in order to determine to which extent every organisation (actor) has contributed to achieving parts of the system functions. Finally, these results are input for the overall system evaluation to identify the performance of the system as a whole in achieving the system goal.

The evaluation was focused on the interdependencies and relations between all actors within the chain of command, and not so much on the performance of every individual actor/organization. The actors/organizations did have dedicated exercise objectives, but these were evaluated by separate evaluation teams. Therefore, emphasis was put on the interaction between crisis management processes on the various levels within the chain of command. An evaluation of operational processes (e.g. medical care for victims, fire-fighting, decontamination) was not the focus of this evaluation and was evaluated by separate evaluation teams. A more detailed overview of the evaluation study and the results is described elsewhere (Van der Sluys-Veer, Kaouass, Van Berlo, Jadoul, Van Zanten, & Van Dorssen, 2008).

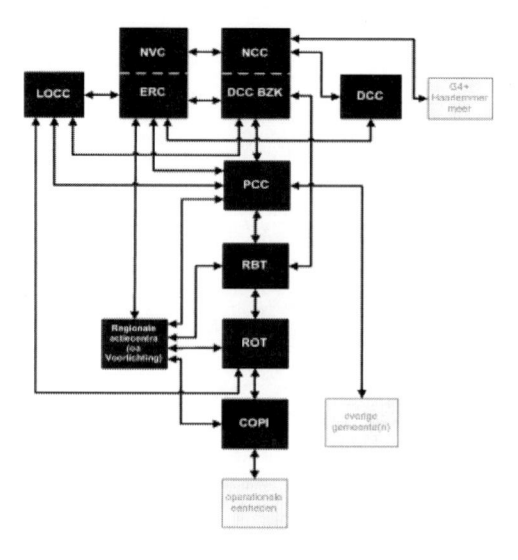

Figure 2. Functional diagram of the Information Management system function

All three system functions (information management, decision-making and communication with the media and public) were described in a uniform manner. For every function, all actors involved were identified, the relations between the actors were indicated, the input-throughput-output was described, and, as much as possible, the criteria for good performance were described. These functional descriptions were based on several official documents, interviews with key personnel within the relevant organizations, and best practices as noted in evaluation studies. Diagrams were constructed to present a visual and comprehensive overview from every system function (see Figure 2 for an example). The descriptions, input-throughput-output and functional diagrams were carefully discussed with relevant stakeholders resulting in a common agreement on these system functions (Van Zanten, Van Dorssen, Slooff, Van der Sluys-Veer, Kaouass, Evers, Van Berlo, & Jadoul, 2007).

For the system functions (or critical process) the following goals were identified. The goal of the Information Management function is that all relevant actors have timely access to required and validated information to adequately make decisions, and have a shared situational awareness, at all levels, of the crisis. The process-related organisational (actor) goals are: uniform information picture, prioritizing and sharing of critical information, and effective use of ICT-support.

The goal of the Decision Making function is that decisions are made at the right moment and in the right way. The process-related organisational (actor) goals are: making decisions based on all available information, following a structured and effective process, and coordination of decision-making between relevant organisations/actors.

The goal of the Communication with Media and Public function is that the communication is timely and uniform/consistent, and gives confidence to the media and public with respect to the actions undertaken by the organisations/actors to manage the crisis. The process-related organisational (actor) goals are: providing consistent and coordinated communication from all organisations/actors, following a structured approach to communication, and offering confidence to the receivers.

For every process-related organisational/actor goal, two to a maximum of five organisational criteria were defined. For instance, 'prioritizing and sharing of critical information' had the following criteria: from every new piece of information it is determined for whom it is relevant, the information is sent to other relevant actors on a regular basis, and information is actively retrieved from other actors.

The criteria for 'coordination of decision making between relevant organisations/actors' were: decisions with an impact on other organisations/actors are communicated as fast as possible, the rationale and implications are explicitly communicated, and the responsibilities as well as authority are known to each other.

With respect to 'consistent and coordinated communication from all organisations/actors', the following criteria were identified: adapting a general communication plan to the specific crisis situation, clarity concerning what will be communicated and what will not (yet) be communicated, and communication is coordinated between the organisations/actors involved.

Finally, for all criteria a set of performance indicators were formulated. These performance indicators were stated in terms of observable behaviour items as much as possible. In this respect a two-step procedure was followed. First, a general description of performance indicators was made. Next, these general performance indicators were translated and adapted to the specific organisations the observers would observe during the exercise, e.g. with respect to specific terminology or procedures. This was done by conducting interviews with key personnel of all 28 organisations/actors. For every individual organisation involved, an organisation-specific set of performance criteria was described within the framework of the three system functions.

In the end, based on the observations (and possibly other additional data) the evaluator could indicate to what extent the process-related organisational goal was achieved by giving a score: 'yes', 'no' or 'partially'. Every score had to be motivated with arguments.

From evaluation format to data collection

The system approach described above resulted in a general description of system functions and goals, organisational goals, organisational criteria and related performance criteria. This description has the form of a table and can be considered as the general evaluation format. Table 1 is an example illustration of the evaluation format, taken from the Information Management system function.

This general evaluation format is in fact applicable to all kinds of organisations within the national crisis management system, irrespective of the specific crisis and/or scenario. This has the advantage that organisations/actors can be compared with each other, but also that progress between exercises can be monitored in a systematic matter.

The evaluation format in itself is not to be used by the observers during the exercise. It gives little room for notes and observations. For this reason, the observers were given a rather general observation format in which they could fill out the name of the actor, the time, the function which was observed, and the observation itself. Table 2 shows an example of this observation format.

In total, 34 observers gathered data from 28 organisations/actors during the Voyager exercise. The observers came from various organisations: TNO, Netherlands Institute of Physical Security, Cap Gemini, Berenschot, the Hague Centre of Strategic Studies, and the Public Order and Safety Inspectorate. Before the exercise took place, all observers received a brief training on how to use both the evaluation format and the observation format. They were given the specific evaluation formats for the respective organisation/actor they would observe. In this way, the observers were aware of the specific performance indicators they could observe. Other supporting tools for the observers were digital cameras to take photographs, copies of situation reports, and a SMS-alerting service to receive notifications of important events in the scenario.

Table 1. Part of the evaluation format of the system function Information Management

Function 1	Information Management		
System goal	All relevant actors have timely access to required and validated information to adequately make decisions, and have a shared situational awareness, at all levels, of the crisis	Name of the organisation/actor:	
Process-related organisation (actor) goals	Organisation (actor) criteria	Performance indicators	Result: Organisation goal achieved?
1. ...	1.1 ...	1.1.1 ...	...
2. Prioritizing and sharing of critical information	2.1 From every new piece of information it is determined for whom it is relevant	2.1.1. In case of incoming information, relations and interdependencies with other actors are acknowledged and monitored	Yes Partly No
	2.2 The information is sent to other relevant actors on a regular basis	2.2.1. Meeting schedules of other actors are taken into account when providing new information 2.2.2. High-priority information is recognized, labeled and handled as such	
	2.3 Information is actively retrieved from other actors	2.3.1. Information is actively retrieved from other actors on a regular basis; not only in case information is missing, but also in a pro-active manner	
3 ...	3.1 ...	3.1.1 ...	...

Table 2. Illustration of general observation format

Actor	Time (hh-mm)	System Function			Observation
		IM	DM	Comm	
NCC	09:15		X		Start of general meeting
...	...				

Immediately at the end of the exercise, the observers made a First Impression Report (FIR). The evaluation staff integrated all these reports into one coherent document used for a debrief of the most important stakeholders the very same day. During the days after the exercise, the observers made a report of their observations and filled out the evaluation formats. All these reports were gathered by the evaluation staff. Additionally, other data were collected, like for instance e-mail correspondence between actors and situation reports of organisations where no observers were available. For every system function as well as at the level of the system of national crisis management, results, conclusions and recommendations were presented.

Evaluation results

In this section, the main results and conclusions of the evaluation study are briefly outlined. A more extensive presentation of the results can be found elsewhere (Van der Sluys-Veer et al., 2008). More detailed results with respect to the quality of the information management process, and more particular with respect to the netcentric nature, are described by Van de Ven, Frinking, Van Rijk, and Essens (2008).

With respect to Information Management it was concluded that the extent to which all relevant actors had a shared view on the various issues and events, could be improved. Further, improvements could be achieved with respect to uniform descriptions of (actual) crisis situations in terms of (technical) parameters, development (scenarios), severity of consequences of events in the area of effect, and relevance and consequences of measures taken. Moreover, between the various management layers (especially regional versus national) and between the various chains of command (crisis management versus combating terrorism) there was no shared perspective on the various threats, events and their relationships and how to best act on these.

An important reason for both bottlenecks was that the integrated information sharing between the various chains, columns and layers was insufficient and not in time, though this was required given the complexity of the decision-making process and the relations between measures taken at all levels.

With respect to Decision making, it was concluded that explicitly defining possible scenarios on both the short and long term, determining the to-be achieved effects, and using this as input for the decision-making process, were hardly done, especially on the national level. In general, the effect of an intended decision is well known. But it can be questioned whether decision makers are fully aware of the operational effects one wants to achieve and which decisions are needed in order to reach these.

Coordinating the decision making process between relevant organizations/actors can be improved. Responsibilities and authorities within one's own organization are recognized, but those of other organizations/actors are not always familiar. An important reason is the lack of criteria determining the up-scaling of a crisis to the national crisis management level (bottom-up versus top-down) and the related responsibilities and authorities. Also the (legal) difference between a crisis and a terrorism scenario leads to perceived indistinctness by the various first responders and crisis managers.

With respect to Communication with Media and Public it was concluded that more effort is needed to guarantee that the right messages are released at the right moment, taking into account the particular circumstances and restrictions, for instance in case of terrorist activities. Preventing contradictory or conflicting media expressions is an important issue, especially in the Netherlands: this country has a complex crisis management structure and relatively autonomous organisations/actors who can organize contacts with the media themselves.

Discussion

For the national crisis management exercise Voyager, a system approach was followed resulting in a general description of system functions, organisational/actor goals, criteria and related performance criteria. An evaluation format and observation format were developed, guided by the principle that both data collection by means of observations and data analysis, should be easily done in a structured, systematic and uniform way. This has the advantage that evaluations of organisations/actors can be compared with each other and that progress between exercises can be monitored in a systematic manner. It is a standardized evaluation format enabling continuous monitoring, improving and benchmarking crisis management organisations. Both the functional descriptions and the evaluation format are adopted by the Ministry of the Interior and Kingdom Affairs of the Netherlands.

The evaluation format appeared to be useful for the observers and evaluators. It was relatively easy to use after a brief observer training. The generic format facilitated the data analysis afterwards. Discussing the format beforehand with the respective organisations/actors in order to define the performance criteria increased the level of acceptance of the results.

Nevertheless, some improvements can be made with respect to using the evaluation and observation format. Acceptance of the evaluated organisations/actors was relatively good, but may increase if the people who define the performance indicators were the same as those who review the observation and evaluation report. Next, it was not possible to conduct an after-action review with the participants immediately after the exercise. If this had been the case, additional and relevant information and experiences from the participants themselves might have been included more often. Finally, there was no opportunity to have a debriefing discussion with all observers. This might have facilitated defining the crosscutting relationships of observations and issues between the participating organisations/actors.

At this moment, the instrument is used by trained observers and evaluators. These individuals are not part of the system and organisations being evaluated. Following Essens et al. (2004), transforming the instrument such that it can be used by personnel of the organisations themselves, or training them to use the instrument has some advantages. In that case, the instrument would be used more often, the organisations would receive a quick overview of the results, and they could determine the conclusions and recommendations more quickly. This way, evaluations can be conducted more often and in a more natural way, rather than always relying on external evaluators.

In line with the previous issue, further research is needed on improving the measurement of team, organisation and system performance. Although there are many advances with respect to the measurement of team behaviour and cognition (Brannick, Prince, Prince & Salas, 1995; Cooke, Salas, Kiekel & Bell, 2004; Essen et al., 2004), there remains a need for more robust, reliable, valid, and diagnostic measurement approaches (Salas et al., 2008).

Finally, the instrument should be used for evaluating different types of teams, in more contexts and with various scenarios in order to establish the scope of organisations that may benefit from this instrument. This would also help get a better grip on the criteria for and norms of effective crisis management teams, organisations and systems performance. This way, useful feedback and recommendations can be formulated increasing both the learning value of (large) exercises and the operational effectiveness of crisis management systems.

References

Brannick, M.T., Prince, A., Prince, C., & Salas, E. (1995). The measurement of team process. *Human Factors*, 37, 641-651.

Cooke, N.J., Salas, E., Kiekel, P.A., & Bell, B. (2004). Advances in measuring team cognition. In E. Salas and S.M. Fiore (Eds.), *Team cognition: Understanding the factors that drive process and performance* (pp. 83-106). Washington, DC: American Psychological Association.

Dyer, J.L. (1984). Team research and team training: A state of the art review. In F.A. Muckler (Ed.), *Human factors review* (pp. 285-323). Santa Monica, CA: Human Factors Society.

Essens, P.J.M.D., Vogelaar, A.L.W., Mylle, J.J.C., Blendell, C., Paris, C., Halpin, S.M., & Baranski, J.V. (2004). *Team Effectiveness: A framework for commanders*. NATO RTO HFM; Soesterberg, The Netherlands: TNO.

Hackman, J.R. (1987). *The design of work teams*. In J.W. Lorsch (Ed.), Handbook of organisational behaviour (pp. 315-342). Englewood Cliffs, NJ: Prentice-Hall.

Kozlowski, S.W.J., & Klein, K.J. (2000). A multilevel approach to theory and research in organizations: Contextual, temporal, and emergent processes. In K.J. Klein and S.W.J. Kozlowski (Eds.), *Multilevel theory, research, and methods in organizations: Foundations, extensions, and new directions* (pp. 3-90). San Francisco: Jossey-Bass.

Rouse, W.B., Cannon-Bowers, J.A., & Salas, E. (1992). The Role of Mental Models in Team Performance in Complex Systems. *IEEE Transactions on Systems, Man, and Cybernetics*, 22, 1296-1308.

Salas, E., Cooke, N.J., & Rosen, M.A. (2008). On Teams, Teamwork, and Team Performance: Discoveries and Developments. *Human Factors, 50*, 540–547.

Van Berlo, M.P.W., Van Dommele, R., Schneider, P., Van de Veerdonk, I., Braakhekke, E., Hendriks van de Weem, N., Van Dijkman, E., & Wartna, S. (2007). Learning to evaluate multidisciplinary crisis-management team exercises. *Proceedings of the 14th TIEMS Annual Conference (The International Emergency Management Society)*, Trogir, Croatia, June 5-8, 2007 (pp. 158-166).

Van der Sluys-Veer, L., Kaouass, A, Van Berlo, M.P.W., Jadoul, M., Van Zanten, P., & Van Dorssen, M. (2008). *Evaluation Exercise Voyager: A system evaluation of critical crisis management processes* (Evaluatie Oefening Voyager: Een systeemevaluatie van kritische processen bij crisisbeheersing, in Dutch). Utrecht, The Netherlands: Cap Gemini (TNO-DV 2008 IN299).

Van de Ven, J.G.M., Frinking, E., Van Rijk, R., & Essens, P.J.M.D. (2008). *Evaluation of Netcentric Information Management during Voyager* (Evaluatie Netcentrische Informatievoorziening tijdens Voyager, in Dutch). Report TNO-DV 2007 C614. Soesterberg, The Netherlands: TNO.

Van Zanten, P., Van Dorssen, M., Slooff, A., Van der Sluys-Veer, L., Kaouass, A., Evers, W., Van Berlo, M.P.W., & Jadoul, M. (2007). *Model descriptions of three critical crisis management processes* (Modelbeschrijving van drie kritische crisisbeheersingsprocessen, in Dutch). Utrecht, The Netherlands: Bureau Berenschot.

On the relationship between physical distance, communication style, and team situation awareness. Preliminary results and implications for collaborative environments

Marco Camilli, Sara Coletta, Alessandra Talamo, & Francesco Di Nocera
University of Rome "La Sapienza"
Italy

Abstract

The concept of Situation Awareness (SA) has captured the interest of many Human Factors professionals and researchers who were interested in the dynamic interaction between human operators and technological systems. However, others have responded to this construct with a great deal of scepticism. Indeed, SA definitions often neglected a careful standardisation of "external" elements, events, interactions, goals and behaviours that may be closely related to different degrees of situation awareness. Recently, researchers have been challenged to shift their focus from individual SA to team situation awareness (TSA) or shared SA (SSA) within teams of operators. Indeed, the increase in bandwidth and the availability of collaborative technological tools led towards a large use of teams in working environments. The study here introduced represents a first step towards a methodological approach to define SA and TSA by taking into account the functionality (or salience) of system's elements and their geometrical features in a Command and Control (C2) scenario. The electronic version of the strategy board game "Risk!" has been used as task in the experimentation. This chapter describes the development of a unique measure of TSA obtained by contrasting individual SA assessment, and a study aimed at investigating the relation between strategies, communication, collaborative system features and TSA.

Introduction

Knowledge and understanding of dynamic environments is supported by what has been defined "situation awareness" (SA). About four decades ago, military research started addressing this topic in order to better understand the pilot's behaviour in the critical context of air battles. The construct of Situation Awareness has been much investigated in several highly specialised fields such as Air Traffic Control (ATC), military Command and Control (C2), Human-Machine Interaction (HMI), military and commercial aircraft piloting, infantry combat, and air battle, therefore confirming the usefulness of SA research in designing complex systems (e.g. aircraft cockpits) and in testing humans behaviour in critical context (e.g. air battle).

However, a formal definition of the construct labelled "Situation Awareness" has been provided only in the late eighties by Endsley (1988). Later on, the same author (Endsley, 1995a; 1995b) presented a more exhaustive description of SA in a special issue of the "Human Factors" journal devoted to SA. The construct was described as the individuals' ability concerning "the perception of the elements in the environment within a volume of time and space (Level 1 SA), the comprehension of their meaning (Level 2 SA) and the projection of their status in the near future (Level 3 SA)" (p. 36). Despite the numerous attempts to define and modelling SA by different perspectives, Endsley's definition is still the most cited. Moreover, many conferences, symposia, meetings, and special issues of scientific journals have been devoted to SA since 1995. Nevertheless, many authors have responded to this construct with a great deal of scepticism, strongly opposing its use in the scientific literature as the only explanation for the quality of the operator's performance (see Flach, 1995). Indeed, SA development appears to follow many applicative requirements neglecting several basic aspects such as a clear definition of the cognitive processes engaged (e.g. mental models and working memory), and a standardisation (or formalisation) of artefacts, elements, events, interactions, goals and behaviours leading to different degrees of SA (see for example, Smith and Hancock, 1995; Pew, 2000; Flach, Mulder, & van Passen, 2004). Generally, the main criticism of the SA concept can be related to the many complementary views to Endsley's approach and the consequent differences in measuring this construct (see Wickens, 2008). Despite these criticisms, SA concept continues to embrace the most recent and relevant issues in Human Factors and Ergonomics (HF/E) research such as teamwork and communication. In fact, the traditional SA literature reported many attempts aimed at defining and assessing the individual SA of the single operator in complex system (e.g. Endsley, 2001). Recent increases in bandwidth and technological tools led towards a large use of teams in working environments (see for example, Artman & Garbis, 1998, or Salmon et al., 2006). The current relevance of teamwork in military and civilian operations challenges SA researchers to focus their studies on behaviours of SA within teams of operators. Endsley (1995a) described team SA (TSA) as "the degree to which every team member possesses the SA required for his or her responsibilities" (p. 31). Consequently, the TSA is gathered by a summation of the individual team members SA. Differently, the concept of shared SA (SSA) is defined as "the degree to which team members have the same SA on shared SA requirements" (Endsley & Jones, 1997; p. 54). The SSA construct is mainly characterised by the fact that each team member could have specific requirements for his / her task. Thus, SSA degrees seem referring to the overlapping of common elements. Other authors have differently defined the SA within teams of operators. For example, Salas et al. (1995) defined TSA as "at least in part the shared understanding of a situation among team members at one point in time" (p. 131). Artman and Garbis (1998) extended team situation awareness argumentation towards the "artefacts, protocols and descriptions of the situation" that indirectly affect the sense-making process of the team members. Also, distributed SA (DSA) models take the collaborative system as the unit of analysis focusing on the interaction and coordination between agents and sub-systems (see Artman & Garbis, 1998 or Stanton et al., 2006). These models are aimed at assessing potential barriers to SA acquisition and maintenance within complex collaborative systems. The DSA approach provides a viable alternative to the individual-oriented perspectives focusing on the use, passage and sharing of information between agents in the network. In other words, it considers the artefacts used by each agent (e.g. desktop computer, documents, laptop, information display) as sources of DSA in collaborative systems (see Salmon et al., 2008 for a recent review on teams situation awareness). Recently, several studies (Cummings, 2004; Bolia, Nelson, Vidulich, Simpson, & Brungart, 2005; Knott, Nelson, Bolia, & Galster, 2006) have addressed these issues in order to enhance understanding of the relations between communication and resulting performance success in teams workplace operations. Other researchers (e.g. Gorman, Cooke, Pederson, Connor, & DeJoode, 2005) investigated the coordination levels and communication failures as precursors of the team situation awareness. Finally, factors including communication devices and collaborative systems are impacting team SA and performance of distributed teams. However, the consequences associated with the introduction of new technology (e.g. new modes of communication and collaborative systems characteristics) are yet to be fully understood.

The studies here presented were aimed at clarifying the relationship between communication and SA within teams and the effects of the physical characteristics of a collaborative platform. For these purposes a shared-collaborative interface (the Risk! game platform) was used as experimental scenario and an instant messaging device (chat-box) allowed communication between distributed and networked team members. A pilot study was carried out in order to:

1. assess possible intrusiveness effects of the freeze technique (used in the SA assessment) on the natural and dynamic task course;
2. investigate the three-levels SA model in terms of the memory effects on the comprehension and projection abilities (i.e. level 2 and 3 SA);
3. test the SA three-levels measure (developed for this specific task) for assessing its capability in discriminating different degrees of TSA.

As it will be reported in the following sections, the pilot study clarified the existence of relations between TSA, communication and performance in teamwork task. The experimental study was firstly aimed at investigating the relationship between SA in teamwork and the physical features of a collaborative system (i.e. iconic representations of the agents involved in the task and their spatial relations). SA-related knowledge in terms of geometrical distances between agents displayed on a map and communication styles within teams in terms of cooperation and coordination were widely discussed taking into account also unexpected individual differences. These latter were addressed in details to specific analyses on information exchange between members of males and females teams. More specifically, in order to analyse the actions performed by the different teams it was adopted qualitative methods developed in the frame of Discourse Analysis (DA). In particular, they were adopted the recommendations by that particular area of DA which includes methods for studying how discourse constitute practices (Taylor, 2001). As Potter (1997) defined it, Discourse Analysis "has an analytic commitment to studying discourse as texts and talk in social practices. That is, the focus is not on language as an abstract entity [...]. Instead is the medium for interaction: analysis of discourse becomes, then, the analysis of what people do" (Potter, 1997, p. 146). Coherently with the theoretical background in which this research takes place, method relies on an important assumption: language is not a description, rather is action and talk is constitutive of contingent reality: "The discourse analyst searches for patterns in language in use, building and referring back to the assumptions she or he is making about the nature of language, interaction, society and the interrelationships between them." (Taylor, 2001, p. 39). In this perspective method is not a neutral technical research instrument; rather it reflects our own theoretical assumption on the nature of social processes. The great deal of data analysed in this study represented a first attempt to integrate different perspectives providing a preliminary framework to assess SA in teamwork task and collaborative systems. The final aim of this research line will be to study the central role of the interactions between systems' geometric features, communication characteristics and team situation awareness.

Experimentation

Task

A computer game of strategy inspired to the popular board game Risk! has been used as task. This experimental scenario presented a map (figure 1) as a collaborative interface and an instant messaging device and it was enjoyable and engaging for participants. The Risk! game was played in turns: each agent could control only his/her own armies on the map during his/her turn of the game. During the turn, each agent could conquer neighbouring territories moving armies like pieces on a chessboard. Also, before ending the turn, each agent could move armies within his / her neighbouring territories in order to improve defensive force. On the map, the territories had not the same functional value. In fact, all territories were linked between them in different functional areas that, when conquered and kept by one agent, allowed him / her to receive additional (or bonus) armies at the following turn (see figure 1). Therefore, the agents of the same team needed to negotiate common strategic behaviours to conquer and keep these areas in order to reach the final shared goal: destroy the enemies and conquer the world map together. No team roles were assigned to participants; thus, they had the same task requirements and they could coordinate their operations as they preferred. This platform allowed to design an effective teamwork task: two participants -distributed in two separated remote control rooms- were assigned to a team and faced a team of two Artificial Intelligence (AI) players. These latter were developed to simulate the behavioural abilities of beginner (lower task load or "easy" condition) and expert (higher task load or "hard" condition) gamers.

Situation Awareness measures

The dynamic course of the task was stopped for the SA assessment: at a given instant of the task, the scenario was frozen and the screen was blanked. Participants were required to fill three different tests to assess their degree of perception (level 1 SA), comprehension (level 2 SA) and projection (level 3 SA) of the task situation. A pilot study was carried out in order to understand the functionality of the SA-tests

physical distance, communication style, and team situation awareness 45

developed for this type of task. The validity of the freeze technique was assessed and several considerations about the capability of the tests were made. Particularly, these measures should provide an effective weight of the correspondent SA level into the overall SA score. Parts of the comprehension and projection checklist-tests used in the pilot study have been changed for this purpose in the experimental study.

Figure 1. Forty-three territories linked between them in 12 specific functional-bonus areas

In the pilot study the three SA-checklists required to participants to:

1. relocate the correct positions of the four armies (i.e. own army, teammate's army and two enemies' armies) on a white map (level 1 SA – perception);
2. report on the behaviours for each army during the last turn of the game by choosing between "conquering work", "defensive work" and "no work" on each functional area;
3. forecast the behaviours of the two enemies' armies in the next turn of the game by choosing between a sample of ten typical strategies used in this type of game (level 3 SA – projection).

In the experimental study the three SA-checklists required to:

1. relocate the correct positions of the four armies (i.e. own army, teammate's army and two enemies' armies) on a white map (level 1 SA – perception);
2. report "attack/no-attack" behaviour for each army in each functional area during the last turn of the game (level 2 SA – comprehension; see table 1);
3. forecast "attack/no-attack" behaviour for each army in each functional area during the next turn of the game (level 3 SA – projection; see table 2).

Table 1. Comprehension-Level 2 SA checklist. Participants ticked "attack/no-attack" behaviour for each army in each functional area during the last turn of the game (in these experiments blue and yellow armies were the human teams).

Step 2: Comprehension of the Situation on the last turn of the game							
Which armies attacked a territory within a functional area?							
Functional areas:	Strategies of the four armies at the least game's turn:							
	Blue army		**Green army**		**Yellow army**		**Pink army**	
Africa	Attack	No attack	Attack	No attack	Attack	No attack	Attack	No attack
	Blue army		**Green army**		**Yellow army**		**Pink army**	
Australasia	Attack	No attack	Attack	No attack	Attack	No attack	Attack	No attack
	Blue army		**Green army**		**Yellow army**		**Pink army**	
Canada	Attack	No attack	Attack	No attack	Attack	No attack	Attack	No attack
	Blue army		**Green army**		**Yellow army**		**Pink army**	
Central America	Attack	No attack	Attack	No attack	Attack	No attack	Attack	No attack
	Blue army		**Green army**		**Yellow army**		**Pink army**	
Europe	Attack	No attack	Attack	No attack	Attack	No attack	Attack	No attack

Table 2. Projection-Level 3 SA checklist. Participants ticked "attack/no-attack" behaviour for each army in each functional during the next turn of the game.

Step 3: Projection of the Situation on the next turn of the game							
Which armies will attack a territory within a functional area?							
Functional areas:	Strategies of the four armies at the next game's turn:							
	Blue army		**Green army**		**Yellow army**		**Pink army**	
Africa	Attack	No attack	Attack	No attack	Attack	No attack	Attack	No attack
	Blue army		**Green army**		**Yellow army**		**Pink army**	
Australasia	Attack	No attack	Attack	No attack	Attack	No attack	Attack	No attack
	Blue army		**Green army**		**Yellow army**		**Pink army**	
Canada	Attack	No attack	Attack	No attack	Attack	No attack	Attack	No attack
	Blue army		**Green army**		**Yellow army**		**Pink army**	
Central America	Attack	No attack	Attack	No attack	Attack	No attack	Attack	No attack
	Blue army		**Green army**		**Yellow army**		**Pink army**	
Europe	Attack	No attack	Attack	No attack	Attack	No attack	Attack	No attack

In the pilot study, the comprehension checklist included three salient strategies that each agent could follow during the game. However, the "no-work" strategy could occur more often in some sessions of the game than in others and the defensive strategy could be difficult to acknowledge (for example, some attacks could be aimed at a defensive strategy by weakening the enemies). The projection-checklist used in the pilot study included ten different possible enemies' behaviours obtained by interviewing expert gamers. This checklist included several complex and salient actions. However, some occurrences could exclude others. Furthermore, some specific behaviour occurred more frequently in some sessions of the game than in others. For this reason, the level 2 and 3 SA-checklists were simplified by taking into account only the most salient and frequent binary behaviour (i.e. attack and no-attack).

Pilot study

Subjects

Eight participants (4 females and 4 males; mean age = 22 years, SD = 0,5) volunteered in this pilot study were grouped in 4 teams of 2 members. All participants were right-handed, with normal hearing and normal or correct to normal vision.

Apparatus

The computer game of strategy and domination "Lux Delux" (Sillysoft: U.S.A.) inspired by the board game Risk! was used as task in this study. The aim of the game was to control armies in order to conquer and hold strategic countries on the map. Lux Delux map was composed of 43 countries grouped in 12 "functional areas" (figure 1). The "functional" label indicated that each army receives additional troops (i.e. increased own force) by conquering and keeping the occupation of those areas. The game engaged four agents grouped in two opposing teams: "human" vs. "artificial intelligence". The names of the AI agents were the same in both levels and the participants didn't know both the real ability and the artificial nature of the opponent-team. A chat-box was used to allow instant messaging communication between the members of every human team. Games were video-recorded and both agents' actions and chat-messages were collected for successive analyses.

Procedure

Each participant completed a training session prior experimentation. Participants sat in separated sound-attenuated rooms and were informed about the experimental scenario and the teamwork situation (i.e. map, control-keys, agents, communications and goals). Each team was requested to play the Risk! game conquering all countries on the map. The AI team was coded (in java language) in order to have the same goal (i.e. occupation of all countries occupied by the human team). Each human team performed two levels of the game difficulty (Easy and Hard). After completing each level of the game, participants were requested to fill the electronic version of NASA-TLX (U.S. Naval Research Lab) for the assessment of mental workload. The scenario was stopped and frozen at the sixth minute of the game for SA assessment. In order to investigate the bias introduced by memory effects in the freeze technique, participants were required to complete the level 2 and level 3 SA-checklists both in a hidden scenario condition (blank screen) and in an overt scenario condition (frozen screen). Obviously, level 1 SA-test (i.e. the perception-test) could not be replicated using the overt scenario.

Data analysis and results

Mental workload
NASA-TLX weighted ratings were used as dependent variable in a repeated measures ANOVA design using "Task load" (Easy vs. Hard) as repeated factor. Results showed a tendency towards statistical significance for task difficulty ($F(1,7)=4.86$; $p=0.06$). The "hard" condition (mean score = 48.50; SD = 25.28) was perceived as more loading than the "easy" one (mean score = 40.21; SD = 21.54).

Freeze technique validity
In order to assess the potential intrusiveness of the SA assessment (freezing the dynamic course of the task) on participants' performance, strategic behaviours (i.e. turn duration, conquered territories, conquered areas, failed attacks, defensive reinforces, defensive displacements, rearward armies, action rapidity, total attacks, total defences, attacks frequency, defences frequency) were used as dependent variables in a series of repeated measures ANOVA designs using "Task phase" (pre-test vs. post-test) as repeated factor. Results showed no main effects of Task phase on the strategic behaviours ($p>0.05$) except for the attack frequency ($F(1,6)=14.13$; $p<0.01$), which was higher in the post-test (mean frequency = 6.14; SD = 2.33) than in the pre-test (mean frequency = 3.45; SD = 2.46). This result was expected, because –when far into the game- participants had to perform more frequent attacks for reaching their goal (i.e. conquer all enemies' territories).

Memory and SA
The mean ratings of the level 2 and level 3 SA-test were used as dependent variables in a repeated measures ANOVA design "Level SA" (Comprehension vs. Projection) x "Scenario" (Hidden vs. Overt). Results showed a significant interaction between the two SA levels and the scenario's conditions ($F(1,7)=6.12$; $p<0.05$). Duncan post-hoc testing showed that, while comprehension ratings were identical in the two scenario conditions ($p=0.99$), the projection ratings were lower in the hidden scenario condition than in the overt ($p=0.05$; table 3).

Table 3. Mean and standard deviation of comprehension and projection scores (percentage of correct responses) separately for hidden and overt scenario condition

	Hidden scenario		Overt scenario	
	Comprehension –SA	Projection -SA	Comprehension –SA	Projection -SA
Mean	61.72	56.33	61.59	63.44
SD	26.9	26.53	21.33	23.6

Team SA

As showed in Figure 2, teams 1-2 reported lower SA ratings than teams 3-4. That allowed a comparison between "low SA" teams (1-2) and "high SA" teams (3-4) that will be addressed in the following section. The team SA overall ratings were categorised by median value and they were used in the following comparisons. Given the small amount of teams and subjects per team, no statistical analyses have been carried out on these data. Nevertheless, such a division (high SA vs. low SA) allowed comparing strategies and communication performed by teams. Future studies with a considerable amount of subjects should be approached using appropriate statistical models (such as generalisability theory) as variance partitioning may be an issue when comparing groups of individuals belonging to teams.

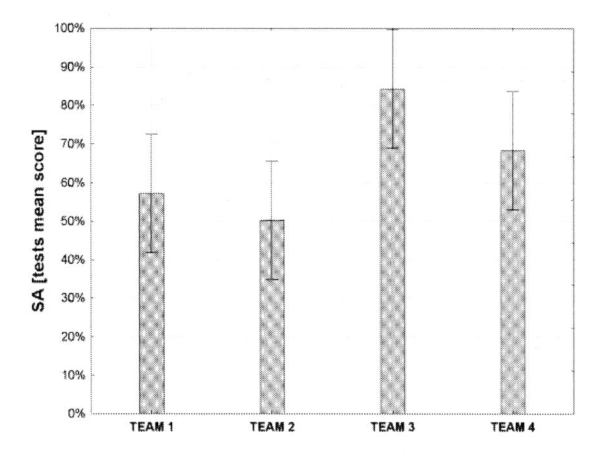

Figure 2. SA overall scores separately for team. Error bars denote 0.95 confidence intervals

Table 4. Mean and standard deviation of strategic and interrogative messages (percentage values on the total messages) separately for "low SA" and "high SA" teams

	"Low SA" teams		"High SA" teams	
	Strategic messages	Interrogative Messages	Strategic Messages	Interrogative Messages
Mean	10	15.55	26.55	3.91
SD	3.61	5.77	17.27	8.8

Strategies and communication

Strategic behaviours and communication features were observed by "team SA" (Low vs. High). Negligible differences were found in the final score obtained in the task, however "low SA" teams spent more time (mean = 30.77 minutes; SD = 3.14) to complete the entire task than "high SA" teams (mean = 9.88 minutes; SD = 2.85). Other noticeable differences were found for turn duration (longer mean duration of the game's turns performed by "low SA" teams than "high SA" teams), frequency of interrogative messages (more frequent interrogative messages used by "low SA" teams than "high SA" teams), and strategic messages (more frequent strategic messages used by "high SA" teams than "low SA" teams). Table 4 shows a comparison between types of messages and degree of team SA. The specific characteristics of the messages exchanged between teams' members will be analysed and discussed in details in a further section of this chapter.

Experiment

Method

Subjects
Sixteen participants (9 females; mean age = 22.8 years; SD = 1.8) volunteered in this experiment were grouped in 8 teams of 2 members. All participants were right-handed, with normal hearing and normal or correct to normal vision.

Apparatus
The same equipment used in the previous study.

Procedure
The same experimental procedure used in the previous study. However, the scenario was frozen at the end of the first (1st Freeze) and third (2nd Freeze) turn of the game in order to have two values of situation awareness during the course of the same task.

Data analysis and results

Mental workload
NASA-TLX weighted ratings were used as dependent variable in a repeated measures ANOVA design using "Task load" (Easy vs. Hard) as repeated factor. Results showed no main effects of the task condition on the TLX scores ($p=0.89$). Participants reported quite similar TLX scores in the "easy" task condition (mean score = 41.81; SD = 25.65) and in the "hard" condition (mean score = 41.12; SD = 25.41). However, gender showed a significant main effect on the TLX scores ($F(1, 14)=10.85$; $p<0.01$): females reporting higher TLX scores (mean = 54.69; SD = 21.5) than males (mean score = 24.48; SD = 18.64). Gender differences will be successively addressed in much detail.

SA in time

The SA tests ratings were used as dependent variable in a repeated measures ANOVA design using the "Freeze" (1st Freeze vs. 2nd Freeze) as repeated factor. Freeze technique has been introduced at the end of the first and third turn of the game. Results showed a main effect of the Freeze on the SA degrees ($F(1, 15)=36.88$; $p<0.0001$). Generally, participants reported a higher SA at the end of the third turn (mean SA = 72.16 percent of correct answers; SD = 9.64) than at the first turn (mean SA = 61.62 percent of correct answers; SD = 9.42).

The SA tests ratings were used as dependent variable in a repeated measures ANOVA design Freeze (1st vs. 2nd) x "SA level" (perception vs. comprehension vs. projection). Results showed a significant interaction between Freeze and SA level ($F(2,30)=11.37$; $p<0.001$). Particularly, participants reported the lowest perception score at the end of the first turn of the game (see table 5). Probably, during the first turn of the game participants had not enough time to carefully observe and perceive all objects represented on the map. For this reason, it has been decided to include into the following analyses only data collected after the third turn of the task.

Table 5. Mean and standard deviation of perception -Level 1 SA- comprehension -Level 2 SA- and projection -Level 3 SA- scores (percentage of correct responses) separately for 1st and 2nd Freeze

	1st Freeze			2nd Freeze		
SA	Level 1	Level 2	Level 3	Level 1	Level 2	Level 3
Mean	48.55	72.64	63.68	73.98	75.91	66.59
SD	14.07	18.76	14.18	17.07	17.06	12.55

Agent-related SA
The SA overall tests scores were used as dependent variable in a repeated measures ANOVA design using the "SA-agent" (Self-related vs. Teammate-related vs. Enemies-related) as repeated factor. Results showed a main effect of the agent on the SA scores ($F(2,30)=14.41$; $p<0.0001$). Duncan post-hoc testing showed that participants reported a higher self-related SA than both teammate-related ($p<0.05$) and enemies-related SA ($p<0.0001$). Also, teammate-related SA was higher than enemies-related SA ($p<0.01$; see table 6).

Table 6. Mean and standard deviation of the SA scores (percentage of correct responses) separately related to the four agents involved in the game

SA	Self-related	Teammate-related	Enemies-related
Mean	81.72	73.46	63.86
SD	12.31	15.16	13.01

Distance by step
The distances between the territories of the four agents were measured in terms of the step that each agent had to cover (moving his/her armies) on the map in order to arrive from his/her own territory to a "target" territory (as in chess). For each participant, the distance (in steps units) was matched to his / her accurate relocation of that territory.

SA by agent
The relocation test scores were used as dependent variable in a repeated measures ANOVA design using the "Agent" (Teammate vs. Enemies) as repeated factor. Results showed a main effect of the agent on the relocation score ($F(1,14)=8.61$; $p<0.05$). On the white map, participants relocated the teammate territories more accurately (mean = 76.87 percent of correct relocations; SD = 19.31) than enemies territories (mean = 60.69 percent of correct relocations; SD = 21.44).

SA by step distance
The relocation test scores were used as dependent variable in a repeated measures ANOVA design "Agent" (Teammate vs. Enemies) x "Step-distance" (Neighbouring

vs. Step 2). Results showed a significant interaction (F(1,14)=16.58; p<0.01) between the agent and distance (see figure 3). Duncan post-hoc testing showed that at the "step 2" distance the teammate's territories were better relocated than the enemies' territories (p<0.01). Moreover, the enemies relocation was better at neighbouring distance than at "step 2" distance (p<0.05). Furthermore, at the "step 2" distance, participants reported a percentage of correct relocations of the teammate's territories (mean = 82.47 percent of correct relocations; SD = 21.83) very close to the percentage of the accurate relocations of their own territories (mean = 82.58 percent of correct relocations; SD = 14.75).

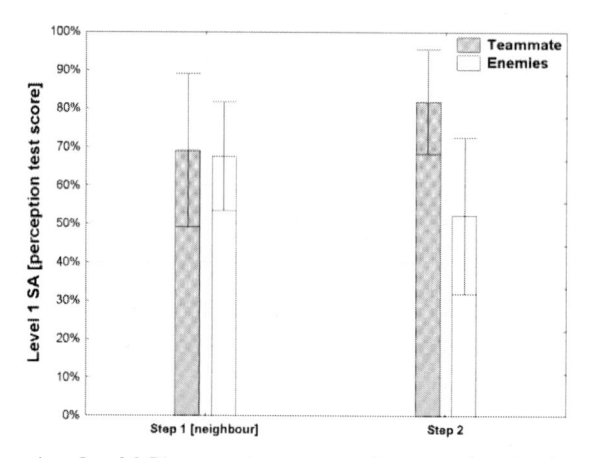

Figure 3. Perception -Level 1 SA- scores (percentage of correct relocations) separately for step distance and for agent. Error bars denote 0.95 confidence intervals

Euclidean distances
The overall SA was investigated as related to the distances between participant's locations and the other agents' locations. All neighbouring territories occupied by the same agent were included in a unique "block" of territories. For each block its smallest rectangle was shaped and its barycentre identified. For each participant the median distance (in centimetres unit) from the other agents was computed taking into account the distances between barycentres. In order to assess the homogenous distribution of the agents' territories on the map, these latter were compared in terms of distances between blocks. The median distances between the blocks were used as dependent variable in a repeated measures ANOVA design "Agent" (Teammate vs. Enemies) x "Task load" (Easy vs. Hard). Results showed an tendency towards statistical significance for the agent x task load interaction (F(1,12)=3.17; p=0.10). Duncan post-hoc testing showed that the participants' territories and their teammates' territories were furthest away in the lower task load condition than in the higher one (p<0.05; table 7). Probably, in the easy task load condition -where the enemies behaviours were more predictable- participants could agree working on different sections of the map. For this reason it has been decided to include into the following analyses only data collected during the high task load condition. In fact, these data could not be affected by this behavioural strategy (enemies are more unpredictable) and the territories on the map were homogenously distributed *per*

physical distance, communication style, and team situation awareness 53

agent. The Euclidean distances were categorised by the median and they were used as independent variable in the following analyses.

Table 7. Mean and standard deviation of the distances between territories (median values in centimetres) separately for agent and for task load condition

	EASY		HARD	
	Teammate	Enemies	Teammate	Enemies
Mean	8.24	7.78	7.43	7.75
SD	1.11	1.45	0.51	0.93

SA by Euclidean distance
The enemies-related SA scores were used as dependent variable in a mixed ANOVA design "Distance" (Short vs. Long) x "Gender" (Females vs. Males). Results showed a not-significant, but yet not negligible effect (F(1,12)=2.49; p=0.14; figure 4) for the distance x gender interaction. Duncan post-hoc testing showed a p=0.12 difference between females and males at short distances. As a matter of fact, males reported higher degrees of SA than females. Even if these results are not statistically significant, it should be considered that the small numbers of teams used in this study may be responsible for this. Therefore, it has been decided to discuss these results.

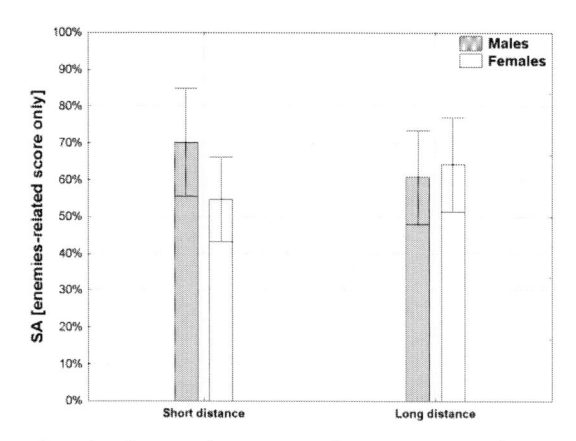

Figure 4. SA enemies-related scores (percentage of correct responses) separately for distance and for gender. Error bars denote 0.95 confidence intervals

The teammate-related SA scores were used as dependent variable in a mixed ANOVA design "Distance" (Short vs. Long) x "Gender" (Females vs. Males). Results showed a tendency towards the statistical significance (F(1,12)=3.17; p=0.10; figure 5) for the distance x gender interaction. Duncan post-hoc testing showed that, at long distances, females reported a higher degree of SA than males (p<0.05).

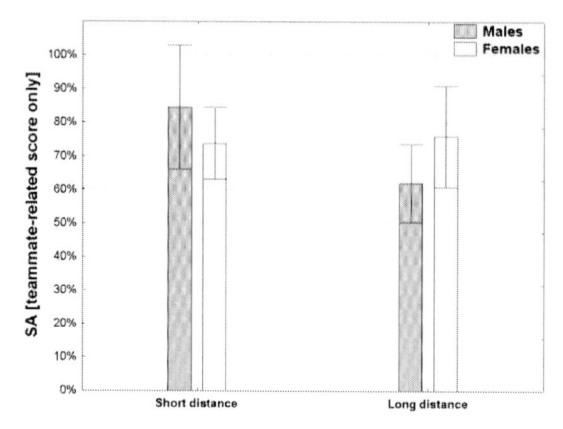

Figure 5. SA teammate-related scores (percentage of correct responses) separately for distance and for gender. Error bars denote 0.95 confidence intervals

Communication

Communication frequencies were used as dependent variable in a mixed ANOVA design "Gender" (Females vs. Males) x "Type of message" (Strategic vs. Interrogative). Results showed a significant interaction between gender and types of messages ($F(1,14)=5.17$; $p<0,05$). Duncan post-hoc testing showed that males more frequently used strategic messages than interrogative messages ($p<0.05$) while females used more frequently interrogative messages than males ($p<0.05$; see table 8).

Table 8. Mean and standard deviation of strategic and interrogative messages (percentage values on the total messages) separately for males and females

	Strategic messages		Interrogative messages	
	Males	Females	Males	Females
Mean	11.59	11.22	4.29	18.04
SD	2.82	3.40	3.05	16.93

Table 9. Mean and standard deviation of strategic and interrogative messages (percentage values on the total messages) separately for males only teams (2), mixed teams (3) and females only teams (3).

	Strategic messages			Interrogative messages		
TEAM	Males only	Mixed	Females only	Males only	Mixed	Females only
Mean	12.03	11.02	11.32	4.49	8.38	20.69
SD	2.39	2.83	4.03	2.04	5.76	20.77

Communication by team composition

Also in this case, sample size and composition of the teams did not allow any statistical comparison. Nevertheless, communication frequencies were compared by "Type of message" (Strategic vs. Interrogative) and "Team composition" (Males-only vs. Mixed vs. Females-only). Females-only teams more frequently used interrogative messages than mixed teams and males-only teams (see table 9).

Qualitative analysis on teams' communication
Chat transcripts have been analysed using Discourse Analysis (DA). Particularly, micro-interactions were selected from a continuum of interactions, which was segmented and clustered by active interpretation. Relevant pieces were not identified in advance, the aim was to select the relevant bits by having participants "define" what was relevant and what was not for implementing their own actions. The qualitative analysis aimed at providing a deeper understanding of the gender issues in performing the task and more specifically in exploring the different performances by groups in terms of the use of interrogative vs. strategic expressions during the game sessions. To this aim it was been run discourse analysis separately on the chat logs of males and females teams (excluding the mixed teams) to see how they differed in terms of interaction.

Experience and competence in gaming as precursors of SA
A first result of DA concerned the competence in playing the game. Despite the equal training, two out of the three teams composed by females showed during the interactions relevant differences in terms of competence in gaming even at a very basic layer, as shown in the excerpt below:

> *Excerpt n.1 [team_07_H_chat, turn 5]*
> Female player 1: Where should I go?
> Female player 2: What's up!?
> Female player 1: Do you remember how to move the armies?
> Female player 2: Nope
> Female player 2: Which ones would you like to move?
> Female player 1: Those placed in South Africa
> Female player 2: I don't know. I've tried as well but did not succeed
> Female player 2: What have you attempted to do?
> Female player 1. No idea. I did something, but couldn't understand how
> Female player 1: I'll try again

As the excerpt 1 shows the interaction by this team included basic questioning about procedures for being able to perform actions in the environment. At this stage interrogative interactions were not included to define a common strategy or a sequence of actions, rather to be able to play the game. The high task load levels reported by female members at TLX questionnaire would corroborate this interpretation. Furthermore, excerpt 1 shows that performing in an efficient way is related to priorities in activities to be performed by the team members. These findings suggest that strategic talk cannot arise till the basic skills for performing actions are not granted. This result is visible in two out of three of teams composed by females. The third team composed by females only showed an example of more expert teammates. In that case they did not need to question about the procedure for playing the game and they showed to be more focused on joining strategic activities.

Monological SA
The qualitative analysis of chat logs of teams composed by males only showed that they behaved in a different way: there was no evidence in male teams of difficulties

in acting or moving the armies, and chatting was used to implement a better performance in terms of effective strategy.

A typical example of interaction in teams composed by males only is showed in excerpt n. 2.

> *Excerpt n. 2 [team _01_H_chat, turn 10]*
> Male player 1: You still need to get the small country up there
> Male player 1: Attack with your 7 and don't care about my troops
> Male player 1: Great!
> Male player 1: As soon as you can, attack the small blue territory
> Male player 1: Put most of your troops there. I am not going to attack you, weirdo
> Male player 1: Go to the right
> Male player 2: Sure sure, I am going
> Male player 2: Chill down
> Male player 2: Otherwise I will send you Bin Laden
> Male player 2: What about me? Should I break in from the left?
> Male player 1: Do you mean from the right?
> Male player 2: Whatever

Excerpt 2 refers to a typical way of acting of male teams to perform a unique and consistent strategy among teammates. In this example "male player 1" strictly drives "male player 2" to what he is supposed to do in order to meet the team goal. This way of acting introduces a particular way of interacting in groups that might be conceived as "monological" (Talamo and Pozzi, accepted). The two members of the team act in a consistent way but are not performing at an equal level. The way of "male player 1" of defining the overall strategy of the team shows that he is simply using "male player 2" as an extension of his own space of activity. "Male player 2" is not required to approve the strategy, nor to introduce any variation but simply to perform actions that are conceived by "male player 1". In terms of SA this kind of team management only needs the establishment of a leader: once one of team members starts leading and demonstrates to keep a clear vision on the actions needed by the team as a whole, the teammate follows indications without discussing them. In the case of "male player 1" and "male player 2" this is clearly visible since turn 1 as shown in excerpt n.3:

> *Excerpt n. 3 [team _01_H_chat, turn 1]*
> Male player 1: You have to fortify there, damn!
> Male player 2: Don't worry

The Dialogical nature of team situation awareness
The chat logs allowed to observing also a different kind of interaction which might be intended as typical of team situation awareness. Two groups (one with female only and one with male only teammates) showed a specific way of interacting that might be called "dialogical". The dialogical interaction is typical of shared activities where the members of a team cooperate to solve a problem. The dialogical interaction showed cohesion by team members in the goal to be achieved and

contemporary coordination in actions to be performed. Excerpt n. 4 shows an example observed in a female group.

> *Excerpt n. 4 [team_04_H_chat, turn 1]*
> Female player 3: Hey
> Female player 4: Let's get organised
> Female player 3: South America
> Female player 4. You get South America
> Female player 4: I get Russia
> Female player 4: Can you leave Australia to me?
> Female player 3: Sure
> Female player 4: Now attack from Australia the green army
> Female player 4: 2 on 1
> Female player 3: I couldn't
> Female player 3: I fortified
> Female player 4: OK
> Female player 3: I am going to take Africa
> Female player 4: What should I do?
> Female player 3: Europe
> Female player 4: OK

Excerpt 4 illustrates very well the dialogical nature of the interaction of this team as reported in the chatting of its members since the very first session. In terms of DA this team showed a relevant difference with what was been observed in the previous examples. "Female player 3" and "female player 4" started the game by allocating zones and tasks in a cooperative way, with no abuse of power by any of team members. The task distribution went on smoothly, with no discussion and with some positive feedback given to each other on reciprocal choices. Generally, this way of proceeding includes two important aspects of team situation awareness: each member is open to the other's proposal; each member understands and agrees on the other's proposal and acts in a synergic way.

Discussion and conclusions

Since the formal definition of "situation awareness" (see Endsley 1995a; 1995b), the interest captured by this construct in the HF/E domain has constantly grown and many SA measures have been developed following different views. The reason for developing novel measurement procedures resides in the need for understanding how several aspects that have been often neglected by literature can affect this construct. Particularly, in this study the effect of distance was assessed. Indeed, distance from and between "objects" in the scene is a key feature of the situation and it should be taken into consideration. A pilot study was firstly carried out in order to assess a possible interference between the freeze technique and the task execution, as well as for testing the experimental platform. Results showed that stopping the task course to assess participants SA did not affect their performance. Other researches (see Endsley, 1996 for a review) reported similar findings when tested freeze technique to assess SA in fighter aviation domain. Also, participants comprehension and projection abilities were assessed both in hidden and in overt scenario situations in order to investigate potential memory effects on SA. Results showed an interesting memory effect in level 2 and level 3 SA. In fact, comprehension was not affected by the lack of visual elements, whereas projection showed a drop in performance. This result suggests that cognitive processes that continuously update changes of the environment could underlie comprehension. Differently, a sort of "on demand" processing may underlie projection. In other words, when we need to make predictions we may benefit from the availability of abundant information. At this stage of this research program, it is difficult to provide clear evidences for this interpretation; future studies will be useful to better understand the memory's effects for each SA ability, and their implications for the operator's performance. As a matter of fact, in many operative contexts (e.g. air traffic control and baggage screening) a temporary display black-out could lead to critical and unsafe situations.

Results of the pilot study also showed that the SA measures that have been used here discriminated between different degrees of TSA, and were related to different teamwork strategies and communications. Salas et al. (1995) suggested that team SA depends on different types of communications (e.g. of mission objectives). The current study seems to confirm this point of view. In fact, high-TSA teams frequently shared information about team performance factors such as strategic activities, individual tasks and team capability. Nevertheless, further studies are needed in order to fully understand the direction of the relationship between SA and communication.

However, the scope of the study reported here was to investigate the role of distance (both in psychological and metrical sense) in the development of SA. Results showed that teammates were better represented than enemies on the map, suggesting a role for "psychological distance" ("close to me" in terms of goals). Thus, team members possessed higher degrees of SA related to the other team member (i.e. task-work SA) than to the other agents. That could be an evidence that SA-related knowledge is distributed between team members through communication, collaboration and coordination (see also Salmon et al., 2008). Also, it was used a distance measure in terms of territories (or step) covered on the map matched to the correct perception (or relocation; level 1 SA) of these territories by participants. Results showed that the neighbouring agents (both teammate and enemies) were better represented on the map, whereas the teammate locations were better reported than enemies locations at the "step 2" distance. Both psychological and physical distances appear to have a role in the development of individual team members' SA. In order to investigate the relationship between distances and the overall SA, the Euclidean distances were taken into account. An interesting (and unplanned) effect found was that of gender. Indeed, task conditions were homogeneous in terms of subjective task load measures. Probably, all participants were well trained to perform this task. However, females perceived the task as more demanding than males. Results showed different effects for females and males also in their representation of the task: males' SA was higher when it was related to agents (enemies and teammate) that were in close proximity, whereas, no matter who was the agent, distance did not affect females' SA.

Despite the role of gender differences in SA is not yet clear, results reported in this study also suggest that males and females differently use communication in collaborative tasks. Collaboration in collective activities is a prerequisite to successful human social life, yet we still know little about how communications takes place and how understanding comes about during such tasks. Of course, the nature of the communications and mutual understanding achieved can be very different. In Bakthin's conception (1981), the monologic (and dialectic) dimension denies the presence of another's consciousness outside the subjective one (for related critical points, see also Lukács, 1923; Ponzio, 1994). Buber (1950, 1979) conceives three potential alternatives for defining the forms of dialogue:

a. the *genuine* dialogue, where each of the participants really has in mind the other as a separate and real existence, and which aims at establishing an authentic mutual relationship;
b. the *technical* dialogue, which is only directed by the need of objective understanding;
c. the *monologue*, disguised as dialogue, in which each one speaks to the other in indirect ways with no intention of sharing.

In this distinction by Buber different types of dialogues are defined accordingly to the specific focus of the interactive activity and with relation to the nature of consideration of the other. The discourse analysis performed on the chat logs of the sessions showed that SA also can follow similar patterns, where some aspects might be related to gender issues (i.e. the expertise in playing the game), but basically relate to the way the team is organised from the very beginning as a monological or a dialogical team.

Generally, these results could provide further guidelines in systems development for teamwork activities. For example, in designing collaborative interfaces, designers should be aware of the role of the distances between objects in maintaining satisfying degrees of related SA. Also, possible individual differences, expertise levels and communication exchanges should be taken into account in making future team performance well predictable. Concluding, these results are encouraging and show three main advancements in the definition of the SA construct: 1) distance (both psychological and physical) can affect team members SA in networked and command control environment; 2) distance interacts with gender in affecting SA; 3) communication characteristics is affected by gender (or expertise) and it is related to SA (however, it is still unclear what is the direction of this relation). In future studies the stability of these results will be tested using a larger sample as well as developing other techniques to assess the team situation awareness (e.g. by matching the responses of the members of the same team). Furthermore, other types of communication (e.g. chat vs. voice) will be used to investigate their effects on TSA and team readiness in command and control environments. Finally, other techniques to measure the individual cognitive load such as eye movements and heart rate variability (HRV) will be used in order to match psychophysiological responses to different degrees of SA. The final aim of this research activity will be to provide a conceptual model of how team situation awareness and operator cognitive states should be represented and displayed in collaborative technologies.

Acknowledgments

The authors thank Davide Corradini and Amelia D'Arco for their help with data collection procedures, and Fabio Magione for his support in adapting artificial intelligence teams to our experimental purposes.

References

Artman, H. & Garbis, C. (1998). Situation awareness as distributed cognition. In *proceedings of ECCE' 98*, Limerick, Ireland.

Bakhtin, M.M. (1981). *The dialogic imagination: four essays by M.M. Bakhtin*. Austin, TX: University of Texas Press.

Bolia, R.S., Nelson, W.T., Vidulich, M.A., Simpson, B.D., & Brungart, D.S. (2005). Communications research for command and control: Human-machine interface technologies supporting effective air battle management. In *proceedings of the 10th International Command and Control Research and Technology Symposium*. Washington, DC: Command and Control Research Program.

Buber, M. (1950). *Urdistanz und Beziehung*. In *Das Dialogische Prinzip*. Heidelberg: Verlag Lambert Schneider (trad. it. Distanza Originaria e relazione. In Il principio dialogico. Milano: Edizioni di Comunità).

Buber, M. (1979). Ich und Du. In *Das Dialogische Prinzip*. Heidelberg: Verlag Lambert Schneider (trad. it. Distanza Originaria e relazione. In Il principio dialogico. Milano: Edizioni di Comunità).

Cummings, M.L. (2004). *Designing Decision Support Systems for Revolutionary Command and Control Domains*. Doctoral Dissertation, Department of Systems and Information Engineering: University of Virginia.

Endsley, M.R. (1988). Design and evaluation for situation awareness enhancement. In *proceedings of the Human Factors and Ergonomics Society 32nd annual meeting* (pp. 97-101). Santa Monica, CA: Human Factors and Ergonomics Society.

Endsley, M.R. (1995a). Toward a theory of situation awareness in dynamic systems. *Human Factors, 37*, 32-64.

Endsley, M.R. (1995b). Measurement of situation awareness in dynamic systems. *Human Factors, 37*, 65-84.

Endsley, M.R. (1996). Situation awareness measurement in test and evaluation. In T.G. O'Brien, S.G. Charlton (Eds.), *Handbook of Human Factors Testing and Evaluation* (pp. 159–178). Mahwah, NJ: Lawrence Erlbaum Associates.

Endsley, M.R. (2001). Designing for Situation Awareness in Complex System. In *Proceedings of the 2nd international workshop on symbiosis of humans, artifacts and environment*. Kyoto, Japan.

Endsley, M.R. & Jones, W.M. (1997). Situation awareness, information warfare & information dominance. (Technical Report 97-01) Belmont, MA: Endsley Consulting.

Flach, J.M. (1995). Situation Awareness: Proceed with Caution. *Human Factors, 37*, 149-157.

Flach, J.M., Mulder, M., & van Paassen, M.M. (2004). The Concept of the Situation in Psychology. In S. Banbury and S. Tremblay (Eds.), *A cognitive approach to situation awareness: theory and application* (pp. 42-60). Aldershot: Ashgate.

Gorman, J.C., Cooke, N.J., Pederson, H.K., Connor, O.O. & DeJoode, J.A. (2005). Coordinated awareness of situation by teams (CAST): Measuring team situation awareness of a communication glitch. In *proceedings of the Human Factors and Ergonomics 49th annual meeting* (pp. 274-277). Santa Monica, CA: Human Factors and Ergonomics Society.

Knott, B.A., Nelson, W.T., Bolia, R.S., & Galster, S.M. (2006). The impact of instant messaging on team performance, subjective workload, and situation awareness in air battle management. In *Proceedings of the 11th International Command and Control Research & Technology Symposium*. Command and Control Research Program: Washington, D.C.

Lukács, G. (1923). *Frühschriften II: Geschichte und Klassenbewusstsein*. Neuwied, Berlin: Luchterhand, 1968 (= Werke. 2.).

Pew, R.W. (2000). The state of situation awareness measurement: heading toward the next century. In M.R. Endsley and D.J. Garland (Eds.), *Situation Awareness Analysis and Measurement* (pp. 33-50). Mahwah, NJ: Lawrence Erlbaum Associates.

Ponzio A. (1994), *Scrittura dialogo alterità*. Firenze: La nuova Italia.

Potter, J. (1997) Discourse Analysis as a Way of Analysing Naturally Occurring Talk. In D. Silverman (Ed.), *Qualitative Research* (pp. 144–160). London: Sage.

Salas, E., Prince, Baker, D.P., & Shrestha, L. (1995). Situation Awareness in Team Performance: Implications for Measurement and Training. *Human Factors, 37*, 123-136.

Salmon, P., Stanton, S., Walker, G., & Green, D. (2006). Situation awareness measurement: A review of applicability for C4i environments. *Applied Ergonomics, 37*, 225–238.

Salmon, P.M., Stanton, N.A., Walker, G.H., Baber, C., Jenkins, D.P., McMaster, R., & Young, M.S. (2008). What really is going on? Review of situation awareness models for individuals and teams. *Theoretical Issues in Ergonomics Science, 9*, 297-323.

Smith, K. & Hancock, P.A., (1995). Situation awareness is adaptive, externally directed consciousness. *Human Factors, 37*, 137-148.

Stanton, N.A., Stewart, R., Harris, D., Houghton, R.J., Baber, C., McMaster, R., Salmon, P., Hoyle, G., Walker, G.H., Young, M.S., Linsell, M., Dymott, R., & Green, D.A. (2006). Distributed situation awareness in dynamic systems: theoretical development and application of an ergonomics methodology. *Ergonomics, 45*, 1288-1311.

Talamo, A. & Pozzi, S. (accepted). The Tension between Dialogicality and Interobjectivity in Cooperative Activities. *Culture and Psychology*.

Taylor, S. (2001). Locating and conducting discourse analytical research. In M. Wetherell, S. Taylor, and S. Yates (Eds.), *Discourse as data: A guide for analysis* (pp. 5-48). London: Sage/Open University Press.

Wickens, C.D. (2008). Situation Awareness: Review of Mica Endsley's 1995 Articles on Situation Awareness Theory and Measurement. *Human Factors, 50*, 397-403.

Training

TECHNOLOGY-ASSISTED SEARCH

In D. de Waard, J. Godthelp, F.L. Kooi, and K.A. Brookhuis (Eds.) (2009). *Human Factors, Security and Safety* (p. 63). Maastricht, the Netherlands: Shaker Publishing.

Transitional information in simplified mine detection tasks

Chris Baber & Robert J. Houghton
EECE, The University of Birmingham
Birmingham, UK

Abstract

There are three sources of information that the military will use in its search for Improvised Explosive Devices (IEDs) and mines. Visual search is used to scan an area for signs of disturbance and metal detectors are used to scan the area for buried objects. One question that arises from this is, how is visual search supported by tools, such as metal detectors? A third source of information relates to the expectations that the searcher might have for the presence of objects. These expectations often arise from the use of 'pattern of life studies' to recognise unusual behaviour in the environment. In this paper, we consider how the integration of these different representations can create problems and how these forms of search require the development of a mental model of the environment, using different forms of representation. The paper presents simple table-top paradigm to explore the differences in representation when search is performed visually versus when it is performed using a metal-detector.

Introduction

The motivation for these experiments came from a comment made by a training officer at the UK MoD Defence Explosives Munitions Search School to the effect that novice uses of mine-detectors tended to 'follow the tool' they were using and to have a limited focus on the environment around them. This implied that the novices were constraining their search to the area in the immediate vicinity of the tip of the metal detector (or 'grudge') rather than planning their search. This point is illustrated by eye-movement recordings we made of a trainee conducting a search using both a metal detector (figure 1) and his hands (figure 2). The direction of gaze (as indicated by the centre of the cross-hair) was typically close to the point at which the hand or the 'grudge' was located, rather than scanning around the area.

Expectation and search

Further discussion with Subject Matter Experts has led to the assumption that search for explosive devices involves a combination of physical search (possibly using some form of detection equipment) coupled with knowledge of likely areas to search. The first set of knowledge concerns the 'pattern of life' surrounding a given area, e.g., if there is a lack of local people in an area (when it is usually busy) then this might be indicative of something being amiss. More often the interpretation of pattern of life factors is far less simple than this example implies, and can be summarised as a feeling that things are 'not right'. The second set of knowledge concerns the physical appearance of the environment, e.g., signs of disturbance of the ground etc., or recognition of vulnerable points, e.g., places which one might expect to encounter booby-traps or other hazards. The notion of what constitutes a 'likely area' changes with experience and it is not for this paper to elaborate on what this knowledge might be. However, it illustrates that search is likely to be informed by a set of hypotheses concerning what one is likely to find and where one is likely to find it (as opposed to random or opportunistic search). From this perspective, scanning only using the tip of the 'grudge' might limit the development of hypotheses. The question of how search is affected by experience and expectations has been studied in the visual domain for many years (e.g., Mourant and Rockwell, 1972, Kundel and La Follette, 1972) but we are not aware of much work that has applied this work to search with tools such as metal detectors.

Figure 1. The Grudge (held in the right hand) is the focus of the soldier's visual attention

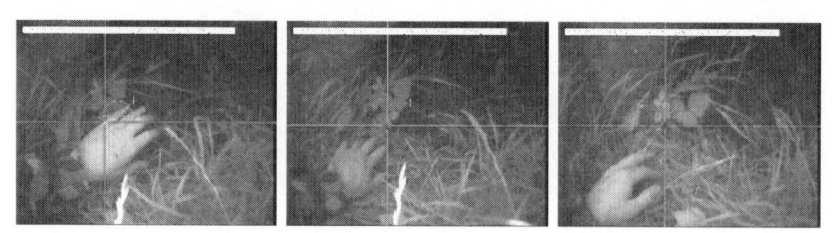

Figure 2. The soldier's visual attention is focused on the left hand as it moves grasses to enable search

Memory and search

Baddeley and Logie's (1999) 'working memory model' suggests that there might be two memory systems involved in visual search: one system being concerned with spatial information, and a separate system being concerned with object identification. Depending on the nature of the search task, one system might be favoured over the other. Thus, Beck et al. (2006), in studies in which search items are modified *during* the search task, have shown that recall seems to be based more on the location of objects than on their identity. The importance of location for a combination of visual search and manual response has been highlighted by Terao et al. (2002), who showed that (gaze) fixation on targets presented sequentially or among distracters enhanced subsequent manual marking of the targets. Parmentier et al. (2005) asked participants to recall sequences of spatial locations and showed that path crossing and path length affected recall and response time. The implication is that path information can be considered analogous to verbal serial memory in that transitional information is required to maintain serial order. From these studies, one might conclude that the location of items was more significant than their identity when performing recall tasks. The reason why such observations might be relevant to search for explosive devices relates to the earlier assertion that search could require a combination of expectations of what is being sought (i.e., identity of both the object and the environment), and where the object is (i.e., location).

Strategies for search for mines

There is a dearth of literature, in the public domain, on the manner in which search is conducted for mines (and improvised explosive devices). Of course, there is good reason why information on this topic might be restricted. However, there are also basic Ergonomics questions that are worth raising, particularly in relation to the development of an appropriate theory to support training and technology development.

Herman and Igelias (1999) note two primary problems with contemporary landmine detection equipment: lack of position feedback for the operator, and lack of a 2D map of detector output. This seems peculiar, given the research into the memory for location in visual search discussed above. In subsequent work, Herman et al. (2001) developed a prototype landmine detector that combines conventional auditory output with a visual display to show a 'map' of this output, and initial trials suggested that this prototype showed improved performance in comparison to conventional equipment.

Staszewski (1999) and Staszewski and Davison (2000) studied the performance of an 'expert' in mine detection. From analysis of this expert's performance, it was found that the expert would use the auditory output from the detector to compose spatial patterns. He would then compare these composed patterns with known patterns that he associated with specific types of mine. This implies that the expert is building a 2D map of the area that contains the landmine, and then interpreting this map against his knowledge of landmine patterns. This work sees landmine detection as a process of translating auditory output into spatial patterns to identify characteristic patterns of specific types of mine. While this does not address the question of how search is managed, it does suggest that the interpretation of auditory output has a spatial component. On the basis of this performance, Staszewski and Davison (2000) developed and implemented a training programme to improve landmine detection performance, and this produced significant improvements (from around 57% to 94%) with only a short (1 hour) period of training.

Aims of the study

We had two aims in conducting this study. First, we wanted to see whether an extremely simple table-top paradigm could be used to capture some of the aspects of search and detection that had been raised by the Subject Matter Experts and previous studies. One motivation for this would be to develop a simple means of introducing trainees to problems relating to the ergonomics of search, prior to more complex and realistic training. The idea is that the table-top paradigm could offer part-task training. Second, we wanted to explore, in a controlled manner, how search performed solely visually differed from search that involved the use of technological assistance. There is a scant literature on how metal detectors are used to support search, and very little that offers a theoretical account as to the relationship between search and technology. Thus, we wanted a simple paradigm that would reflect the cognitive demands of search in a manner that could illustrate the relationship between strategy and expectations.

Approach

Previous laboratory studies, e.g., Cerutti et al. (2000), have used computer-based training with cursor movement being used to scan a search area and auditory feedback to indicate targets. We wished to have a paradigm which could be as convenient as the computer-based study but could include some physical activity in a real environment. The approach in our initial studies used a handheld metal-detector to detect metal targets hidden under identifiable markers arranged in a grid. The advantage of such an approach is three-fold: (i.) it provides a grid pattern against which to check our first assumption, that naïve search might be linear, but that more experienced search might focus on likely locations; (ii.) it provides discrete visual cues that can be tested through recall after the search; (iii.) it provides an opportunity to search with either a metal-detector or (by lifting the markers) visually. The number of items that constituted the search set was kept to three or four. This was primarily because previous research (e.g., Kumar and Jiang, 2005; Luck and Vogel, 1997) had shown that participants were able to retain up to four items when performing search tasks, even if they differ on several properties, such as colour, shape, orientation. We have decided to use playing cards as the markers in these studies. The use of cards to support the study of cognitive ability has a long tradition in cognitive psychology, e.g., Wason (1971).

Experiment 1: Increasing knowledge of the world and visual vs. auditory search

The first experiment required participants to repeat a search task several times, with additional information being given to them on the 'most likely' location for the targets at the start of each search. In this way, one would assume that increasing knowledge of the layout of the environment (from repeated performance of the task) coupled with additional information ought to lead to an increase in performance, but that when the additional information becomes too constraining then bias might creep into performance. The question to be addressed in this experiment was whether a similar pattern of results would occur when the search was guided by visual or auditory feedback.

Fourteen undergraduate students participated in the experiment (9 male, 5 female; average age 19 years). None of the participants had participated in similar studies, nor had they any experience of conducting similar search tasks. This would allow us to consider learning effects through repeating several trials. The experiment was run with two groups: visual and metal detector.

Sixteen playing cards were laid face up on the table. Under each was placed either a coin (three each of: 2p, 10p, 20p) or a plastic counter (so that all cards were raised to a similar height). Participants were instructed that their goal was to pick three cards which would yield the maximum payout (as there were three 20p coins, this ought to 60p for each trial). In each trial, the participant was presented with 'intelligence' on the likely position of the coin. Each trial began with the instructions "There are coins under some of the cards. Your task is to search the cards until you are satisfied that you have found three coins which will give the maximum payout." This 'intelligence' was expanded over the course of the trials as follows:

Trial 1: The coins could be under any card.
Trial 2: The coins are more likely to be under any card whose face value is greater than or equal to 9.
Trial 3: The coins are more likely to be under any card whose face value is greater than or equal to 9, and is not a queen.
Trial 4: The coins are more likely to be under any card whose face value is greater than or equal to 9, is not a queen and not directly above a queen (in the layout).
Trial 5: The coins are more likely to be under any card whose face value is greater than or equal to 9, is not a queen, not directly above a queen (in the layout) and not black.

Participants in the 'visual' search condition were required to lift a playing card, look at the coin underneath and then replace the card. Participants in the 'metal detector' search condition were required to use the handheld metal detector to test cards as to whether or not there was a metal object (coin) underneath it. This means that they had no feedback on the value of the coin during the search. The layout of cards, and placement of coins, was kept constant over all five trials in order to encourage learning of their position. The study produced three sets of results: eye-tracking (using the ASL Mobile-Eye wearable eye tracking system); numbers of cards checked; amount of payout discovered. All searches were timed to take up to 2 minutes and, while participants tended to search more quickly in the later trials, there was no difference in time taken.

Results

Visual search tracking

As this is only a short paper, we will present two stills from the eye-tracking of a participant using the metal detector (figure 3). In the first condition, the fixation point (shown by the solid circle with the cross in its centre) was adjacent to the metal detector. The eye-tracker showed that the fixation point was typically one card ahead of the metal detector. This has interesting implications for the point raised at the start of this paper. In the fifth condition, the fixation point is clearly separate from the tool; the participant would scan the entire set of cards and consider the rules presented to them prior to testing the card.

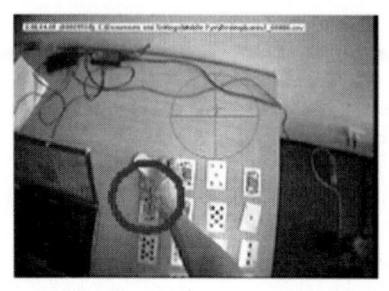

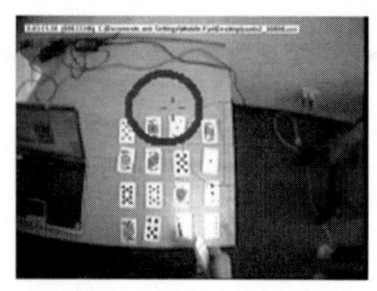

Figure 3. Fixation (shown by thick circle) relative to location of metal detector for low knowledge (left) and high knowledge (right) conditions

The results illustrate a difference between experienced and inexperienced search and can be used to illustrate how knowledge / expectations change search strategy. The less information the participant had, the more likely their search was to be linear, i.e., to be closely coupled to the movement of the metal detector, and the more information they had the more likely they were to follow a more informed strategy. As mentioned above, this interpretation is by no means original or novel. However, it is interesting to note that a common finding from visual search also applies to tool-based search and, as we shall see in the discussion, raises questions as to how search is trained.

Number of cards checked

Table 1 shows that there are differences in the number of cards checked in each condition, but only in the initial trials.

Table 1. Number of cards checked in each condition

Instructions	Visual	Metal
under any card	15.2	22.3
more likely to be under card >=9	6.8	13.1
more likely to be under card >=9 and not queen	5.2	9.3
more likely to be under card >=9, not queen and not directly above queen	7.0	7.3
more likely to be under card >=9, not queen and not directly above queen and not black	4.5	4.4

A two-way Analysis of Variance (ANOVA) of condition (visual vs. metal) and instructions (level 1 to 5) showed significant main effects of condition ($F(1,50) = 15.01$, $p = 0.0003$) and instruction ($F(4,50) = 37.42$, $p = 0.0001$). Post-hoc analysis (using Tukey tests) of condition reveals that there is only a significant difference with the first instructions ($p<0.0005$), and that the subsequent trials, although clearly involving more tests for the metal detector, failed to reach significance in this study.

Payout

The number of cards checked suggests that the detector leads to more checking during the initial trials. This could be due to the relatively low physical cost of moving the detector to the card, in order to perform the check (as opposed to lifting the card) or could be due to the memory load involved in keeping track of where targets have been found. As trials progress, so the number of checks evens out between the conditions. It is not possible to determine whether the decrease in selections over the five sets is due to participants learning and remembering the location of specific cards or whether it is due to their using the instruction set to reduce their choice.

We also consider the average amount of money found on each trial. This was calculated as the proportion of the average amount of money found per condition relative to the maximum amount defined by the rules. Figure 4 shows this analysis graphically and shows a similar pattern of results to the search data, i.e., improving performance over the first three trials, with some difference between the initial conditions. It is, perhaps, particularly interesting to note the difference between visual and detector for the second condition; while the visual condition tends to have a linear improvement over the first three trials, the detector performance drops on the second condition. A possible explanation was that the visual condition simply involved the participant remembering the specific locations of the three coins (and so having to check fewer cards as the experiment progressed) but the detector condition was less able to either retain this information or associate a specific location with a target.

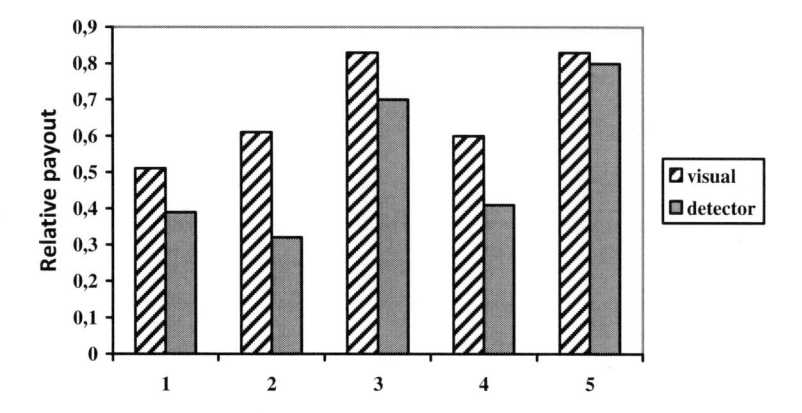

Figure 4. Relative payout in each condition

Conclusions

The higher sampling for the initial conditions with the metal detector can be interpreted in several ways. If there is a higher recency effect for auditory spatial information, then it is possible that such information is quite fragile and requires continual reinforcement in order to establish a reliable trace. This could mean that the information provided by the tones were not processed sufficiently to enable recall. Another explanation was that the relationship between the position of the metal detector and the beep it emitted were not fully reconciled during the initial trials. In other words, the participants tended to follow a scan path that progressed along rows, and to recheck when a beep occurred. This would mean that a set search path was being followed, with interruptions for checking beeps, but that this was not an efficient strategy for learning the location of the targets. As the trials progressed, so participants tended to reduce the number of checks and also to follow a more directed search, although this still tended to employ more checks than the visual condition.

Experiment 2: Memory for location versus memory for target

In this experiment, in addition to identifying the target locations, participants were asked to perform one of two tasks. In both cases, participants were asked to turn away from the table and look at a sheet printed with a grid of boxes in the same arrangement as the cards. One group of participants was asked to point to the locations of the selected targets, the other group was asked to look at the grid and state the names of the cards for the selected targets. Twelve people took part in this study (9 male, 3 female). All were members of the University of Birmingham's Officer Training Corps and none had participated in experiment one. This experiment involved a repeated measures design, in which all participants performed search using metal detector or visual search.

As with the first experiment, 16 playing cards were arranged on the table-top and nine coins placed under the cards. Participants were instructed to find either the maximum or minimum payout that could be made using four of the coins hidden under the cards. The order in which searching was performed was counter-balanced so that either visual search or metal detector was performed first, and either 'high' or 'low' value was sought first. Finally, after reporting their selection, participants were asked to turn away from the table and then report either the location of their selected coins, or the identification of the playing cards covering their selected coins. This means that, rather than continuing to find the same coins (as in experiment one), there was some uncertainty as to which coins to report and what form the report would take.

Results

Number of cards checked

The number of cards checked varied as a result of the search condition, e.g., the metal detector had an average of 19.5 cards per trial and the visual search had an average of 16.4 cards per trial. A two-way ANOVA (condition x trial) did not show any main or interaction effects. The number of cards selected varies across trials, but this variation is not dependent on the technology. This result implies that participants were trying to apply a similar search strategy.

Recall of selected cards

In addition to identifying specific cards, participants were also required to recall the identification of the playing card under which their selected coins were hidden. Participants only need to recall four cards, but there was a difference between conditions (the metal detector conditions had a mean recall of 2.63, and the visual search conditions had a recall of 3.5).

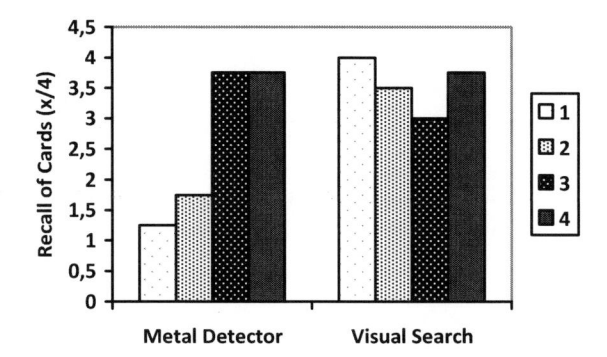

Figure 5. Mean recall of playing card name across trials under each search condition

A two-way ANOVA (condition x trial) showed main effects of condition ($F(1,24) = 8.4$, p=0.0079) and trial ($F(3,24) = 3.257$, p = 0.0391) and a significant interaction effect ($F(3,24) = 7.829$, p=0.0008). This shows some effect of practice, through the effect of trial, but a strong effect of condition (which explains the interaction effect). These results show that the visual search conditions led to superior recall of the identification of the playing cards in comparison with the metal detector conditions, at least in the initial conditions.

Recall of location of selected coins

Participants were also asked to identify the location of the coins. The visual search conditions had a mean recall of 3.85 and the metal detector conditions had a mean recall of 3.5. Two-way ANOVA yielded no significant main effects, which implies that the two conditions had the same performance of recall of location.

Recall of location of all coins

Performance of recall of the location of all coins, was an inverse of the results of the recall of the cards identification. The visual search group correctly identified a mean of 6.1 of the 9 coins, while the metal detector group correctly identified a mean of 7.5 of the 9 coins.

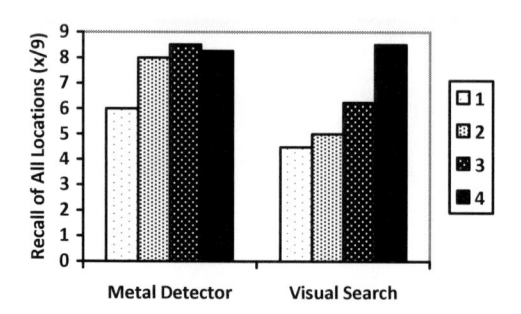

Figure 6. Mean recall of all locations of coins across trials under each search condition

In other words, there was a small, but non-significant effect of condition for recall of location of all coins in favour of the metal detector conditions ($F(1,15) = 6.913$, $p=0.19$). There was also a main effect of trial ($F(3,15) = 5.594$, $p=0.0089$) but no interaction effects.

Conclusions

The results of this study point to the separate search strategies proposed by Baddeley and Logie (1999) in that there is evidence to suggest a spatial search that is similar for the recall of specific items but superior for the metal detector in the recall of all items, and an object search that is superior for visual search.

Discussion

By way of conclusion to the paper, it is possible to use a very simple paradigm to illustrate repeatable differences in search strategy with visual search or metal detector, and with the amount of information provided to the searcher. The implication is that these differences might be less to do with the manner in which the scene is scanned, and more to do with the manner in which information is processed. This suggests that, as argued by Staszewski, the translation from auditory beeps to a spatial pattern is key to both search and recall; that, as argued by Baddeley and Logie, there are two memory systems involved in search; and that search using metal detector or visual search tends to support one or other of these memory systems.

Experiment one showed how additional knowledge can (up to a point) enhance search, and that too much information can constrain search. It also showed that the metal detector condition tends to have more sampling of the scene. Experiment two showed how different memory systems can be supported by difference means of searching the scene. While both media support the discovery of the coins (which might be seen as the primary motive of the search), there were marked differences in terms of the recall of all items or recall of cards' identity.

From these two, admittedly simple studies, we suggest that searching for objects in a scene that can be described according to a grid (such as a mine field), then converting auditory information to a spatial pattern can enhance performance (which agrees with the proposals of Staszewski considered above). However, when searching for objects which lie in a different type of scene (i.e., one which is not so easy to grid), then there is a need to combine feedback from the search with additional information. These studies imply that the combination of this information is not as straightforward as one might anticipate, and that search using a metal detector might bias the search away from visual cues in the scene.

References

Baddeley, A. & Logie, R.H. (1999). Working memory: the multiple component model. In P. Shah and A. Miyake (Eds.) *Models of Working Memory* (pp. 28-61). Cambridge: Cambridge University Press

Beck, M.R., Peterson, M.S., & Vomela, M. (2006). Memory for where, but not what, is used during visual search, *Journal of Experimental Psychology: Human Performance and Perception, 32,* 235-250

Cerutti, D.T., Chelaru, I.M., & Staddon, J.E.R. (2000). Detecting hidden targets: a procedure for studying performance in a mine-detection-like task. In A.C. Dubey, J.F. Harvey, J.T. Broach and R.E. Dugan (eds.) *Detection and Remediation Technologies for Mines and Mine-like Targets V, Proceedings of SPIE, 4038.* (pp. 102-109). Bellingham, WA: SPIE

Herman, H. & Iglesias, D. (1999). Human-in-the-loop issues for demining, In A.C. Dubey, J.F. Harvey, J.T. Broach, and R.E. Dugan (Eds.) *Detection and Remediation Technologies for Mines and Mine-like Targets IV, Proceedings of SPIE, 3710* (pp. 797-805). Bellingham, WA: SPIE,

Herman, H., McMahill, J.D., & Kantor, G. (2001). Enhanced operator interface for hand-held landmine detector. In A.C. Dubey, J.F. Harvey, J.T. Broach, and R.E. Dugan (Eds.) *Detection and Remediation Technologies for Mines and Mine-like Targets VI, Proceedings of SPIE, 4394* (pp. 844-851), Bellingham, WA: SPIE

Kundel, H.L. & La Follette, P.S. (1972). Visual search patterns and experience with radiological images, *Radiology, 103,* 523-528

Kumar, A. & Jiang, Y. (2005). Visual short-term memory for sequential arrays, *Memory and Cognition, 33,* 488-498

Luck, S.J. & Vogel, E.K. (1997). The capacity of working memory for features and conjunctions, *Nature, 390,* 279-281

Mourant, R.R. and Rockwell, T.H. (1972). Strategies of visual search by novice and experienced drivers, *Human Factors, 14,* 325-336

Parmentier, F.B.R., Elford, G., & Mayberry, M. (2005). Transitional information in spatial serial memory: path characteristics affect recall performance, *Journal of Experimental Psychology: Learning, Memory and Cognition, 31,* 412-427

Staszewski, J.J. (1999). Information processing analysis of human land mine detection skill, In A.C. Dubey, J.F. Harvey, J.T. Broach and R.E. Dugan (eds.) *Detection and Remediation Technologies for Mines and Mine-like Targets IV, Proceedings SPIE 3710* (pp. 766-777). Bellingham, WA: SPIE.

Staszewski, J.J. & Davison, A. (2000). Mine detection training based on expert skill. In A.C. Dubey, J.F. Harvey, J.T. Broach, and R.E. Dugan (Eds.) *Detection and Remediation Technologies for Mines and Mine-like Targets V, Proceedings of SPIE, 4038* (pp. 90-101). Bellingham, WA: SPIE

Terao, Y., Andersson, N.E.M., Flanagan, J.R., & Johansson, R.S. (2002). Engagement of gaze in capturing targets for future sequential manual actions, *Journal of Neurophysiology, 88,* 1716-1725

Tremblay, S., Parmentier, F.B.R., Guerard, K., Nicholls, A.P., & Jones, D.M. (2006). A spatial modality effect in serial memory, *Journal of Experimental Psychology: Learning, Memory and Cognition, 32,* 1208-1215

Wason, P.C. (1971). Natural and contrived experience in a reasoning problem, *Quarterly Journal of Experimental Psychology, 23,* 63–71

Cognitive and personality variables of operators: individual characteristics and process control performance

Dina Burkolter, Annette Kluge, & Matthias Brand
University of Duisburg-Essen
Duisburg, Germany

Abstract

Two studies analysing the relationships between operator characteristics and process control performance were conducted in order to replicate and extend earlier research by Burkolter et al. (2007, 2009). The first study (N = 41) examined the link between general mental ability (GMA), cognitive flexibility and conscientiousness on the one hand, and performance in a simulated process control task on the other. GMA correlated significantly with diagnostic performance and knowledge, while cognitive flexibility and conscientiousness were not related to performance. The second study (N = 50) incorporated working memory capacity, set-shifting and decision-making as well as the personality variables Need for Cognition (NFC) and perfectionism. Working memory capacity and set-shifting were related to diagnostic performance, while decision-making (risky decisions) was associated with both system control and diagnostic performance. Moreover, NFC and declarative knowledge were significantly correlated. Confirming earlier results, GMA exhibited relations to diagnostic performance and declarative knowledge. Practical implications as well as implications for future research are discussed.

Introduction

The present studies aim to continue and extend previous research results, which addressed relationships between individual characteristics and performance in early stages of skill acquisition (see Burkolter et al., 2007, 2009; Kluge et al., submitted).

Altogether, these studies form part of a comprehensive research programme that links cognitive requirements of process control tasks to training. When designing training, it is necessary to analyse the cognitive requirements of a task (e.g. through cognitive task analysis) in order to derive training objectives and criteria (Burkolter et al., 2007). However, the amount of learning and the training results might be limited by individual characteristics. Therefore, the overriding research question focuses on which aspects of process control performance are affected by training, and how training can best support learning and performance with respect to individual characteristics. Our underlying assumption is that process control performance is composed of a dynamic interplay between cognitive processes and

In D. de Waard, J. Godthelp, F.L. Kooi, and K.A. Brookhuis (Eds.) (2009). *Human Factors, Security and Safety* (pp. 77 - 88). Maastricht, the Netherlands: Shaker Publishing.

individual characteristics. Therefore, before developing training scenarios and methods, it is necessary to investigate the elements of a process control task that are affected by person-related variables and address the question of which of these are trainable.

Our earlier research showed that general mental ability (GMA) was positively related to system control performance and declarative knowledge (knowledge about the system and the relationships between parameters) as well as to diagnostic performance of novel faults, i.e. not previously encountered system faults. These findings confirmed the results of earlier studies (e.g. Schmidt & Hunter, 1998; Morris & Rouse, 1985). Furthermore, cognitive flexibility was found to be a good predictor of process control performance. Cognitive flexibility as a construct distinguishes between a reductive and an expansive cognitive style. Individuals with high cognitive flexibility have a flexible and situation-adaptive assembly of knowledge, while those with low cognitive flexibility prefer simplicity and rigid prescriptions from memory (Spiro et al., 1996). Interestingly, the effect of cognitive flexibility was found to occur in different directions. Operators with high cognitive flexibility performed best on diagnostic performance, whereas operators with low cognitive flexibility performed best on system control. GMA, cognitive flexibility and declarative knowledge accounted for about 30% of the variability of system control. A second study (Kluge et al., submitted) with a similar sample showed comparable results, and corroborated the impact of cognitive flexibility on process control performance. Finally, regarding personality traits, no correlation was found between conscientiousness and process control performance (Burkolter et al., 2009), despite the fact that a meta-analysis by Barrick and Mount (1991) had suggested such a relationship.

Although these findings demonstrated the relevance of operator characteristics in terms of explaining variance in performance, they also uncovered the need to (possibly) confirm previous results and address individual differences more specifically. We therefore sought to extend this type of research by investigating the interplay between GMA, cognitive style and task-related personality traits in different samples, for instance with regard to age and experience. These differing samples were selected based on the insight that in order to support operators most effectively, training needs to be tailored according to the trainees' education, experience and abilities.

Moreover, in addition to the basic constructs such as GMA, cognitive style and personality traits, there is a need to examine more specific cognitive variables and personality traits, which might explain a higher amount of variance in complex task performance than the basic concepts. Based on empirical findings summarised below, this study began by addressing Need for Cognition (NFC), perfectionism, working memory capacity, and set-shifting performance as a promising set of individual characteristics that are expected to have an impact on process control performance: NFC refers to "the tendency for an individual to engage in and enjoy thinking" (Cacioppo & Petty, 1982, p. 116). In a recent study by Day et al. (2007), NFC was shown to be related to performance of a complex skill. Facets of perfectionism such as concern over mistakes and personal standards were found to be related to decision-making (Brand & Altstötter-Gleich, 2008), and working memory capacity was associated with performance in complex systems characterised by a high amount of interconnectivity (e.g. Wittmann & Hattrup, 2004). Set-shifting (or categorisation) as a sub-component of executive functions, is assumed to be related to process control performance, as it has been found to be relevant for problem-solving (Hockey, 1990). Finally, decision-making under uncertainty was included in the study as an important factor in dealing with complex systems (e.g. Wickens et al., 1998).

The setting

Two studies were conducted in order to investigate the relationships between cognitive and personality variables and process control performance. Both studies involved a simulated process control environment called Cabin Air Management System (CAMS), which has already been employed in a number of previous studies (for details, see Sauer et al., 2000). By employing a simulation, we chose a compromise between experimental control and realism, under the assumption that the simulation represents the essential characteristics of a real-world environment (Gonzalez et al., 2005; Sauer et al., 2008). CAMS simulates the automated life-support system of a spacecraft, but its underlying principles correspond to a process control task. Accordingly, the two main tasks of the operator are to maintain five parameters in a predefined zone and to intervene in the case of a system fault. System control performance was measured as the duration with which one or more of the parameters lay outside of the predefined zone, converted into percentages. Diagnostic performance included the two measures diagnostic errors (percentage of incorrect fault diagnoses) and diagnosis time (required time for correct fault diagnosis in seconds). Knowledge was assessed using a written questionnaire that contained items relating to the parameters, processes and relationships in CAMS (declarative knowledge) as well as items regarding fault descriptions, fault symptoms and repair steps (procedural knowledge). The knowledge test was composed of multiple-choice items as well as open questions (for details see Sauer et al., 2008; Kluge et al., submitted).

The procedure was similar in both studies. Cognitive and personality variables were assessed before the training session either through questionnaires or using computer-based tests. Participants in both studies were trained in the use of CAMS for several hours, including an introduction by means of multimedia-based instructions and a main training session on the control and diagnosis of five system faults. Training was carried out in groups of typically five to ten participants, supervised by one instructor.

Study 1

The goal of the first study was to replicate earlier findings and examine the relationship between operator characteristics and performance more deeply in different samples, i.e. by inviting participants of different ages and with different levels of education.

Participants and methods

Forty-one engineering students (four female) from universities of applied sciences in Switzerland took part in the study. The mean age of the participants was 24.7 years (SD = 4.0), with ages ranging from 19 to 35 years. They were paid approximately € 60 (100 CHF) for participation.

At the beginning of the training session, which lasted five hours in total, different questionnaires were administered (approx. 30 minutes). Cognitive flexibility was measured using the 'Cognitive Flexibility Inventory' (Spiro et al., 1996), GMA using the 'Wonderlic Personnel Test' (Wonderlic Inc, 2002), and items measuring conscientiousness were extracted from Saucier's (1994) 'Big Five Markers'. Following this, participants were introduced to the CAMS task through multimedia-based instructions, and were provided with training and time to practise the process control task. Results are reported for the test (70 min) that was conducted two weeks after the first session. Knowledge was assessed using a paper-and-pencil questionnaire.

Results

Descriptive statistics on all performance variables as well as personality variables are provided in Table 1. While there were significant relationships between GMA and performance, cognitive flexibility and conscientiousness were not significantly associated with process control (see Table 2).

Table 1. Descriptive statistics on performance and personality variables

Variables	M	SD
Process control performance		
System control failures (%)	27.05	9.36
Diagnostic errors (%)	53.66	21.89
Diagnostic speed (s)	330.07	66.91
Knowledge (max. 14 pts)	6.76	1.97
Personality variables		
GMA[1] (max. 50 pts)	25.61	5.65
Cognitive flexibility (–4 to +4)	0.77	1.07
Conscientiousness (1-9)	6.71	1.09

Notes: N = 41, [1]General mental ability

Table 2. Correlations between personality and performance variables

Performance measures	GMA[1]	Cognitive flexibility	Conscientiousness
Process control performance			
System control failures	–.28	–.27	–.31
Diagnostic errors	–.47**	–.16	–.21
Diagnostic speed	–.43**	–.12	–.11
Knowledge	.33*	.15	.03

Notes: N = 41; *$p < .05$, **$p < .01$; [1]General mental ability

GMA was significantly correlated with both measures of diagnostic performance (diagnostic errors and diagnostic speed, see Table 2) as well as acquired knowledge.

The negative correlations indicate that as a tendency, the higher the participants' GMA scores, the better their performance. Neither cognitive flexibility nor conscientiousness was significantly related to process control performance.

Summary

In part, the present results confirmed earlier findings discussed in the Introduction. In line with previous research, GMA was significantly correlated with diagnostic performance and knowledge. Similarly to other work environments (Schmidt & Hunter, 1998), GMA also supports performance with complex systems. As in the previous study, conscientiousness was not significantly related to process control performance. However, there were also differences compared to the previously obtained results, for instance regarding cognitive flexibility. While previous research showed a relationship between cognitive flexibility and process control performance, these findings were not confirmed in the present study.

Study 2

The purpose of the second study was to include additional variables in order to assess their incremental predictive qualities in terms of process control performance. The supplementary variables were derived from studies from neuropsychology and cognitive science, especially research concerning executive functions. Working memory capacity, set-shifting performance and decision-making, assessed with computer-based tests, were included as cognitive variables. As personality variables, NFC and perfectionism were additionally incorporated. The cognitive variables as well as the personality variables were shown in earlier research to be related to performance in complex systems.

Participants and methods

Fifty engineering students (26 female) from a university in Germany took part in the study. Their mean age was 23.6 years (SD = 2.9), with ages ranging from 20 to 34 years. Participants were paid 100 EUR to attend the experiment, which was held over three sessions.

At the first session, person-related variables were assessed (approx. one hour). Working memory capacity was assessed using the 'n-back' task with one-digit numbers, presented in a random sequence. In this n-back task, the participant monitors a series of numbers and is required to indicate whenever a number is presented that was presented two or three trials previously (see Owen et al., 2005). Ten stimulus blocks (two practice blocks and eight experimental blocks) with 24 stimulus trials each were given, alternating between 2-back and 3-back conditions (only 2-back data were analysed here; see Schoofs et al., 2008). Set-shifting performance was measured by the 'Modified Card Sorting Test' (Nelson, 1976), in which the participant is required to sort a deck of cards according to a particular rule. Feedback regarding whether the correct rule has been applied is given after each card has been sorted. To assess the participants' decision-making, the 'Game of Dice Task' (Brand et al., 2005) was employed. In this game, a fictitious starting capital has to be increased within 18 throws of a die by choosing numbers or a combination of numbers that are related to different winning probabilities (i.e. gains or losses). All of the assessments mentioned were carried out using computer-based tasks. Furthermore, different questionnaires on personality were administered: Perfectionism was measured with the German version of the 'Multidimensional Perfectionism Scale' (Altstötter-Gleich & Bergemann, 2006), 'Need for Cognition' was examined using the German version of the 'Rational-Experiential Inventory' (Keller et al., 2000), cognitive flexibility with the 'Cognitive Flexibility Inventory' (Spiro et al. 1996), GMA with the 'Wonderlic Personnel Test' (Wonderlic Inc, 2002), and conscientiousness by Saucier's (1994) Big Five Markers.

At the second session, participants were introduced to the experimental task and provided with training and some time to practise (approx. four hours). The training was followed by a first immediate testing session and a second testing session one week later (both 70 min). Knowledge was assessed at the second testing session (Sauer et al., 2008; Kluge et al., submitted). Results are reported from the second testing session one week later.

Results

Compared to Study 1, participants in Study 2 scored similarly on GMA but differed with regard to cognitive flexibility (see Table 3). They also differed in terms of their absolute process control task performance. The overall pattern of results revealed significant relationships between GMA as well as NFC and process control performance (see Table 4) as well as significant correlations between working memory capacity, set-shifting performance, decision-making and process control (see Table 5).

Table 3. Descriptive statistics on performance, personality and cognitive variables

Variables	M	SD
Process control performance		
System control failures (%)	44.78	15.76
Diagnostic errors (%)	45.33	25.65
Diagnostic speed (sec)	333.31	63.37
Knowledge (max. 33 pts)	15.75	5.59
Personality variables		
GMA (max. 50 pts)	25.78	6.28
Cognitive flexibility (–4 to +4)	–2.39	0.38
Conscientiousness (1-9)	5.95	1.30
Need for cognition (1-7)	5.27	0.55
Perfectionism[2] (1-6)	1.90	0.95
Cognitive variables		
Working memory: Correct (%)	81.77	15.30
Set-shifting: Correct (max. 48)	40.50	7.11
Decision-making: Risky decisions (max. 18)	1.20	1.87

Notes: N = 50, [1]General mental ability; [2]subfacet: doubts about actions

GMA was significantly correlated with diagnostic performance (diagnostic errors, see Table 4) and system knowledge. Working memory capacity was significantly related to both measures of diagnostic performance. This negative correlation indicates as a tendency that participants scoring high on the working memory test committed fewer diagnostic errors and needed less time to diagnose a fault state correctly.

Table 4. Correlations between personality variables and process control performance

Performance measures	GMA[1]	Cognitive flexibility	Conscientiousness	Need for cognition	Perfectionism[2]
Process control performance					
System control failures	−.20	.13	−.11	−.14	−.01
Diagnostic errors	−.28*	−.16	−.08	−.21	.19
Diagnostic speed	−.14	−.18	−.15	−.21	.20
Knowledge	.43**	.12	.15	.40**	−.27

Notes: N = 50; *$p < .05$, **$p < .01$; [1]General mental ability; [2]subfacet: doubts about actions

Cognitive flexibility was not significantly related to process control performance, but the tendency was similar to results obtained in earlier research (see Introduction). Regarding NFC, a significant relationship with system knowledge was found, but not with the other performance measures. Again, conscientiousness was not associated with the performance criteria. Finally, perfectionism was not significantly related to process control performance, except for a marginally significant relationship between knowledge and the perfectionism facet 'doubts about actions' ($r = -.27$, $p = .058$).

Table 5. Correlations between cognitive variables and process control performance

Performance measures	Working memory: Correct	Set-shifting: Correct	Decision-making: Risky decisions
Process control performance			
System control failures	−.14	−.08	.32*
Diagnostic errors	−.34*	−.36*	.30*
Diagnostic speed	−.30*	−.23	.30*
Knowledge	.16	.40**	−.23

Notes: N = 50; *$p < .05$, **$p < .01$

Regarding set-shifting performance, there was a significant relationship with diagnostic performance in terms of diagnostic errors as well as knowledge. Significant correlations were observed between decision-making and both system control and diagnostic performance (diagnostic errors and diagnostic speed, see Table 5). Thus, operators making more risky decisions tended to perform lower in terms of process control.

Hierarchical regression analyses were carried out in order to reveal whether, and how much more, variance in performance could be explained by several variables together compared to a single variable alone. The regression analyses were conducted with the variables that correlated most highly with the performance measures chosen as predictors. Thus, two regression analyses were conducted with diagnostic errors and diagnostic speed as criteria and working memory capacity as well as decision-making as predictors (see Table 6). Working memory and decision-making were both significant predictors of diagnostic performance. Working memory accounted for 11% of variation in diagnostic errors and 9% of variation in diagnostic speed. Overall, 22% of the variation in diagnostic errors and 20% of variation in diagnostic speed was explained by the two predictors working memory capacity and decision-making.

Table 6. Summary of hierarchical regression analyses with diagnostic errors and diagnostic speed as criteria

Predictors of diagnostic errors	B	SE B	β	Significance
Step 1				
Working memory	−0.56	0.23	−.34	.019
Step 2				
Working memory	−0.62	0.22	−.38	.007
Decision-making: Risky decisions	4.52	1.76	.34	.014
			$R^2 = .11$ for step 1. $\Delta R^2 = .11$ for step 2	
Predictors of diagnostic speed				
Step 1				
Working memory	−1.23	0.59	−.30	.041
Step 2				
Working memory	−1.40	0.56	−.34	.017
Decision-making: Risky decisions	11.09	4.54	.33	.018
			$R^2 = .09$ for step 1. $\Delta R^2 = .11$ for step 2	

A third hierarchical regression analysis was performed with acquired knowledge as a criterion and GMA and NFC as predictors (see Table 7). GMA accounted for 19% of the variation in knowledge alone, but including NFC made it possible to explain a further 8%, so that in total, 27% of the variation was accounted for by these two predictors alone.

Table 7. Summary of hierarchical regression analysis with knowledge as a criterion

Predictors of knowledge	B	SE B	β	Significance
Step 1				
GMA	0.39	0.12	.43	.002
Step 2				
GMA	0.30	0.12	.34	.015
Need for cognition	3.02	1.36	.30	.032

Notes: $R^2 = .19$ for step 1. $\Delta R^2 = .08$ for step 2

Summary

There were several results that go beyond earlier findings, such as the positive relationship between NFC and knowledge. Furthermore, GMA and NFC were significant predictors of knowledge, and accounted for 27% of variance in knowledge acquisition. Confirming earlier results, Study 2 showed that GMA was related to the correct diagnosis of system faults as well as the amount of acquired knowledge.

As for the cognitive variables, working memory capacity and decision-making (i.e. risky decisions) were related to diagnostic performance, and moreover, they were significant predictors of diagnostic performance. 20% of the variation in diagnostic performance could be explained by working memory and decision-making. In view of all possible influences on performance (e.g. experience or work design), the amount of variance explained by just a few person-related variables is considerable.

As in Study 1, there were no significant relationships either between cognitive flexibility and the performance measures or between conscientiousness and performance, even though conscientiousness has been shown to be a strong predictor of job performance for a range of occupations (Barrick & Mount, 1991) in general. However, this does not seem to be the case in terms of process control.

Discussion

The goal of the conducted studies was to continue and extend research on the relationships between individual characteristics and process control performance in order to gain more insights for personnel selection and training design. One result emerged as the most stable across the different samples: GMA has consistently been shown to be related to process control performance, especially knowledge, independently of the sample. This finding is in line with research by Schmidt and Hunter (1998) or Morris and Rouse (1985).

The inclusion of executive functions such as working memory capacity and set-shifting performance revealed additional interesting results. Working memory capacity and set-shifting performance were found to be of relevance for diagnostic performance and knowledge acquisition, but not for system control performance. Future research should investigate the relationships between executive functions and process control further by integrating other executive functions such as goal-setting, selection and implementation of strategies or control of actions (see Hockey, 1990). Various selection criteria such as problem-solving abilities or communication skills are suggested by the IAEA (2002), but the assessment of executive functions such as set-shifting performance or working memory capacity is not among the recommendations. It might be of interest for personnel selection in process industries that the computer-based tools employed to assess working memory capacity and set-shifting performance are relatively easy and quick to administer. Furthermore, knowledge on individual characteristics might be employed to tailor training appropriately to trainees, for example with CBT, which can be matched to different learning and cognitive styles by presenting information accordingly (cf. Russell, 1997).

The study findings give rise to the suggestion of examining the relationships between individual characteristics and performance closely regarding age or experience of individuals. Significant relationships between process control performance and cognitive flexibility were found for apprentices, but not for students (although apprentices and students were comparable regarding GMA, see also Burkolter et al., 2007). A possible explanation for this finding could be that the cognitive style may be relevant for apprentices to guide their learning and application process, while this might be less relevant for students, who may be more aware of the advantages and disadvantages of their cognitive styles and might have acquired compensation strategies. Although cognitive styles are seen as fixed attributes, individuals can nevertheless "develop strategies to maximise their strength and minimise weaknesses once aware of their own style" (Russell, 1997, p. 208). However, as age and education (apprentices vs. students) are confounded in this case, it cannot be determined whether age or education in particular might have influenced the observed difference.

To put the findings in an appropriate context, it has to be considered that the studies involved engineering students and a simulated process control task instead of experienced operators in a real work setting. This might limit the validity of the findings for process industries. Particularly in light of the fact that our first studies regarding operator characteristics and the employed task showed differences between samples, further research with more experienced operators from process industries is needed in order to investigate under what conditions findings also apply to more experienced workers. However, even experienced operators start out by learning the fundamentals and building up a generic mental model on how a plant works, upon which, for example, they later build their situational model (Vicente et al., 2004). The situational model then forms expectations regarding how the plant is going to act or react. Studying performance in the early stages of skill acquisition thus holds special importance for supporting operators.

References

Altstötter-Gleich, C. & Bergemann, N. (2006). Testgüte einer deutschsprachigen Version der Mehrdimensionalen Perfektionismus Skala von Frost, Marten, Lahart und Rosenblate (MPS-F). [Statistics and validity of a German version of the Multidimensional Perfectionism Scale of Frost, Marten, Lahart and Rosenblate (MPS-F)]. *Diagnostica, 52,* 105–118.

Barrick, M.R. & Mount, M.K. (1991). The Big Five personality dimensions and job performance: A meta-analysis. *Personnel Psychology, 44,* 1–26.

Brand, M. & Altstötter-Gleich, C. (2008). Personality and decision-making in laboratory gambling tasks – evidence for a relationship between deciding advantageously under risk conditions and perfectionism. *Personality and Individual Differences, 45,* 226–231.

Brand, M., Fujiwara, E., Borsutzky, S., Kalbe, E., Kessler, J., & Markowitsch, H.J. (2005). Decision-making deficits of Korsakoff patients in a new gambling task with explicit rules: associations with executive functions. *Neuropsychology, 19*, 267–277.

Burkolter, D., Kluge, A., Sauer, J., & Ritzmann, S. (2009). The predictive qualities of operator characteristics for process control performance: The influence of personality and cognitive variables. *Ergonomics, 52*, 302-311.

Burkolter, D., Kluge, A., Schüler, K., Sauer, J., & Ritzmann, S. (2007). Cognitive requirement analysis to derive training models for controlling complex systems. In D. de Waard, G.R.J. Hockey, P. Nickel, and K.A. Brookhuis (Eds.), *Human Factors Issues in Complex System Performance* (pp. 475–484). Maastricht: Shaker Publishing.

Cacioppo, J.T. & Petty, R.E. (1982). The need for cognition. *Personality and Social Psychology, 42*, 116–131.

Day, E.A., Espejoa, J., Kowollka, V., Boatmana, P.R., & McEntire, L.E. (2007). Modeling the links between need for cognition and the acquisition of a complex skill. *Personality and Individual Differences, 42*, 201–212.

Gonzalez, C., Vanyukov, P. & Martin, M.K. (2005). The use of microworlds to study dynamic decision making. *Computers in Human Behavior, 21*, 273–286.

Hockey, G.R.J. (1990). Styles, skills and strategies: Cognitive variability and its implications for the role of mental models in HCI. In D. Ackermann and M.J. Tauber (Eds.), *Mental models and human-computer interaction 1* (pp. 113–129). North-Holland: Elsevier.

Keller, J., Bohner, G. & Erb, H.-P. (2000). Intuitive und heuristische Urteilsbildung - verschiedene Prozesse? Präsentation einer deutschen Fassung des «Rational-Experiential Inventory» sowie neuer Selbstberichtskalen zur Heuristiknutzung. [Intuitive and heuristic decision making – different processes? Presentation of a German version of the 'Rational-Experiential Inventory' and new self report scales for heuristic use]. *Zeitschrift für Sozialpsychologie, 31*, 87–101.

Kluge, A., Ritzmann, S., Burkolter, D., & Sauer, J. (submitted). The interaction of drill and practice and error training with individual differences.

Morris, N.M. & Rouse, W.B. (1985). Review and evaluation of empirical research in troubleshooting. *Human Factors, 27*, 503–530.

Nelson, H. (1976). A modified card sorting test sensitive to frontal lobe defects. *Cortex, 12*, 313–324.

Owen, A.M., McMillan, K.M., Laird, A.R., & Bullmore, E. (2005). N-back working memory paradigm: A meta-analysis of normative functional neuroimaging studies. *Human Brain Mapping, 25*, 46–59.

Russell, A.J. (1997). The effect of learner variables cognitive style on learning performance in a vocational training environment. *Educational Psychology, 17*, 195–208.

Saucier, G. (1994). Mini-Markers: A brief version of Goldberg's unipolar big-five markers. *Journal of Personality Assessment, 63*, 506–516.

Sauer, J., Burkolter, D., Kluge, A., Ritzmann, S., & Schüler, K. (2008). The effects of heuristic rule training on operator performance in a simulated process control environment. *Ergonomics, 51*, 953–967.

Sauer, J., Wastell, D.G., & Hockey, G.R.J. (2000). A conceptual framework for designing microworlds for complex work domains: A case study of the Cabin Air Management System. *Computers and Human Behavior, 16*, 45–58.

Schmidt, F.L. & Hunter, J.E. (1998). The validity and utility of selection methods in personnel psychology: Practical and theoretical implications of 85 years of research findings. *Psychological Bulletin, 124*, 262–274.

Schoofs, D., Preuss, D. & Wolf, O.T. (2008). Psychosocial stress induces working memory impairments in an *n*-back paradigm. *Psychoneuroendocrinology, 33*, 643-653.

Spiro, R. J., Feltovich, P.J., Coulson, R.L. (1996). Two epistemic world-views: prefigurative schemas and learning in complex domains. *Applied Cognitive Psychology, 10*, 51–61.

Vicente, K.J., Mumaw, R.J., & Roth. E.M. (2004). Operator monitoring in a complex dynamic work environment: a qualitative cognitive model based on field observations. *Theoretical Issues in Ergonomic Science, 5*, 359–384.

Wickens, C.D., Gordon, S.E., & Liu, Y. (1998). *An introduction to human factors engineering*. New York: Addison Wesley.

Wittmann, W.W. & Hattrup, K. (2004). The relationship between performance in dynamic systems and intelligence. *Systems Research and Behavioural Science, 21*, 393–409.

Wonderlic (2002). *Wonderlic Personnel Test*. Libertyville: Wonderlic Inc.

The use of simulator training for train drivers in Europe – an overview and new approaches in 2TRAIN

Christian Maag & Marcus Schmitz
Center for Traffic Sciences
University of Würzburg
Germany

Abstract

In recent years, driving simulation has become one of the most effective methods for the training of train drivers. Nevertheless, the usage of simulators for training often does not take advantage of all possible capabilities. The European project 2TRAIN aims at creating software modules for already existing train driving simulators (Rule-based Expert System, Virtual Instructor, and Assessment Database). These add-on modules allow a more objective, transparent, and detailed assessment of the competence of train drivers.

To get an overview of the existing simulation systems in Europe 18 railway undertakings were interviewed and asked to give detailed information concerning their training concept and technology (research sample: 138 simulator systems).

One of the main findings was that the training concepts of different companies as well as the technical characteristics and capabilities of the simulation systems are very diverse (e.g. real cab replica, small driver desks, software interfaces). The paper presents several aspects of this diversity including the application fields of different types of simulators and a comparison with other training methods. Shortcomings and problems in the application of the current simulator technology are reported. Finally, recommendations for a further technical improvement of simulation technology and training concepts are given.

Introduction

The establishment of safe, competitive, and interoperable railways in Europe requires particular attention to the qualification of train drivers. In order to strengthen a European harmonisation in rail traffic, it is appropriate to advance common training concepts as well as common training technology. Developments concerning these two aspects are the scientific and technological objectives of the rail research project 2TRAIN funded by the European Commission under the Sixth Framework Programme (http://www.2train.eu).

In D. de Waard, J. Godthelp, F.L. Kooi, and K.A. Brookhuis (Eds.) (2009). *Human Factors, Security and Safety* (pp. 89 - 99). Maastricht, the Netherlands: Shaker Publishing.

The railway systems of all European countries differ in signalling systems, engine technologies, rule books, legal conditions, and training structures. As a consequence of this diversity, a complete harmonisation of training technology and training contents is unachievable. Nevertheless, as Europe grows together and cross-border operations increase there is a strong need to adjust and coordinate the education and training of train drivers concerning general driving and operational abilities as well as particular crisis management competencies. Taking into account experiences made in the past, 2TRAIN aims at developing European best-practice solutions for an efficient, safety-enhancing, and cost-effective use of modern technologies for the training as well as for the ongoing competence and performance assessment of train drivers (Schmitz & Maag, 2007).

Important benefits can be obtained by the use of computer-based systems and simulators in training (Rail Safety & Standards Board, 2007). These computer-based technologies – and especially simulation – facilitate the establishment of common training efforts for train drivers in Europe and enable an enhancement of training efficiency by using realistic and interactive training exercises (European Commission, 2007). In contrast to training in the real environment, the training session can be composed of different technical failures and abnormal operational situations of which many cannot be trained in a real setting. Furthermore, it is possible to replicate scenarios at any given time. Many safety-related and economic advantages of simulation have led to the wide dissemination of this training method. In the near future, railway companies will even intensify the usage of simulators for the training and assessment of train drivers.

A fundamental objective of 2TRAIN is the benchmarking of training concepts and training technologies already in use in different European railway companies. The benchmarking process aims at describing the present situation of computer-based training in different companies and the settings under which the training is applied. The benchmarking methodology and selected key findings are presented and discussed in the following sections (see also Schmitz & Maag, 2008).

Methods

Benchmarking process

For the purpose of benchmarking the state of the art concerning training technology (e.g. simulator hardware, architecture, and software) and training contents was analysed by reviewing the relevant publications. Subsequently, data about training systems of European railway undertakings were collected with a special focus on simulation and training contents related to driving under abnormal and irregular conditions. By these two steps, the current status and quality of train driver training was evaluated and the requirements for future computer-based training concepts were specified.

Figure 1 illustrates the stages of the benchmarking process in the frame of 2TRAIN.

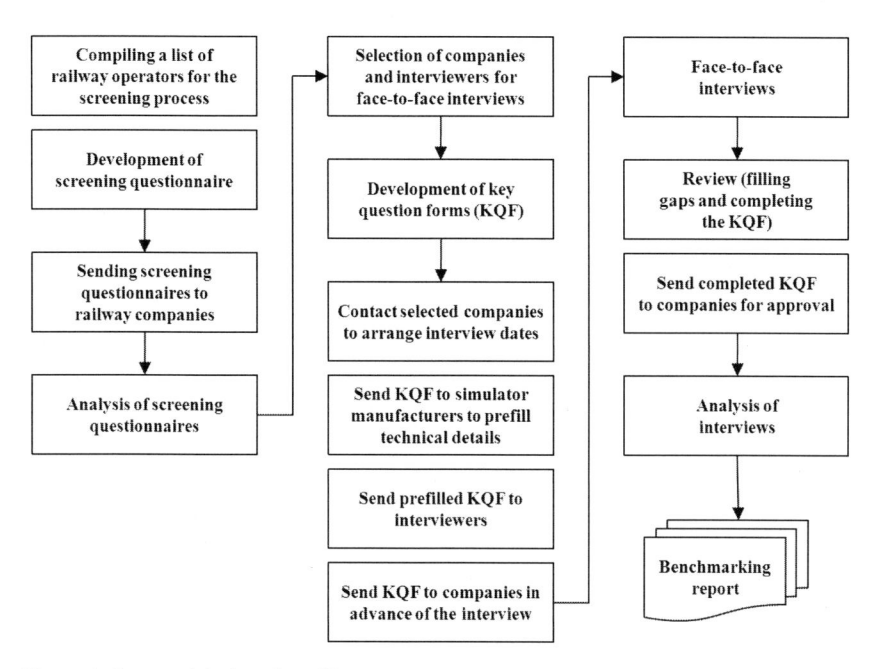

Figure 1. Stages of the benchmarking process

The data gathering itself was a central part of the benchmarking and followed a two-step approach. The first step consisted of a screening questionnaire sent to 75 companies in more than 20 European countries. This screening questionnaire aimed at giving a preliminary look on the usage of simulation and other e-learning tools at European railway companies. Based on the results of this first questionnaire, a sample of 18 railway companies was selected for the detailed face-to-face interviews.

All questionnaires were reviewed in order to complete missing information. The completed questionnaires were sent to the interviewed companies for approval before the benchmarking results were analysed. The whole process ended in the compilation of the final benchmarking report (Schmitz & Maag, 2008).

Sample

The sample of the 18 interviewed railway companies consisted of Ceske Drahy (CD), Societe Nationale des Chemins de Fer Luxembourgeois (CFL), Deutsche Bahn (DB), First ScotRail (ScotRail), Irish Rail, MAV Magyar Allamvasutak, Metro de Madrid (Metro), Norges Statsbaner (NSB), Red Nacional de los Ferrocarriles Espanoles (renfe), S-Bahn Berlin, Societe Nationale des Chemins de fer Belges (SNCB), Societe Nationale des Chemins de fer Francais (SNCF), Southern Railway (Southern), SouthWestTrains, Stuttgarter Strassenbahnen (SSB), Transports Metropolitans de Barcelona (TMB), Trenitalia, and Verkehrs-Aktiengesellschaft Nuernberg (VAG) (Figure 2).

Figure 2. Map of interviewed railway companies

Most of the interviewed companies have many years of experience in running training simulation and other e-learning tools. The sample represents the European railway area as the selected companies employ more than 600 000 people. This number includes more than 77 000 train drivers. Not only major railway undertakings but also metro and tram operators of different European countries were part of the research sample.

The study was based on the analysis of 138 simulator systems developed by eight different manufacturers: Citef, Corys T.E.S.S., EADS, Indra, Krauss-Maffei Wegmann, Lander, OKTAL, and Sydac.

Benchmarking questionnaire

A detailed key question form (KQF) was prepared with the objective of obtaining an comprehensive overview of (1) training contents for train drivers, (2) tools used for training, and (3) organisational aspects of training. As different interviewers with different professional background conducted the benchmarking interviews, the objective of the KQF was to guarantee a standardised interview procedure and to enable a common analysis of the gathered data. One part of the KQF was compiled out of multiple-choice questions asking for the training methods used for specific training topics. Another part of the KQF required the respondent to estimate the amount of current training on several topics by means of different training methods. Furthermore, the respondent had to give information about the amount of simulator usage in the course of initial training, advanced training, and competence check. The last part of the KQF covered general and organisational aspects of the training concept, e.g. the development of training exercises, relevant legal and internal regulations, assessment schemes, evaluation of training, and encountered problems and drawbacks.

Results

In the following section, the results of the benchmarking interviews are presented. The first part describes the main fields of simulator usage and compares computer-based training with conventional training methods. The second part deals with the expectations and experiences of the interviewed companies related to the training technology. The last section focuses on shortcomings of the existing technology, on feedback concerning the day-to-day use of simulation technology, and on planned future improvements.

Five different types of simulators were differentiated in the frame of the study: (A) software interface of cab equipment, (B) part-task trainer, (C) partial cab, (D) full replica cab without motion system, and (E) full replica cab with motion system. Most of the 138 analysed systems were partial cab simulators and simulators equipped with a full-replica cab and motion system (Figure 3). Several companies have procured different simulator types in order to cover full-mission simulation on the one hand and cost-effective partial-cab simulation on the other hand.

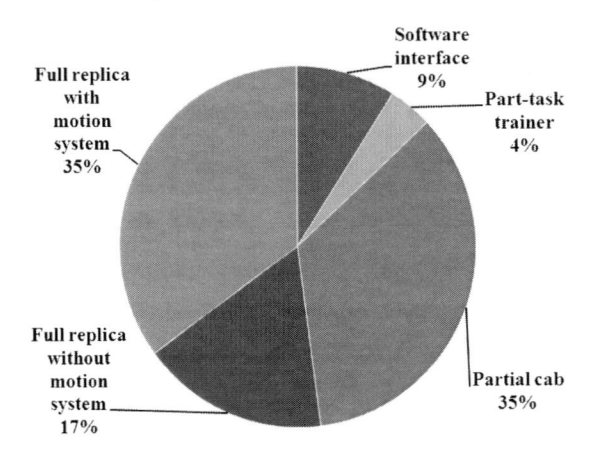

Figure 3. Distribution of the different analysed simulator types at N=17 railway companies

Usage of simulator training

The usage of driving simulators in comparison to other training methods (classroom or real cab) in concrete phases and for specific topics of training is rather inconsistent. Simulators are only used for 7% of the overall initial training amount. Concerning advanced training, the companies cover in average 25% of the complete training with simulation. 26% of all competence checks are realised by means of a simulator (Figure 4).

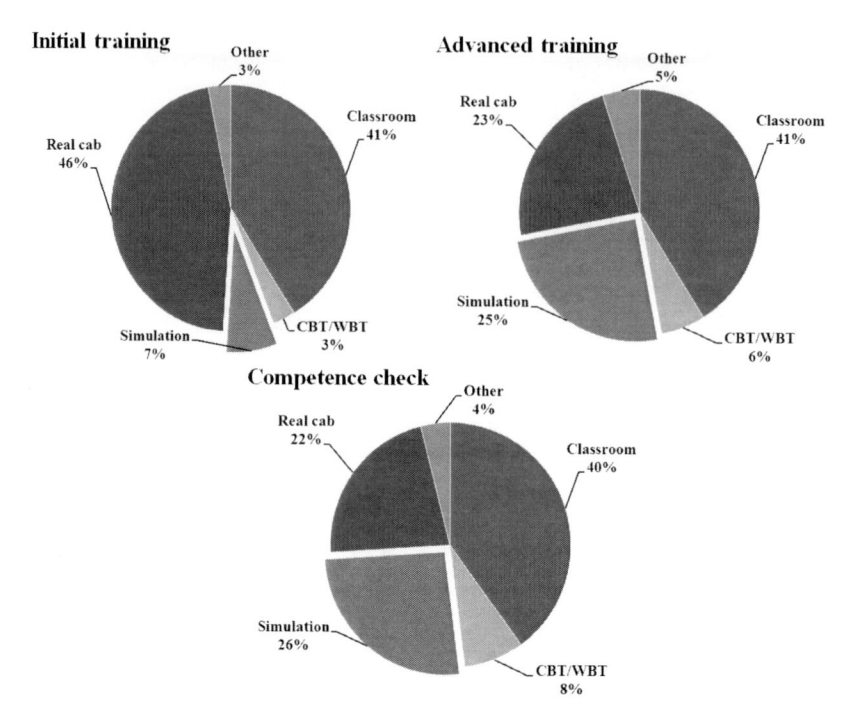

Figure 4. Usage of simulation and other training methods for initial training, advanced training, and competence check (CBT/WBT. Computer/Web-Based Training)

These findings are a result of the advantages of simulator-based training, e.g. the possibility to simulate emergency situations and driving under degraded conditions. Conventional training methods as classroom or training on the real cab still play an important role, especially for those areas that cannot be fully substituted by a simulator (e.g. because of legal reasons or internal directives) like route knowledge or rolling stock training (Figure 5).

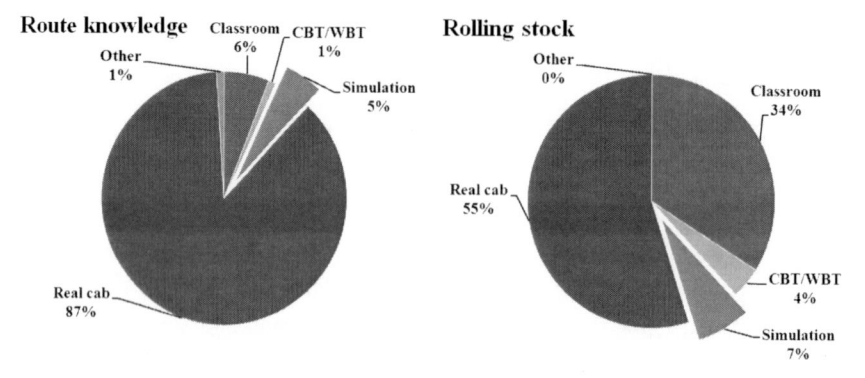

Figure 5. Usage of simulation and other training methods for the training of route knowledge and rolling stock

In average, about 40% of all simulator sessions are allocated for initial training. Therefore, initial training is the main topic the simulator is used for. Besides that, simulator training is an instrument for advanced training (~30%) and competence check (~24%). The remaining 6% are allocated to topics like acquisition of route knowledge, training on rolling stock or ATC/ATP systems.

The average number of hours a train driver spends on a simulator within the scope of initial training is approximately 30 hours with some companies considerably above average (up to 160 hours per trainee). Simulators are used for advanced training only to a limited extent in terms of hours per year (one to eight hours per trainee per year) due to a high number of drivers in many railway undertakings and the limited training hours in the simulator. Another reason is that advanced training in the simulator is not obligatory in all companies. Nevertheless, on average nearly every fourth hour of advanced training is realised as simulator exercise. Simulators are also used for competence check, i.e. for a regular performance assessment of the train driver. Ten out of 17 companies use the simulator for this training topic with a range of half an hour to three hours per trainee per year.

Concerning the usage of simulators, there are only small differences between the major railway operators (e.g. SNCF, DB) and the smaller metro and light rail operators (e.g. VAG Nuremberg). Whereas railway operators use simulators in a significant amount for competence checks, using the simulator for these checks is not very common at metro and light rail operators. Nevertheless, the specific training objectives are more important for the used training approach than the mere size or operational area of a company.

Training simulation technology – the users' view

The 18 companies were asked to rate the importance of technical features of a training simulator on a five point scale (1 = "not important", 5 = "very important"). The most important features with average ratings above 4.5 were the realistic train behaviour and the possibility of changing settings during an exercise (e.g. the aspect of a specific signal). Average ratings above 4.0 were given for an objective assessment system, real cabin controls, realistic tracks and signalisation, realistic visualisation and sound, as well as different driving and weather conditions. Average ratings above 3.0 were stated for the existence of a full replica cab, for the modelling of all subsystems (e.g. electric, pneumatic system), for the possibility of simulating different rolling stock, and for compliance with compatibility standards (e.g. High Level Architecture). The lowest average ratings below 3.0 were given for the existence of a motion system, the ability to connect different simulators for combined exercises, and the representation of road traffic, passengers and pedestrians. It has to be considered that the scoring on some of these items was greatly dispersed. For example, the motion system was of high importance for those operators running a simulator with motion system but was nearly unnecessary to other companies.

During the interviews, the operators were also asked to evaluate the existing training technology on a six point scale (0 = "simulator does not have the characteristic", 1 =

"not realistic", 5 = "very close to reality") regarding the fidelity of the simulator system. As a result, the simulation of road traffic, pedestrian and passenger behaviour were the relatively low-rated items. The mathematical train model, the signalisation system and the similarity between simulator and real train cabin (including the sensation of movement) were the best rated items. Medium-high marks received the simulation of other trains (apart from the ego train), the ability to simulate failures, and the visual/sound system.

Finally, the day-to-day usage of the simulators should be rated (0 = "not available", 5 = "maximal customer satisfaction"). It was found that the operators give good marks for the improvement of training-learning capacity due to simulators and for the amount of information given to the instructor during the exercise. The relatively low-rated items were the quality of the final exercise report and the existing possibilities to create new exercises. A recording system, the time needed to load the simulator exercise, and the general manageability and ease of use received medium marks.

Shortcomings and future planning

The last part of the interviews dealt with evaluation approaches, and participants' comments concerning shortcomings of the existing training technology as well as improvements planned for the future.

Companies were asked to report evaluation studies that were conducted to clarify if the training is interesting and motivating for the trainee and if it achieves the expected results in terms of knowledge transfer, behavioural change, skills and competence development. Besides the fact that some companies have introduced a regular feedback instrument asking for the trainees' acceptance of the computer-based training sessions, none of the interviewed companies evaluates the training effects in a systematic and standardised way.

Each company stated certain insufficiencies regarding the operation, maintenance and development of simulator training. The answers can be divided into (1) technical and economical shortcomings, (2) shortcomings related to the assessment capabilities, (3) feedback given by the train drivers, (4) feedback given by the labour unions, and (5) the intention of future improvements.

The given answers imply that it is difficult to update and keep the simulator at the level of the real system (according to changes affecting the infrastructure or the rolling stock), because implementing changes in the simulation is often cost-intensive and time-consuming and can only be done by the manufacturer. The simulation of a generic driver's cab and a generic infrastructure was suggested as a reasonable solution to deal with that problem and to achieve a higher flexibility in developing new exercises.

Other comments were related to the quality of post-run analyses, because often there is not enough feedback and data to support the instructor's assessment. At some training simulators an automated objective assessment system is totally missing. As a consequence, the use of objective performance markers for the assessment of the train driver and a connection of the assessment database with the company's information system were proposed.

Concerning the trainees' feedback, some train drivers asked for more sessions on the simulator. The fidelity of the simulation system should be as high as possible. Due to a missing full motion system, some existing drivers were not keen on simulators as for example the sensation of braking is not realistic.

The feedback given by labour unions and staff councils was inconsistent. At some companies an agreement with the staff council has been settled not to use the simulator for the regular competence check of train drivers. At other companies, unions do not allow the detailed performance check by using objective assessment data. Sometimes the labour unions insist that simulator sessions are treated as additional training and never reduce the amount of conventional training. Other companies reported that the acceptance of simulation as training method is quite high. Good experience was made with the involvement of the labour unions from the beginning of the simulator project (e.g. as part of the simulator steering group).

Asked for future improvements and planning some companies argued for an intensified use of simulation in the course of initial training, for training sessions focused on the distraction of drivers in critical phases, and for implementing cross-border training on partial-cab simulators. Other companies would like to increase the number of simulated engine types and realise a higher number of small-desk simulators.

Conclusions

2TRAIN benchmarking

The objective of 2TRAIN is to give best-practice guidelines for an efficient, safety-enhancing and cost-effective use of modern technologies for the training of train drivers as well as for the ongoing competence and performance assessment. The objective of the benchmarking process is to provide a comprehensive overview of the training concepts used in different railway companies in different European countries.

The benchmarking results are based on a structured research process and interviews with 18 European railway undertakings (including light rail and metro companies). The results show that the technical features and capabilities of simulators are extremely diverse as each company is focused on different training objectives. Some companies combine full-mission simulators with cost-effective partial-cab solutions whereas others are only focused on a single type. All in all, the analysis highlights the important role of simulation for the training and assessment of train drivers. Nevertheless, the interviewed companies state some shortcomings that hinder an optimal use of the existing training technology. Major points of criticism are the missing possibility to create new training exercises and scenarios as well as the assessment capabilities of the simulators, in particular the quality of the simulator reports.

An optimal benefit from simulator training could be achieved, if it is embedded into the overall competence management of the company. In addition, the simulator exercises should be tailored to the intended training objectives. Furthermore, it is important that the train drivers become familiar with the simulator technology by using this training tool regularly in the course of initial training, advanced training, and competence check. Standardised scenarios and procedures for all training centres of a company lead to a transparent training and assessment. The usage of simulator data supports an objective and detailed assessment of the train drivers' performance.

2TRAIN approach and technical development

Based on the benchmarking results, the requirements for future computer-based training concepts have been specified, taking into account the information derived from the questionnaires and from the discussions with the representatives of the interviewed companies. Concerning the technical approach of 2TRAIN, a universal simulator for all European countries is not realistic due to the current individual company standards in training technology and the companies' investments in simulation technology already made in the past. Instead, a Common Data Simulation Interface (CDSI) is developed within 2TRAIN to allow a standardised data recording and to provide the opportunity to implement simulator add-on systems. These add-ons are a Rule-based Expert System (ExSys), a Virtual Instructor (VI), and an Assessment Database (AssDB) (Figure 6).

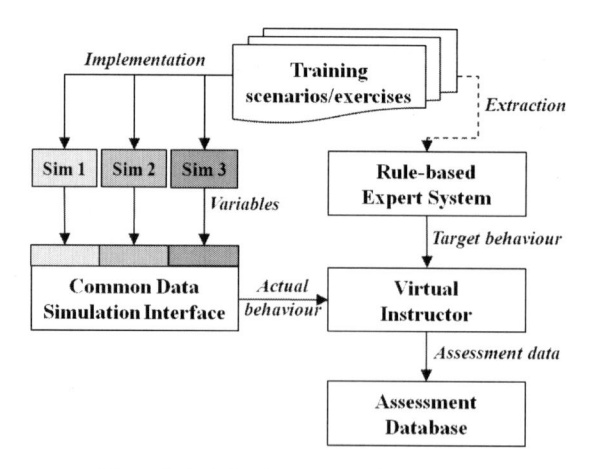

Figure 6. Illustration of the technical development work in 2TRAIN

In order to realise a detailed assessment of the train driver's performance during a simulator exercise, the simulator data are compared with the data from the ExSys. The ExSys defines and stores the target behaviour (i.e. correct behaviour) of the train driver for different scenarios. The VI is responsible for running the assessment procedure online. The VI receives the data about the actual behaviour of the train driver during the simulator exercise from the CDSI and compares it with the target behaviour and assessment rules from the ExSys. The resulting assessment data (e.g.

driving deviations, errors, reaction times) are stored in the AssDB. The VI also provides online information to the train driver on how to improve his performance. This is realised by messages presented to the train driver (via the visual system of the simulator) and by showing documents or multimedia chapters (via a PC screen placed in the driver's cab). The assessment data that are stored in the AssDB in a standardised form serve as a basis for generating an assessment report at the end of the training session. Further input to the AssDB is information that derives from other training methods (e.g. CBT/WBT). By this, the AssDB is expanded to be a competence management system.

These technological developments of 2TRAIN increase the efficiency of simulator training, the transparency of performance assessment, and contribute to get highly qualified train drivers that are able to handle all kinds of operational situations in a safe and appropriate manner.

References

European Commission (2007). Directive 2007/59/EC of the European Parliament and of the Council of 23 October 2007 on the certification of train drivers operating locomotives and trains on the railway system in the Community. *Official Journal of the European Union, L315,* 51-78.

Rail Safety & Standards Board (2007). *Good practice guide on simulation as a tool for training and assessment* (RS/501, Issue 2, June 2007). London, UK: Rail Safety & Standards Board.

Schmitz, M. & Maag, C. (2007). 2TRAIN – TRAINing of TRAIN drivers in safety relevant issues with validated and integrated computer-based technology. In *Proceedings of the 15th International Symposium EURNEX - Žel 2007, Volume 2* (pp. 206–211). Žilina, Slovakia: Eurnex-ZEL 2007.

Schmitz, M. & Maag, C. (Eds.) (2008). *Benchmarking report on computer-based railway training in Europe* (2TRAIN project report). Würzburg, Germany: Center for Traffic Sciences at the University of Würzburg.

Evaluating the effectiveness of a human factors training –Focus on situation awareness

Nicki Marquardt, Swantje Robelski, Gwen Jenkins, & Rainer Höger
Leuphana University of Lueneburg
Lüneburg, Germany

Abstract

This paper presents the evaluation of a human factors training specifically designed for employees in manufacturing within the automotive industry. The central objective of this training program was to increase the workers situational awareness about potential errors that can occur during the production process. According to Endsley (1995) situation awareness encompasses three levels: the *perception* of elements in the environment, the *comprehension* of their meaning for the current situation and the *projection* of current states into the future. Various tools were used to improve the workers three levels of situation awareness. Specifically, a flicker paradigm and a structure formation technique were adjusted and used to train the perception, comprehension and projection of errors at various stages within the production process. Seventy employees, all of which were working in a production unit for gearbox manufacturing, participated in the training program of this study. All workers and their direct supervisors had to answer general questionnaires and take task-specific tests for measuring the workers degree of situational awareness one month prior to and after, as well as six months after the training session. The results showed a significant increase of the workers situational awareness after the training. At the end of this paper recommendations for future research in the field of human factors training are made.

Introduction

Today, industrial workplaces are often very dynamic and the electronic and automation systems are highly complex. Such workplaces impose a considerable burden on workers because mistakes on their part could have severe consequences. Endsley (1995) developed a model of situation awareness that explains which requirements and conditions have to be given for humans to attain a state of awareness that would lead to making appropriate decisions and actions in a given situation. According to Endsley "Situation Awareness is the perception of the elements in the environment within a volume of time and space, the comprehension of their meaning, and the projection of their status in the near future" (1995, p 36). Due to not being bound to a specific field of work, the theory of Situation Awareness is applicable to various industries including production and manufacturing.

In D. de Waard, J. Godthelp, F.L. Kooi, and K.A. Brookhuis (Eds.) (2009). *Human Factors, Security and Safety* (pp. 101 - 110). Maastricht, the Netherlands: Shaker Publishing.

The first level of situation awareness implies that the operator is capable of perceiving critical factors in his current environment by means of attention and information processing. The subsequent level also interprets those factors within the given situation by including information on individual goals and expectations (Endsley, 1995). The accuracy of the interpretation largely depends on how well the critical elements are perceived and how suitable the underlying memory structures such as mental models, schemata and scripts are. The highest level of situation awareness consists of understanding what can happen with this system in the future (Endsley, 1995). In short, level one and two serve as a basis for the third and final level. When an individual has reached the latter, he is fully aware of the situation and has the necessary knowledge critical for further actions such as decision making and performance. Situation awareness (SA) makes system errors more accessible and enables individuals to make better predictions of dynamic events and take necessary actions. SA can be seen as a unique product of incorporated external information, working memory processes and long-term memory stores which form an internal representation of the environment (Endsley, 2000).

However, various sources of error, such as incomplete knowledge of situational elements or erroneous knowledge regarding elements and their meaning, have a substantial effect on the development of Situation Awareness (Endsley, 1995). These errors can occur on all three levels of situation awareness. Level 1 is prone to perception errors which could lead to misconceiving crucial situational factors. Failures on level 2 can be attributed to the incapability of integrating the operator's goals or interpreting perceived information. These failures could result from incomplete models, incorrect choices of mental models or incongruous goals (Endsley, 1995). Finally, when looking at the third level, environmental dynamics can make projecting the current situation to the future more difficult. Additionally, workplace characteristics such as automation reliability can disturb the development of Situation Awareness.

Within aviation and other industries, many of the most common errors involve the loss of situation awareness among different individuals or teams (Jones & Endsley, 1996; Robertson, 2001). Therefore, situation awareness is a central part of every training program which is designed for human error prevention like crew resource management (CRM) or human factors training (Burke, Wilson & Salas, 2005; Flin & O'Connor, 2001; Safety Regulation Group, 2006). Consequently, human factors trainings which were first implemented in aviation have been transferred to other high risk industries like offshore oil and gas, medicine, nuclear power or merchant navy (Flin & O´Connor, 2001; Glavin & Maran, 2003; O'Connor, Campbell, Newon, Melton, Salas, & Wilson, 2008; Salas, Wilson, Burke, & Wightman, 2006). Still, a lot of high risk industries such as the automotive industry have not yet quite recognized the importance of introducing human factors trainings as part of their safety culture or corporate error management concept. One important thing is that the design of such a program i.e. its trainings methods and exercises fit the needs of the organization in question and its members (Flin & O'Connor, 2001). Until now, many human factors training methods for increasing situation awareness – e.g. high fidelity simulators – are mainly focussed on aviation work environments. As a consequence, necessary human factors training programs as well as effective situation awareness training methods for other non-aviation industries like automotive are still in need to be designed.

The aim of this study was, therefore, to develop, implement, and evaluate a human factors training program specifically designed for the automotive industry. As a result, new situation awareness training approaches were created. The expectation was that this human factors training would increase the workers situational awareness about potential errors that can occur during a production process.

Method

Participants

Seventy employees of the automotive company Volkswagen in Kassel, Germany participated in this study. All employees were shift workers working in a production unit for gearbox manufacturing. Accordingly, all participants belonged to the "sharp end" of the organization (Reason, 1997). In contrast to people at the "blunt end" of a system or an organisation (e.g. managers, engineers) members at the "sharp end" (e.g. operators, pilots, workers) are more likely to make active errors that directly lead to accidents. Due to the precondition of the survey being anonymous, as requested by the firms work council, personal data such as age, sex and qualification could not be collected. The workers from this department were selected specifically because here deficiencies in their situational awareness could lead to fatal errors and incidents during the process of production. Moreover, undetected incidents could result in the production of failing gearboxes which in turn could induce car accidents with casualties.

Design & procedure

A pre-post training design was used for the evaluation. Thus, the participants' SA was measured and SA behavioural assessments, made by their direct supervisors, were inquired one month prior to (t_0) and after (t_1), as well as six months after (t_2) the human factors training. Due to company restrictions, there was no opportunity to compare these measurements to a control group, as would be mandatory according to traditional experimental designs. The data were collected in a separate quiet room during the working shifts of the company's three-shift operation. Each group consisted of 15 workers.

Training

The human factors training involved a two days course, including interactive tutorials and lectures on situation awareness, communication, teamwork, and stress management. However, this study only focused on situation awareness. Hence, only descriptions of training methods and results relevant for SA are reported here. Several training instruments were used to improve the workers three levels of task specific SA. Therefore, two methods derived from cognitive psychology were adapted to train *perception, comprehension* and *projection* of errors at various stages within the production process.

Method to increase SA level 1 "perception"
The first level of SA, *perception* of elements in the environment, was trained with a method called flicker paradigm (Rensink et al., 1997). Within the flicker paradigm an original image of a scene continually alternates with the same image modified in some way (e.g. an item is altered or is removed). A brief blank slide is interposed (for about 80 ms or more) between the two images. The detection of the change requires a series of alternations and depends on the effectiveness of visual search strategies which in turn is affected by task specific SA. Figure 1 shows the temporal relationships of the flicker paradigm. The display times typically vary between 200 – 600 ms, while the duration time of the blank does not exceed 80 – 100 ms. The sequence alternates between the original and modified image until the observer responds.

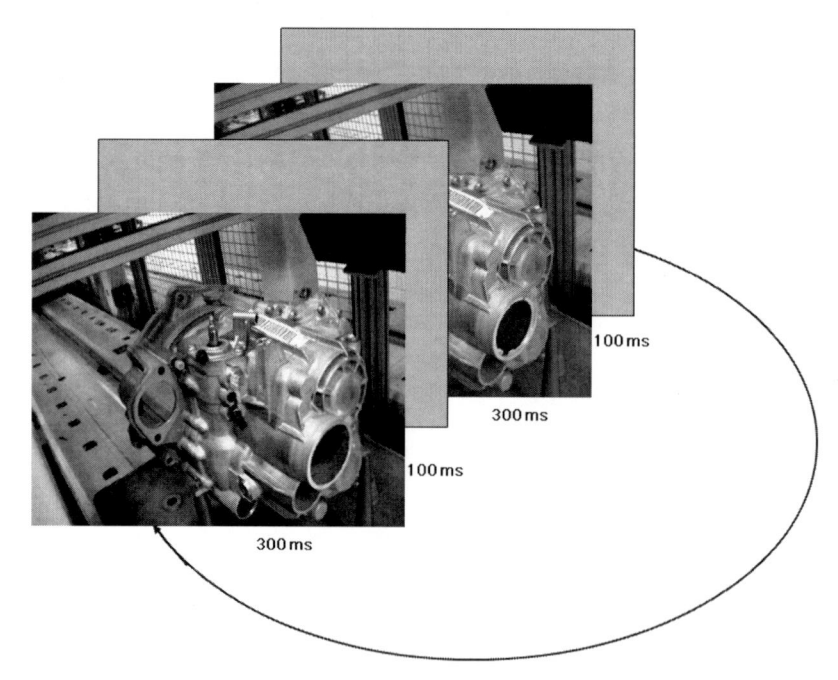

Figure 1. Example of the flicker paradigm and a photograph of a gearbox used in the training. The snap ring was removed from the modified image.

In a first step, photographs of the gearbox production process from nine workstations were taken. The intention was to grasp a wider range of error prone situations as well as various levels of hazardous potential. The photos vary from flashing error displays to incorrect and missing components at several stages of the gearbox production process. The selection of the workstations was based on a survey on errors caused by flaws in SA.

In a second step, the photos were modified using digital image processing and prepared for the flicker paradigm. The modified version of the original image was created by changing or removing items relevant for an appropriate perception of the error-causing situation. Gearbox engineering specialists from this department selected the relevant items.

In a third step, the photographs as well as the flicker paradigm were presented on an IBM-compatible personal computer and a projector within the classroom. Nine photographs and their modified variants were used. The participants had to detect in which way two photographs differ from each other. The original and modified images were displayed up to 300 ms each whereas the interposed blanks had a duration of 100 ms each. The alternation of the original and modified image was repeated for 60 seconds. Participants who found the mistake of the error-causing item were asked to write it on a card. After all the photographs were presented, they were asked to post their cards on a display board and compare their own perceptions with those of the other participants. This was done in order to sensitise the workers for flaws in their perception concerning their work. The workers were supposed to realize that their perception of errors and hazards is dependent on the mental models underlying their task specific SA.

Method for increasing SA level 2 & 3 "comprehension & projection"
The second and third level of SA, *comprehension* of the perceived elements and *projection* to their future states, were trained with the structure formation technique (Scheele & Groeben, 1984). Firstly, a huge poster with a layout of the gearbox production process with nine marked workstations was presented to all participants. Secondly, participants had to visualise their mental models of error causation (comprehension of elements) and its consequences (projection of their future states) for each of the nine workstations by writing contributing factors for and likely consequences of these errors on cards and posting them on a display board.

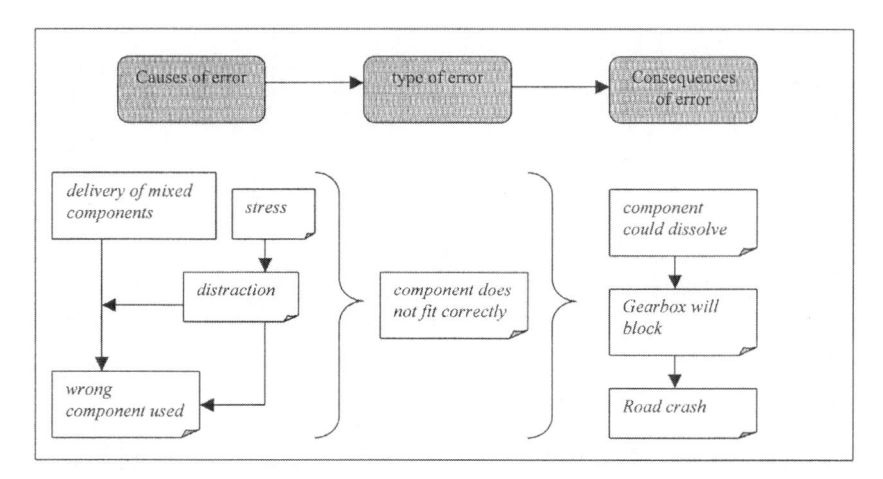

Figure 2. A mental model of causes and consequences for an error type at a specific workstation visualized by the structure formation technique

Next, every participant presented his visualized mental model of causes and consequences for each error type (see figure 2) in front of the class. In addition, all the participants discussed each of the visualized mental structures according to the correctness of error causation and likelihood of error consequences for the production process, the product itself, the customer or the whole company. The exchange of differing experience, knowledge and perspective of error causes and consequences is supposed to support the alignment of the workers mental models. Consequently, an increase of shared mental models as well as distributed situation awareness (Burke et al., 2005; Endsley, 1995) across the workforce is expected.

SA assessment methods

SA self report measure
The first SA assessment tool was a questionnaire that incorporated five items. The items were developed with the help of academic literature, accidents reports and work experience of human factors practitioners. The items captured several behavioural aspects of flaws in SA. Examples are "In my work environment I always recognize dangers in time" or "I am very precise in the work I do - even in stressful situations". For all items a five-point Likert scale with the anchors "agree" and "disagree" was applied.

SA behaviour rating
The second SA measure was a SA behaviour rating by the workers direct supervisors. Eleven foremen, in charge as direct supervisors, rated their workers SA relevant behavioural deficits on a one-item rating scale "How do you assess workers deficits in SA relevant behaviour?" A five-point Likert scale with the anchors "Slight SA deficits" and "Great SA deficits" was applied.

SA performance test
An additional, more objective indicator for SA, a task specific SA performance test, was developed. Task specific questions were generated for assessing the workers SA according to potential errors in the production process. The task specific performance test contained the layout of the gearbox production process with nine marked workstations. Every worker had to answer three open-ended questions, each representing definite indicators for SA for each of the nine marked workstations: "What type of errors can occur at this stage of the production process?" (*Perception of errors*), "What could be the cause of this error?" (*Comprehension of error causing elements*), "What could be the consequences of this error?" (*Projection of consequences for an error*).

Results

Data reduction

Seven participants attended the pre-training but not the first post-training data collection of the SA performance test and the SA self-rating questionnaire. Moreover, fourteen participants filled out the SA performance test at the first but not at the second evaluation. Thirteen of the workers answered the self-rating questionnaire one month after but not six months after their training. In all, 49 participants answered the SA performance test and 50 participants filled out the SA self-rating questionnaire one month prior to and after, as well as six months after the training session. Therefore, there were 49-50 participants left for the training effectiveness analysis.

The department's gearbox engineering specialist rated the qualitative data format of the task specific SA performance tests. Next, the qualitative data were transferred into a quantitative format. This was done by averaging the number of entries for each of the four questions by the number of marked workstations. Hence, this resulting mean for each of the four questions was a quantitative measure of the individual task specific situation awareness.

SA self report measure

The participants' SA self report ratings revealed a decrease of flaws in SA. Using repeated-measures ANOVA to examine the serial trend, *lack of SA* ($F_{(2, 48)} = 9.41$, $p < .05$; $\eta^2 = .28$) decreased significantly after the training. As can be seen in figure 3, the scale *lack of SA* (Cronbach's $\alpha = 0.77$) decreased significantly from 2.19 (t_0) to 1.97 (t_1) and to 1.94 (t_2).

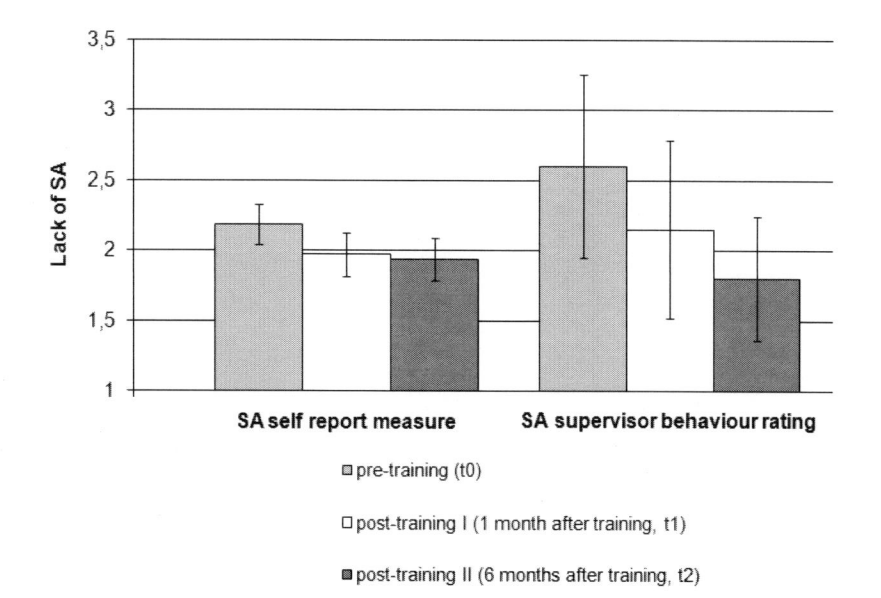

Figure 3. Results of the self report measure and the supervisor's behaviour rating for lack of SA. Error bars denote .95 confidence intervals

SA behaviour rating

SA behaviour ratings by the eleven supervisors also showed a decrease in SA relevant behavioural deficits. As can be seen in figure 3 again, the one-item rating scale (test-retest-reliability $r_{t1-t2} = .67$) decreased significantly ($F_{(2, 8)} = 12.62$, $p < .05$; $\eta^2 = .76$) from 2.60 (t_0) to 2.15 (t_1) and to 1.80 (t_2).

SA performance test

The SA performance test revealed a significant increase in task specific SA. Although the increase was smaller after the second post-training measurement (t_2), it was still highly significant for all SA performances indicators. As can be seen in figure four, the first task specific SA performance indicator, *perception of errors* (test-retest-reliability $r_{t1-t2} = .77$), increased significantly ($F_{(2, 47)} = 5.48$, $p < .05$; $\eta^2 = .19$) from 4.27 (t_0) to 6.55 (t_1), and to 5.59 (t_2). The second indicator, *comprehension of error causing elements* (test-retest-reliability $r_{t1-t2} = .83$), also increased significantly ($F_{(2, 47)} = 5.23$, $p < .05$; $\eta^2 = .18$) from 4.29 (t_0) to 6.69 (t_1), and finally to 6.02 (t_2). The third indicator, *projection of consequences of an error* (test-retest-reliability $r_{t1-t2} = .70$), *again* increased significantly ($F_{(2, 47)} = 5.21$, $p < .05$; $\eta^2 = .18$) from 3.53 (t_0) to 5.16 (t_1), and finally to 5.06 (t_2). Figure 4 shows results of the SA performance test. Higher mean values of known errors, causes and consequences of errors across the nine selected workstations, measured one month as well as six months after the training, illustrate the increase of task specific situation awareness. It should be noted that there were no significant differences between the two post evaluations (t_1 and t_2) across all SA measures (self report, behaviour rating, performance test) of the training evaluation.

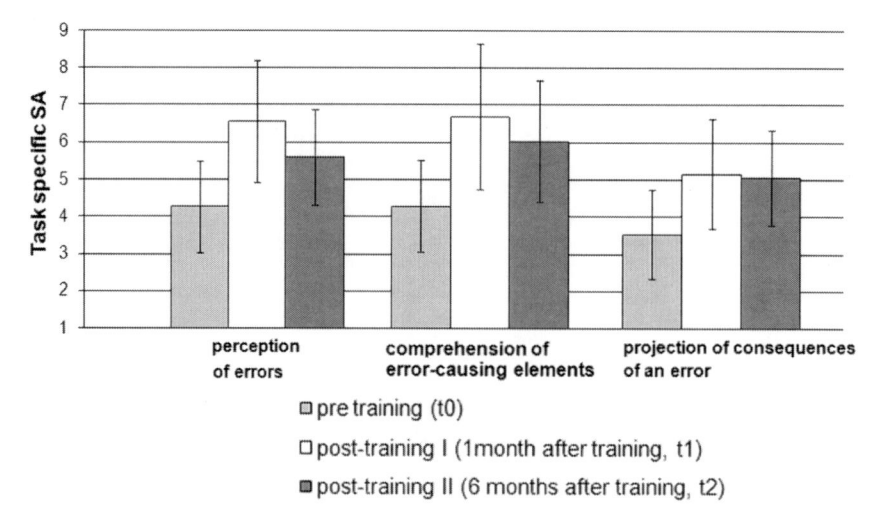

Figure 4. Results of the SA performance test on task specific SA. Error bars denote .95 confidence intervals

Discussion

This was the first application of human factors training specifically designed for the automotive industry. Before, human factors trainings were mainly implemented in civil and military aviation, health care and the oil industry (Flin & O'Connor, 2001;

Salas et al., 2006). Also, measures such as the flicker paradigm and the structure formation technique from cognitive psychology were adapted to industrial and task specific SA training for the first time. Special emphasis was put upon using real examples – e.g. photos of error prone steps within the production process – from the workers actual workplace and their specific tasks. The training was specifically focused on the three levels of situation awareness (perception, comprehension, projection). The workers' situational awareness showed significant changes and small, medium as well as large effect sizes within the three measures (self report $\eta^2 = .28$, supervisor behaviour rating $\eta^2 = .76$ and SA performance test $\eta^2 = .18 - .19$) for both evaluations. As a result, the increase of the workers' situational awareness and therefore their awareness of potential errors within their workplace can be attributed to the impact of the human factors training. In addition, a mixture of methods for evaluating the training program – e.g. qualitative and quantitative data from questionnaires and tests – was used. Furthermore, for the evaluations various point of views like SA self report by the workers, SA behaviour ratings by their supervisors and objective SA ratings with a task specific test were taken into consideration. Finally, the longitudinal design of this study, using three different points in time for evaluation (more than six months apart) allowed the examination of serial trends.

Despite the strengths and implications for human factors training research, some limitations of the current study, however, are noteworthy. Unfortunately, a limitation of this human factors training and evaluation program is that the results are prone to be influenced by other components of the training such as communication or teamwork. This means that confounded variables cannot be easily detected. Furthermore, even this evaluation has the same problem as many other evaluations of human factors trainings for it also lacks a control group for comparison (Edkins, 2005). However, a comparison of error rates to those of a control group is planned for the near future.

References

Burke, C.S., Wilson, K.A. & Salas, E. (2005). Teamwork at 35,000 feet: enhancing safety through team training. In D. Harris and H.C. Meier (Eds), *Contemporary issues in human factors and aviation safety* (pp. 155–180). Aldershot: Ashgate.

Edkins, G.D. (2005). A review of the benefits of aviation human factors training. In D. Harris and H.C. Meier (eds), *Contemporary issues in human factors and aviation safety* (pp. 117–131). Aldershot: Ashgate.

Endsley, M.R. (1995). Toward a Theory of Situation Awarenss in Dynamic Systems. *Human Factors, 37*, 32–64.

Endsley, M.R. (2000). Theoretical Underpinnings of Situation Awareness: A Critical Review. In M.R. Endsley, and D.J Garland (Eds.). *Situation Awareness Analysis and Measuremen*. Mahwah, NJ: Lawrence Erlbaum Associates.

Flin, R., & O´Connor, P. (2001). Applying crew resource management on offshore oil platforms. In E. Salas, C.A. Bowers, and E. Edens (Eds.), *Improving teamwork in organizations: applications of resource management training* (pp. 217-234). Mahwah, N.J: Lawrence Erlbaum.

Glavin, R.J. & Maran, N.J. (2003). Integrating human factors into the medical curriculum. *Medical Education, 37*, 59-64.

Jones, D.G., & Endsley, M.R. (1996). Sources of Situation Awareness Errors in Aviation. *Aviation, Space, and Environmental Medicine, 67*, 507–512.

Reason, J.T. (1997). *Managing the risks of organizational accidents*. Aldershot: Ashgate

Rensink, R.A., O'Regan, J.K. and Clark, J.J. (1997), To See or not to See: The Need for Attention to Perceive Changes in Scenes. *Psychological Science, 8*, 368-373.

Robertson, M.M. (2001). Resource management for aviation maintenance teams. In E. Salas, C.A. Bowers and E. Edenes, (Eds.): *Improving teamwork in organizations. Application in resource managment training* (pp. 235-263). Mahwah, N.J.: Lawrence Erlbaum.

O'Connor, P., Campbell, J., Newon, J., Melton, J., Salas, E., & Wilson, K.A. (2008). Crew Resource Management training effectiveness: A meta-analysis and some critical needs. *The International Journal of Aviation Psychology, 18*, 353-368.

Safety Regulation Group (2006). *Crew Resource Management (CRM) Training. Guidance for Flight Crew, CRM Instructors (CRMIS) and CRM Instructor Examiners (CRMIES). (Civil Aviation Publication, CAP 737 of the UK Civil Aviation Authority)*. Norwich, UK: TSO. (Retrieved 15.03.2007 from: http://www.caa.co.uk/docs/33/CAP737.PDF)

Salas, E., Wilson, K. A., Burke, C.S., & Wightman, D.C. (2006). Does crew resource management training work? An update, an extension, and some critical needs. *Human Factors, 48*, 392-412.

Scheele, B. & Groeben, N. (1984). *Die Heidelberger Struktur-Lege-Technik (SLT). Eine Dialog-Konsens-Methode zur Erhebung Subjektiver Theorien mittlerer Reichweite*. Weinheim/Basel, Switzerland: Beltz.

Aviation

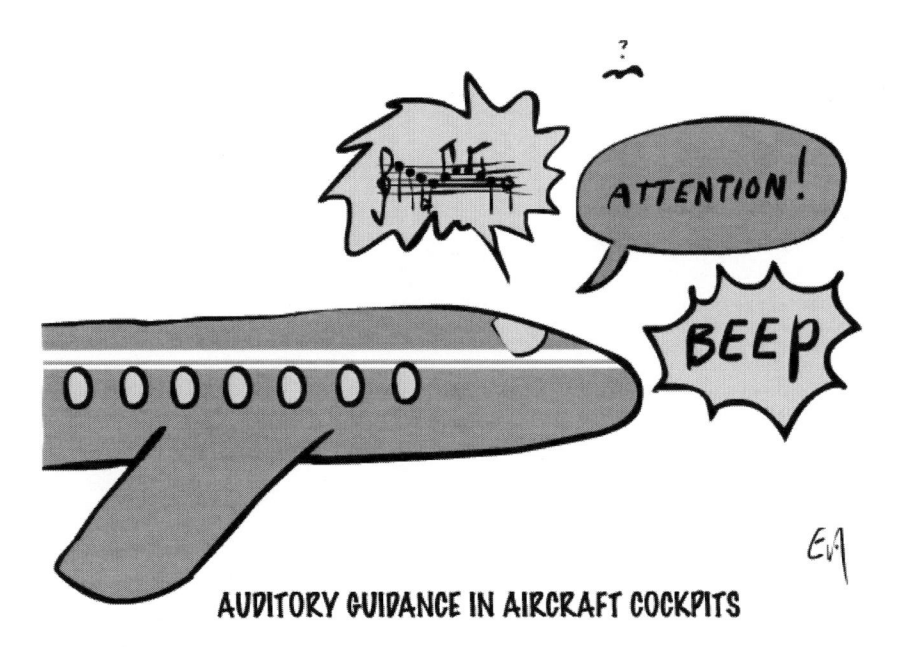

In D. de Waard, J. Godthelp, F.L. Kooi, and K.A. Brookhuis (Eds.) (2009). *Human Factors, Security and Safety* (p. 111). Maastricht, the Netherlands: Shaker Publishing.

Sound design for auditory guidance in aircraft cockpits

Bernd-Burkhard Borys
University of Kassel
Kassel, Germany

Abstract

Auditory information from the situation around us, available in open nature is lost in vehicles. To maintain the three-dimensional picture and situational awareness, pilots in an aircraft cockpit need to interpret two-dimensional data provided on flat screens. The visual information prevails; the auditory channel is used mainly for alerting or getting attention. In an ongoing research project we evaluate means to provide information using audio signals. For these audio signals, carrying information about state, direction, and amount of deviation explicit sound design is essential. Even when using voice commands, where the information is put into the words chosen, parameters exist that can support or oppose to this semantic information, like pitch (low to relax, high to alert), personality (robotic for regular automatic updates, human for unexpected trustworthy information). The paper shows, how an existing flight instrument is mapped to an audio display and what principles from psychology and music are applied to sound design.

Introduction

Since 1999, the Human-Machine Systems Engineering group of the University of Kassel conducted research in auditory displays for processes and vehicles. This research originated from the multi-media control room concept (Borys & Johannsen, 1997): in the clean, quiet, and safe control rooms in modern plants, information in mainly presented visually, and other modalities – noise, vibration, temperature, smell – is locked out. However, these sensations provide valuable information about a plant's status for an experienced operator and modern computer equipment, mainly from media and gaming applications, can bring back some of this experience in a safe and controlled manner. Later concentrating on audio displays, we evaluated how parameters known from music may be used for controlling industrial applications (Johannsen, 2004). This included short tunes to inform people about intended movements of a service robot (Johannsen, 2001), sound signals positioned in space for collision warnings in aircraft cockpits (Gudehus, 2002), and obstacle warning for the driver of a large construction vehicle (Carl, 2006).

An ongoing research project supported by the German Research Foundation DFG 2006 to 2008 evaluated several ways to support aircraft pilots with auditory signals in tasks, where quick and reliable perception of directions is necessary – collision warnings, approach and landing, and taxi guidance. A set of verbal auditory signals was designed, consisting of spoken commands, articulated by different speakers in different pitch, speed, and rhythm and presented using either a conventional audio system with one central speaker or a multi-speaker directional audio environment. The current paper will concentrate on the sound design of the verbal sounds. Other technical aspects of this project have been illustrated earlier (Borys, 2008, 2007, 2004) while evaluation of data captured during experiments is still not completed.

Guiding aircraft approach and landing – by audio displays?

In the following, guiding an aircraft during approach will be taken as a scenario for auditory support. During an instrument (Instrument Landing System, ILS-) approach of a modern, glass-cockpit aircraft, two needles integrated into the primary flight display in front of the pilot show vertical and lateral deviation from the glide path. Unless the technical equipment of aircraft and airport provides guidance to the runway until touch-down, the pilot needs to see the runway below a specified height above ground and to decide, whether a safe landing is possible or not (decision height – DH), take over control and land the aircraft manually. Thus, and because head-up displays are not common in civil aviation, for the last approx. 90 metres (300 feet) of descent the pilot must share visual attention between the instrument and the outside view. The experiments accomplished during this research should show, whether directional or non-directional audio guidance gives sufficient support to land an aircraft. An important work package during this research was sound design.

Sound design for verbal approach guidance

Sonnenschein (2001) lists different sound qualities: Rhythm, intensity, pitch, timbre, speed, shape, and organization. These form the dimensions, along which parameters for sound design can be picked. Introducing slight changes in only one dimension produces sets of similar sounds, changing several of these dimensions results in different and unconnected sounds.

The two important physical parameters describing a human voice are pitch (the frequency of the movements of the vocal folds) and the formants (the peaks in the spectrum of a voice sound). Based on these two parameters, a voice itself can express a personality and a physical state. Sonnenschein (2001) links personality to association of a voice to well known existing or standardized persons, e.g., the cowboy John Wayne, the gentleman Carry Grant, the sinister computer Colossus. Physical state is detected from the frequency distribution over the formants (e.g., a robust, self-confident voice with energy equally distributed over all formants, a hollow, tired, depressed voice with energy contained mainly in the first formant). In search of voices for a voice guidance and alerting system, we could start as if selecting actors for a movie.

The human voice is flexible in several more dimensions and in using rhythm, stress, and intonation, speech is able to add additional information to the pure text message. This makes the difference between question (*gear down?* Is it already?), information (*gear down*. It is down.), and command (*gear down*! Perform the necessary action now!). (As readers see: written language uses signs or marks to provide the prosodic information and is even capable of expressing different levels of urgency – *gear down!!*)

However, to use all these capabilities of human speech to transfer information in a voice-based guidance and alerting system we would need to record the speech of several well-trained professional speakers. This would be too expensive, at least in a small research project and for that reason we generated the voice commands with the speech synthesizer. Several freeware text-to-speech synthesizers turned out not to be flexible enough for the intended purpose, finally the synthesizer included in Image Line's Fruityloops software (Image Line, 2008) was used. This synthesizer converts text entered into the command window into speech and store this in a sound file with 16 bit resolution and 48 kHz sapling rate. It does not give control over all aspects of speech, e.g., no direct control over formant power distribution, and only limited control over prosody, but provides several pre-defined personalities (male, female, child, large, small, human, robotic), control over pitch, rate, and total power. Control over prosody is not documented but seems to be possible as the examples below will show.

Three different voice personalities have been chosen for the experiment, the male human voice *Anton*, the female human voice *Berta*, both based on predefined human voices, and the mechanical and monotone voice *Robby*, starting from a pre-defined voice *Colossus*, but shifting pitch to make it sound more friendly. Berta's appearance is female, because its basic pitch (F in the third octave) is one octave higher than Anton's and Robby's (E and C, both in the second octave) and it is speaking with a rate of 130 slightly faster than Anton's 120. Robby speaks with a rate of 150 even faster, this is necessary because it is used for short call-outs and must not interfere with lateral and vertical guidance information.

Design example

The Instrument Landing System (ILS) provides the information necessary to guide an aircraft to land at the beginning of the runway. Its core is the glide path, a line connecting an entry point (Final Approach Fix) 40 km before and 1 500 m (5 000 ft) above the runway with the touch-down point on the runway. Radio transmitters mark the glide path through the air and receivers on the aircraft calculate deviations using the signals received. The track, its projection onto the ground extends the runway centreline. In the cockpit, the vertical and the lateral deviation is shown on the ILS instrument, either mechanically with a vertical and a horizontal needle or electronically with two crossing lines on a screen.

Four principles used in the ILS instrument need to be noted here: The conventional ILS indicator provides (a) deviation as a command: Pointer to the right means correction towards the right side necessary, it shows (b) the deviations constantly, it shows (c) vertical and lateral deviation in parallel, and it shows (d) the deviations on a continuous linear scale. Finally, the amount of deviation is indicated by the distance of the pointer from the centre of the display which (e) provides the display with a balanced, thus, friendly appearance when everything is correct.

An auditory display should follow a similar philosophy. Although this is not completely possible, it may provide guiding principles for the design:

(a) Command indicator: Following the first design principle for the visual indicator, the audio display must provide a command, like *turn to the left* instead of *deviation to the right*. As commands must be short, the words selected in this example for the unbiased basic commands are *up, down, left*, and *right*. Modifications described later will generate a vocabulary of 14 commands.

(b) Continuous commands: Voice commands take some time, e.g., one third of a second for the word *up*, half a second for the words *left* or *right*. On the other hand, a pilot will also not monitor the single ILS display constantly, but include it in a scanning cycle. Using short words and repeating these in a short cycle will be sufficient to provide authentic deviation information.

(c) Parallel presentation: While visual information is perceived in parallel, auditory information is perceived sequential. When presenting two spoken commands in parallel we cannot assure that both are understood. The solution selected is to use a short sequence of two commands, starting, because near the ground more critical, with the vertical correction, immediately followed by the lateral, this followed by a longer silence. Two voice personalities, Berta for vertical and Anton for horizontal deviation, underline the difference between these two. As Berta's pitch is one octave higher it sounds also more urgent, again stressing the importance of keeping the correct altitude.

This design for apparent parallel presentation implements the three Gestalt principles of (1) proximity, (2) similarity, and (3) completeness (Sonnenschein, 2003):

(1) Proximity is the property that in music creates a melody from single notes by timing of the auditory events. Anton and Berta belong together, because these two different voices provide their statements close together in time, separated from the next command by a longer break. If necessary, more related information can be added without violating proximity by adding it to the command sequence before the break, any unrelated information must be separated in time. The break after one cycle is noticeable, but short, following the guideline that a "critical warning shall be repeated with no more than a 3 sec pause between messages (…)" (MIL-STD-1472F cited in Ahlstrom & Longo, 2003).

(2) Similarity forms a stream from the unconnected sounds. After the break, Anton and Berta continue the same message, approach guidance, because the same human voices speak again using the same vocabulary, speed, and tone. The underlying principle is that when a sound is noticed, always a source of this sound is imagined, even if this source is invisible and unknown. When a following sound with similar characteristics is heard, though after a longer break, it is associated with the same source unless strong evidence stands against it. Reversing this principle provides a guideline on how to implement unrelated commands: These must differ in several characteristic parameters, for example by using a completely different voice. Consequently, altitude call-outs, that start counting down the remaining height when

crossing 500 feet (152 m) above ground, have been implemented using the voice Robby. This voice is dissimilar from the approach guidance commands, because Robby's voice is non-human, faster and of lower pitch. Robby's count-down provides a second, non-interfering message stream.

(3) Completeness makes a sentence from single words. One deviation message, started by Berta and continued by Anton is complete, because Anton's voice is lowering at the end. The necessary control over prosody is gained by ending Anton's command with a full stop: "on track." or "left.".

(d) Continuous scale: Using a limited number of command words, it is not possible to provide a continuous scale, specially because the number of commands must be kept low, as reaction time increases with vocabulary size. The result is a set of six commands and a modification scheme for increased deviations. The set consists of two commands for the correct vertical and horizontal position (not really commands, but the messages *on glide* and *on track*) and four basic commands (*up, down, left, right*), that indicate the directions of the corrective action in case of deviations. When deviation increases, the correction command is repeated two and three times (e.g., from *up* over *up...up* to finally *up...up...up*).

In all sound design, the intended urgency of a message must match the perceived urgency of a sound. Factors influencing perceived urgency are speed, pitch, loudness, and the number of repetitions of a sound (Haas & Edworthy, 1996). The repetition described above already generates an impression of urgency. In the cockpit, there is no room to increase loudness; however, we can manipulate pitch and speed. The voice Anton uses its original pitch of E in the second octave in the *on track*-message. In the *left*- and *right*-command, pitch raises a minor second to F and for the repetitions additional minor seconds to F# and G. The same method is used for the *down*-commands: Raising Berta's normal pitch F in minor second intervals to F#, G, and G#. More weight is given to the *up*-command, as being too low is dangerous: In this case, the pitch raises in major seconds to G, A, and B. Manipulating pitch will not break the similarity principle introduced earlier, as similarity is much less depending on pitch.

Manipulation of speed is more complicated as the commands are already short. The solution is to start the third repetition earlier as expected, similar to a syncope in music: As an example, the word *left* takes 554 ms, the repetition starts already after 539 ms during the /t/-sound of the first word, and the second repetition starts even 115 ms earlier, 963 ms from the beginning.

Taking *left* and *up* (with increased urgency) as examples, the Table 1 shows the resulting design for manipulating perceived urgency (*right* and *down* show the same characteristics in pitch and timing as *left*).

(e) Friendly when correct: Similar to the balanced appearance of the visual display with both pointers in the centre, the voice message for correct position (*on glide* and *on track*) both sound more friendly and calming and indicate less urgency than the correction commands – two vowels, the long "o" in *on*, no repetitions, basic (lowest) and constant pitch. Table 2 compares the resulting voice design with the conventional instrument or display.

Table 1. Pitch and Timing for Increasing Urgency

Direction	Deviation	Command	Voice Personality	Pitch	Timing in ms from beginning
Lateral	No	*On track*	male (Anton)	Normal: E	-
	To the right	*Left*		Raised: F	-
	More to the right	*Left...left*		Rising: F...F#	0...539
	Much to the right	*Left...left...left*		Rising: F...F#...G	0...539...963
Vertical	No	*On glide*	female (Berta)	Normal: F	-
	Below	*Up*		Raised: G	-
	More below	*Up...up*		Rising: G...A	0...425
	Much below	*Up...up...up*		Rising: G...A...B	0...425...719

Table 2. Conventional ILS and Voice Guidance

Visual ILS Screen Display or Instrument	Voice Approach Guidance
Conventional screen display with two pointers	Auditory display with two voice personalities
Pointer indicates necessary correction	Voices command necessary correction
Vertical and lateral deviation shown in parallel with two pointers	Sequential commands for vertical and lateral deviation by grouped by proximity, presented by two voice personalities
Constant indication (scanned in intervals)	Intermittent commands, linked by similarity
Continuous linear scale	Few different commands
Balanced, when correct	Friendly, when correct

Directional audio

The commands described above use words to indicate a direction. Using two ears and sensing differences in phase and amplitude, humans are capable of directional hearing. Direction-dependent filtering by the pinae adds clues, also in horizontal direction (Blauert, 1974). Presenting a command from the direction to correct to instead of speaking the direction is expected to trigger faster corrections.

One method to expand audio from one single source to more dimensions is to use headphones and to provide each ear with a separate signal while manipulating amplitude, phase, and overall spectrum using head-related transfer functions. As the position of the pilot's head is quite fixed in the cockpit, another feasible way to create the impression of sound distributed in space is to use several speakers. This method is familiar from entertainment, starting with two speakers for stereophonic radio, more for cinema, finally speaker arrays that produce a sound wave as if originating from the desired position (Theile, 2007). The performance of multi-speaker settings is comparable with head-related transfer functions (Wenzel et al., 1991) without the need of head trackers, headphones, and specific audio processing equipment.

For directional audio experiments, eight speakers were used, four in a plane above, four below the pilots head, in each plane one speaker before, one behind, and two left and right. Using two speakers in parallel generates a phantom sound source between these speakers on an apparent position controlled by the amplitude difference (Sengpiel, 2008). With the sound coming from the direction to turn to, the direction needs not to be mentioned in the command any longer, reducing the commands to *glide* and *track*. Directional verbal commands opposed to centralized commands and screen display is the third display mode used in the evaluation experiments.

Evaluation

A fixed-base cockpit mock-up provided the environment to evaluate the design with airline pilots. The scenario used in the experiments included four consecutive segments, beginning north of Frankfurt Airport during descent to the initial approach fix at Taunus VOR. The second segment was a five minutes (30 km) level flight to the south and a left turn on localizer intercept, followed by an instrument approach in clouds and landing on the right of two parallel runways (07R) as the third segment, and finally the fourth segment – taxi to the parking position. One or two collision warnings have been triggered in the second segment, approach guidance was given in the third and taxi guidance in the fourth segment. The simulator provided only those controls necessary in the scenario – stick and throttle, controls for flaps, speed brakes, gear, thrust reverser, wheel brakes, and front wheel steering. Microsoft's Flight Simulator 2002 running on a PC simulated a Boeing 777 aircraft. A flat screen in the cockpit showed the instrument panel, a video beamer the outside vision on the wall in front of the simulator. A second PC connected as observer to the flight simulator, using its multi-player facility, and either showed, as a traditional cockpit display on a second flat panel screen, in sequence a TCAS collision warning display, the ILS approach guidance instrument, and an airport map with taxi instructions, or, while blanking out the display, controlled the auditory information. For more technical details see Borys (2008).

Six experimental sessions with airline pilots have been accomplished. Independent variable was the display type with three modes – conventional screen, verbal centralized, verbal 3D. Dependent variables are answers to questionnaires and the data recorded during the flights. The pilots received a 17-page description of the simulator, the instrument approach chart and two pages of instructions comparable to the clearances received from the controller during this approach prior to the experiments by mail. Each pilot had to perform nine flights during one day, after filling a first questionnaire about general opinion on auditory displays. During the first three flights, the pilots should become familiar with the experimental environment, the visual and auditory displays, and three more questionnaires used after a collision warning, after landing, and after taxi. These elicited opinions about the voices heard, situational awareness, workload, and judgment of current performance. Flights four to six have been used to evaluate the two verbal designs, and, for comparison, the conventional instruments, using questionnaires and recording flight data. The display modes are assigned randomly to the flights, the sequence was repeated in flights seven to nine after a longer break (usually lunch), resulting in 6 to 7 hours of experiments.

First results and conclusions

The experiments have been performed and evaluation of the data captured is in progress. The answers to the questionnaires and comments received show, that pilots accept verbal guidance (more often stated as helpful than hindering), although only as an addition to the conventional instrumentation, while the amount of verbal information should be as low as possible and not interfering with radio and crew communication. Speech quality was acceptable, although the voices, especially the female, should sound more naturally. For the directional audio, lateral deviation was significantly better detected than vertical and awareness of the own position was significantly better with conventional pointer instruments than in audio settings. Recorded tracks show oscillations in altitude and vertical speed. The track data is currently under further examination, it is, however, already obvious that audio guidance alone, as implemented for the experiments described, is not suited for approach.

Microsoft's Flight Simulator was accepted as a research vehicle by the pilots. However, two of them examined the behaviour of the simulated aircraft farther and detected faults in the relation of pitch, power and vertical speed. Means for fine-tuning the aircraft model are necessary for future experiments.

Acknowledgement

The work described was supported by Deutsche Forschungsgemeinschaft (German Research Foundation) from April 2006 to December 2008.

References

Ahlstrom, V. & Longo, K. (2003). Chapter 7: Alarms, Audio, and Voice. In *Human Factors Design Standard*. Atlantic City: Federal Aviation Administration Technical Center.
Blauert, J. (1974). *Räumliches Hören*. Stuttgart: Hirzel.
Borys, B.-B. (2001). Hardware, Software, and Experiments for Auditory Displays. In M. Lind (Ed.), *Human Decision Making and Manual Control*. Lyngby: DTU.
Borys, B.-B. (2004). Evaluation of auditory displays supporting aircraft approach and landing. In J. Stahre (Ed.), *Automated systems based on human skill and knowledge* (pp. 129–134). Oxford: Elsevier.
Borys, B.-B. (2007). Microsoft's Flight Simulator and DirectX to Drive Auditory Displays in a Cockpit Mock-up. Poster. *Annual Conference of the Human Factors and Ergonomics Society Europe Chapter*, Braunschweig.

Borys, B.-B. (2008). Ein Cockpit-Mockup zur Evaluierung auditiver Flugführungsanzeigen (A Cockpit Mock-up for the Evaluation of Auditory Flight Instruments). *i-com, 7* (1), 30–33.

Borys, B.-B. & Johannsen, G. (1997). An experimental multimedia process control room. In K.-P. Holzhausen (Ed.), *Advances in multimedia and simulation* (pp. 276-289). Bochum, Germany: Fachhochschule Bochum.

Carl, L. (2006). *Auditive Anzeige zur Kollisionswarnung auf einem Bagger.* [Auditory Display for Collision Warning on an Excavator]. Diploma Thesis. Kassel: Universität Kassel.

Gudehus, T.C. (2002). *Drei-dimensionale akustische Anzeigen zur Signalisierung von Kollisionswarnungen im Flugzeugcockpit.* [Three-Dimensional Acoustic Displays to Signal Collision Warnings in the Aircraft Cockpit]. Diploma Thesis. Kassel: Universität Kassel.

Haas, E. C. & Edworthy, J. (1996). Designing urgency into auditory warnings using pitch, speed, and loudness. *Computing & Control Engineering Journal, 7* (4), 193–198.

Image Line (2008). Fruityloops FL Studio. Overview. Sint-Martens-Latem, Belgium: Image Line Software BVBA. Retrieved from http://www.fruityloops.com/documents/what.html.

Johannsen, G. (2001). Auditory displays in human-machine interfaces of mobile robots for non-speech communication with humans. *Journal of Intelligent & Robotic Systems, 33* (2), 161–169.

Johannsen, G. (2004): Proceedings of the IEEE, Special issue on engineering and music – supervisory control and auditory communication. Piscataway: IEEE

Sengpiel, Eberhard (2008). Forum für Mikrofonaufnahme und Tonstudiotechnik. Retrieved from http://www.sengpielaudio.com/.

Sonnenschein, D. (2001). *Sound design. The expressive power of music, voice, and sound effects in cinema.* Studio City: Michael Wiese Productions.

Theile, G. (2007). Neue Anwendung der Wellenfeldsynthese. Binaural Sky. *FKT Fernseh- und Kinotechnik, 10,* 539–542.

Wenzel, E.M., Wightman, F.L., & Kistler, D.J. (1991). Localization with non-individualized virtual acoustic display cues. In S.P. Robertson, *Reaching Through Technology* (pp. 351–359). New York: Addison-Wesley.

Wittenberg, C. (2001). *Virtuelle Prozessvisualisierung am Beispiel eines verfahrenstechnischen Prozesses.* Düsseldorf: VDI-Verlag.

Barriers and accidents: the flight of information

Thomas G.C. Griffin, Mark S. Young, & Neville A. Stanton
Brunel University,
Uxbridge, UK

Abstract

Situation awareness remains a contentious topic within the human factors community. This paper investigates the application *of information networks* (derived from the Event Analysis of Systemic Teamwork, EAST, methodology (Stanton et al., 2008) to *aviation incidents*. The information network is a novel way of expressing who, within a system, owns what information, and how is it being communicated around the network. Information networks capture the information present within a system at any given time. Through comparison within and across networks the passage of information into, around and out of a system can be studied. Accident data within general aviation have been accrued and generic information networks have begun to be developed for different stages of flight. This paper looks at a generic landing network and investigates the integration of probability data to establish causality within the network. Manipulation of the networks based on conditional probability theory is a possible predictor of error migration routes following the imposition of constraints or barriers on a system. More than just illustrating the information present within a system space, these networks offer the opportunity to identify where barriers can be placed to improve the flow of information. In this way, the networks envelop the various aspects of communication and information-ownership that are central to distributed situation awareness.

The evolution of information networks

The Event Analysis of Systemic Teamwork methodology (EAST; Stanton et al., 2008) offers a network model based approach to understanding complex multi-causal, multi-agent and multi-linear accident sequences. Originally developed to examine the role of actors and groups of actors within complex socio-technical systems, EAST has been applied to a number of domains, including rail (Walker et al., 2006), aviation (Griffin et al., 2007), and, 'command, control, communications, computers and intelligence' (C4i) scenarios in the military (Stanton et al., 2006).

In order to apply the EAST methodology to a particular system, three network types are created for 'snapshots' in time, and can then be animated over a time period to illustrate the changes within the networks. The three network types used within this methodology are Task, Social and Information (figure 1). For a full description of the creation and population of each of these networks, the reader is directed to Stanton et al. (2008).

In D. de Waard, J. Godthelp, F.L. Kooi, and K.A. Brookhuis (Eds.) (2009). *Human Factors, Security and Safety* (pp. 123 - 129). Maastricht, the Netherlands: Shaker Publishing.

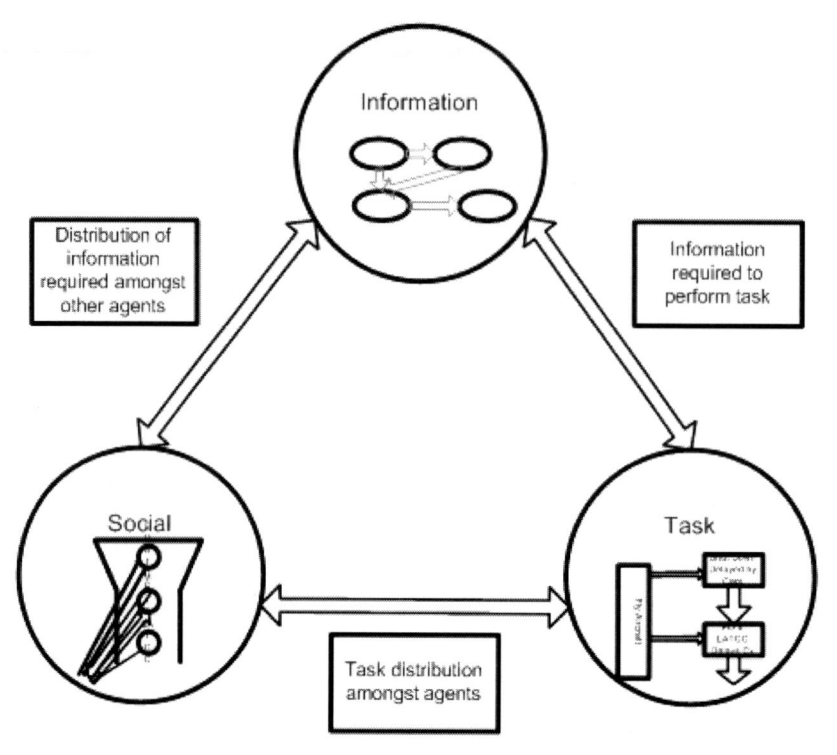

Figure 1. Interrelationship between Task, Social and Information networks (adapted from Stanton et al., 2008)

Network models of accidents and incidents to date (e.g., Griffin et al., 2007) have allowed us to illustrate and understand more clearly the information space in which decisions were made and interactions between agents (human and non-human) took place. The methodology allows identification of the causal factors central to an event, and is being modified to allow for the use of statistical analysis to highlight the major risks. In this paper, the methodology is adapted to allow the information networks to be populated with conditional probability data for the first time.

The information network is a novel way of expressing who, within a system, owns what information, and how is it being communicated around the network. Where appropriate these networks can be annotated to show conscious ownership of information nodes by specific agents within a system and making it possible to track information flow. Information networks can also be compared with 'ideal' alternatives. This ability to illustrate information propagation through a system clearly supports the application of such networks to the study of situation awareness and in particular team situation awareness (Salmon et al., 2008)

Figure 2 represents an information network of a generic landing network created from aviation accident and incident data. One hundred aviation incidents during the landing stage of flight were harvested from the Air Accidents Investigation Branch

(AAIB) database for general aviation aircraft between 2005 and 2007. General aviation (GA) is defined, in the UK, as any aircraft in use excluding military or commercial air transport. For the purposes of clarity, the one hundred accident scenarios used involved single-engine piston fixed wing light aircraft, e.g., the Cessna 152 or Piper Warrior. Each of the AAIB formal report transcripts were studied and, using no additional external information so as to limit bias, a taxonomy and a database of factors reported in the accident was developed.

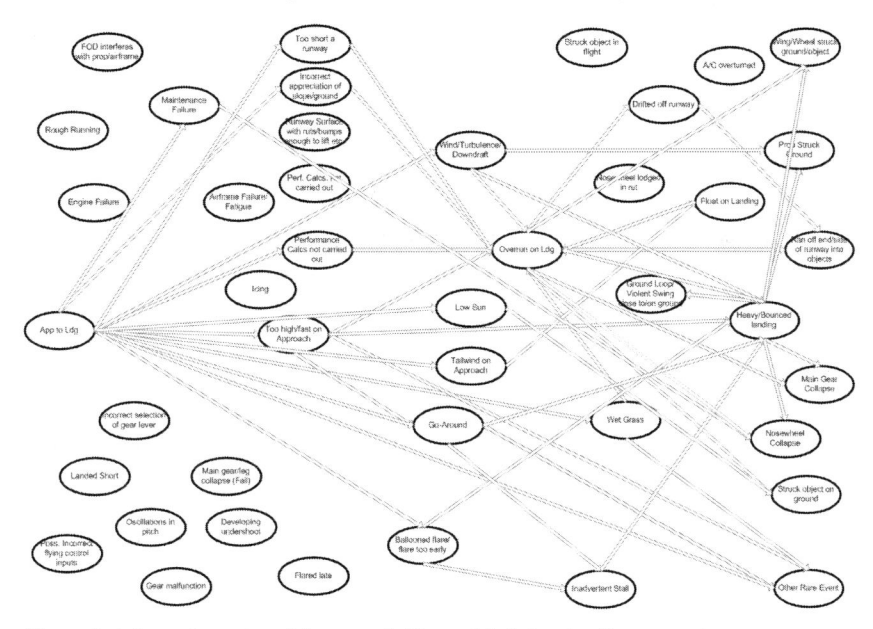

Figure 2. Information network for generic General Aviation landing scenario

Originally, information networks were modeled using knowledge objects taken from a Hierarchical Task Analysis for the nodes with propositions such as 'has' or 'causes' forming the links. However, for this study, the taxonomic labels, e.g., "too high/fast on approach", are entered as nodes into the information network and the links between nodes are drawn up to represent viable relationships, thus representing possible incident pathways. Due to the nature of these nodes, it is possible to consider some as 'influencing factors', others further along the information network as 'intermediate outcomes' or 'further influencing factors' and finally as 'outcomes'. More importantly though, each node represents information objects that are present within the system at that time. Further to this, an event cannot occur without the information being present in the system to indicate that event (even if that information goes unnoticed). The links, or arrows, between nodes represent the flow through this information network resulting in a specific outcome.

For the purposes of this preliminary study many of the information nodes which only occurred once in one hundred times were either grouped together or moved into the node "other rare event". This allowed for a more manageable network that still retains the potential to extend the networks to include hundreds or even thousands of accident pathways and additional nodes, limited only by computing power as will be seen later in the paper. To further limit the calculations required at this stage, a typical 'influencing factor', "too high/fast on approach" was selected. Only the associated nodes were kept, and all others removed from the network. The 'influencing factor' "too high/fast on approach" occurs seventeen times out of the one hundred accidents. The nodes associated with this network pathway are: "approach to landing", "too high/fast on approach", "heavy landing", "heavy landing & overrun", "overrun" and "nosewheel collapse". There are also the associated nodes of "go-around" (an approach that fails and results in another attempt), "decision to land" and "other rare event" included for completeness.

Probabilities and possibilities

Conditional probabilities

The frequency of occurrence of each node was recorded and from this, conditional probabilities given the occurrence of the preceding node(s) were calculated using adjacency matrices. Figure 3 shows the simplified information network for a "too high/fast on approach" landing scenario. The probability of each node occurring in the one hundred accidents, non-conditionally, is given on each link. The additional node of "decision to land" was created in order to allow statistical computation and comparison to the "go-around" node. By developing a series of equations in MATLAB (a numerical computing programming language) and populating them with the conditional probability data, it is possible to use the software to manipulate the values and calculate the effect on occurrence probabilities of each node in a Bayesian style.

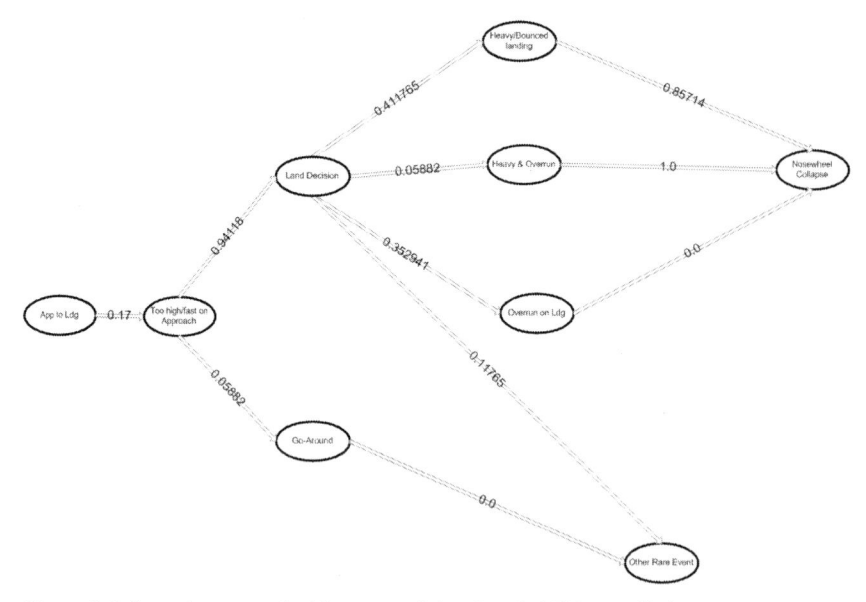

Figure 3. Information network with non-conditional probabilities applied

These modified networks essentially allow a level of prediction of the effect on the whole network of manipulating event probabilities. The next section introduces the idea of barriers within a network, and how this may provide a new way of looking at safety improvements combined with the methodology outlined here.

Barriers

Barriers are ubiquitously linked with safety in the literature and yet their use within aviation, outside physical barriers (e.g., engineering related or physical safety barriers), appears to be limited. Within network models of a system, it would be possible to change probabilities of occurrence of events, or probabilities of conscious ownership of information, by inserting (or removing) barriers into the links between nodes. It is these arrows, or links, that result in the flow of information within a system and any action to affect them should manifest itself as changes in outcome or changes in the probability of an outcome. Barriers, in this sense, can be seen as safety measures, i.e., the quality of a barrier defines the increased metaphorical 'distance' it creates from a negative outcome in this instance.

However, if probabilities must add up to one (i.e., the event will occur on 100% of occasions) and the barriers are affecting the probability of any node occurring within a network, then the value of the probability difference cannot simply be lost, so we must therefore look at alternatives to this. An adiabatic process, in relation to thermodynamics, is one where there is no heat lost or gained from/to the system (e.g., a gas or fluid) despite changes occurring within. In a similar vein, the probabilities within our aviation network system also cannot be simply lost or gained to exceed a probability value of one leaving an adiabatic style information system. If these probability values cannot simply disappear then the only option is migration of probability around the network and, given that any network model developed is aimed to be as comprehensive as possible, this may give some scope as to predicting the migration of the values around the system. Ideally, of course, the simulations should allow us to identify what barrier implementation or changes would allow for maximizing the positive, or safe, outcomes within a system.

Returning to our mathematical programming software, the author is now in the process of developing a Bayes' network type methodology of identifying all possible outcomes (i.e., effects on probability of related nodes) from modifying a particular link's probability with the addition of a barrier. This would identify possible information, and hence error, migration routes created when attempting to mitigate the accident pathways described in the information network. The benefits of developing a semi-automated mathematical program are that the calculations for a simple network such as that described in this paper are extensive yet once developed it is hoped the system could work on a much larger scale.

The evolution of a project

The potential for use of these networks does not stop at GA though. Commercial aviation holds great repositories and databases of abnormal operation information and accident/incident data that could be used to populate information networks in order to learn about salient information and communication issues. An information network model using probabilities in this way, it is anticipated, would combine very well with current safety management systems in use through aviation today. By exploiting the ability of the networks to give some insight into possible migration effects post-barrier implementation, it may be possible to provide cost-benefit analysis of future changes to an aviation system.

The analysis detailed in this paper uses only negative-outcome data (i.e., 100% of flights used resulted in an accident) and this currently limits the migration issues relating to a "safe outcome"; a central element of this work. The addition of positive-outcome data is not without its challenges but commercial aviation certainly has the equipment and ability to record and observe vast amounts of normal operation data should it prove useful to do so.

Central then to this methodology is the ability to construct relevant and comprehensive information networks for the aviation system. These information networks allow a visualisation and comparison of the propagation and flow of information within a system, providing a higher level of understanding of a team, or of distributed, situation awareness (Stanton et al., 2006a).

As an ongoing project there is much to be refined and honed in order to ensure a high quality of results. A study using a multi-axis flight simulator based at Brunel University has been developed to validate the results gained from the conditional probability manipulations for the General Aviation landing scenario outlined in figure 3. The aim of this study is to place pilots in the landing scenario mapped out above and then attempt to manipulate the network-barriers by controlling pilot behaviour. A full methodology and results will be produced and disseminated once the study is complete. It is hoped this study will begin to validate the use of conditional-probability networks to understand more fully the development and migration of error in General Aviation incidents.

References

Griffin, T.G.C., Young, M.S., & Stanton, N.A., (2007) A complex network approach to aviation accident analysis. *Contemporary Ergonomics 2007.* Aldershot, UK: Ashgate Publishing

Salmon, P.M., Stanton, N.A, Walker, G.H., Jenkins, D.P., Baber, C., & McMaster, R. (2008) Representing situation awareness in collaborative systems: A case study in the energy distribution domain. *Ergonomics, 51*, 367-384

Stanton, N.A., Walker, G.H., Houghton, R.J., & Salmon, P. (2006) *Work Package 1.1.4: Generic Process of C4I Activities.* Alvington, Yeovil: Aerosystems International

Stanton, N.A, Stewart, R., Harris, D., Houghton, R.J., Baber, C., McMaster, R., Salmon, P., Hoyle, G., Walker, G., Young, M.S., Linsell, M., Dymott, R. & Green, D. (2006a) Distributed situation awareness in dynamic systems: theoretical development and application of an ergonomics methodology. *Ergonomics, 49*, 1288-1311

Stanton, N.A., Salmon, P.M., Walker, G.H., Baber, C., & Jenkins, D.P. (2008) *Modelling Command and Control*. First Edition. Aldershot, UK: Ashgate Publishing

Walker, G.H., Gibson, H., Stanton, N.A., Baber, C., Salmon, P., & Green, D. (2006) Event Analysis of Systemic Teamwork (EAST): A Novel Integration of Ergonomics Methods to Analyse C4i Activity. *Ergonomics, 49,* 1345-1369.

PAMELA, a portable solution for workflow support and human factors feedback in the aircraft maintenance environment

Martine Hakkeling-Mesland & Koen van de Merwe
National Aerospace Laboratory NLR
Amsterdam
The Netherlands

Abstract

This paper describes the NLR PAMELA (Personal Aid for Maintenance Engineers in Line operAtions) concept and the graphical user interface design of some aspects of it, developed in the EC co-funded projects TATEM and HILAS. The PAMELA concept is based on the use of nomadic computer devices to provide workflow support to aircraft maintenance engineers. It includes a set of functions for potential efficiency improvement, error reduction and the collection of performance data and human factors information. Earlier research in this area yielded a better understanding of the issues related to documentation use and task support requirements. The structure of the concept is modular, thus allowing for supplementary functions to be added.

The TATEM project focuses on the process orientation of the future maintenance system and the HILAS project on the human element in the lifecycle of aviation systems. These distinct project scopes allowed for different yet complementary viewpoints and functionalities to explore and develop under one coordinated concept. Due to the generic character of the EC projects, the project results are not expected to be *the* solution for one specific maintenance organisation, but sets of elements can be taken out for further customisation by individual end-user parties.

Introduction

Previous research has identified a number of human factors issues in the aircraft maintenance environment, of which a large number were related to the use of procedures and task documentation, as described by Van Avermaete and Hakkeling-Mesland (2001). Many of the results pointed out that a better form of task-support was needed in order to improve both safety and efficiency. Two main indicators as derived from these studies, which were based on a number of surveys done in maintenance organisations across the EU, supported this conclusion: the high figures for non-compliance to task procedures (up to 34% of the total number of cases), and the omnipresent use of so-called 'black books', illegal and personal note books, containing notes for later reference. Deficiencies in the content of procedures (e.g. there are easier and quicker ways) and the cumbersome accessibility of the information with respect to the actual way of working were often cited as motivation for this behaviour. Although a strong commitment to safety was observed, so was a double standard that allowed the official and actual way of working to exist side by side.

When unofficial ways of working take place, this clearly has an impact on safety. Safety systems (e.g. sign-off for having completed the work according to procedure) are being by-passed and the organisation will not be able to monitor what actually happens. In addition, operational feedback on the mismatch between the prescribed and actual way of working will not become available. This would potentially be relevant information for the improvement of technical and organisational procedures and processes. Ironically, the high sense of safety that was experienced among maintenance personnel was found to contribute to this situation occurring. Aircraft Maintenance Engineers (AMEs) were found to be inclined to take on additional tasks on their own discretion for safety reasons (e.g. to perform additional inspections of problematic parts). This would add to an already high workload situation and lead to (perceived) high time pressures. Based on the same sense of safety, it is therefore expected that AMEs will not be unwilling to follow the task instructions to the letter, if the required information for the task is provided in a more efficient manner, at the right time. A functional feedback and improvement system, effectively using comments from the work floor to adapt procedures and processes, potentially would add to this. Nomadic computer devices could facilitate both information consultation and feedback provision and as such would contribute to an overall objective of continuous safety and efficiency improvement.

The PAMELA concept is based on the use of nomadic computer devices. Its main goals are to provide workflow support to AMEs and simultaneously capture human factors and performance related data and information. The use of nomadic devices will directly improve the on-the-job availability of the task information. In addition, a task oriented interface design is applied, which, combined with context based linkage of information, is expected to improve the accessibility of the information. The PAMELA concept comprises the following three basic elements: First, access to the maintenance information to perform the task. Second, additional functions relevant for task execution, such as links to relevant support departments and automation of administrative actions. And third, functionality for data collection and provision of feedback. The modular set-up of these functions allows for future extensions to be added.

The PAMELA concept is being further developed in the ongoing EC co-funded projects TATEM (Technologies And Techniques for nEw Maintenance concepts) and HILAS (Human Integration into the Life cycle of Aviation Systems). Both projects have a specific outlook on the future maintenance system and potential improvements. The TATEM project focuses on new technologies for health managed aircraft and the process orientation of the future maintenance system whereas the HILAS project focuses on the human element in the lifecycle of aviation systems. In both projects a selection of user interface design elements were developed that, although based on distinct project scopes, fit in with the overall PAMELA concept.

Description of the PAMELA concept

PAMELA is meant as a personal aid. Based on previous research, the hardware of the PAMELA concept comprises a suite of devices appropriate for the various working conditions that occur in aircraft maintenance, such as a PDA, a small laptop or a large screen display. Input mechanisms are either a hard or soft keyboard, stylus (touch screen) or voice control. The hardware should be ruggedized to outlast the harsh maintenance environment. Ideally, in the future situation, an AME should be able to pick up the hardware device most appropriate for the task to be executed. Application of the same interface design philosophy regarding user interaction and 'look and feel' for each of the devices will allow for smooth conversions between devices used for different tasks. The equipment is connected through a wireless network link with a data kernel containing the up to date versions of all relevant information.

Main users of PAMELA are AMEs directly involved in the execution of maintenance tasks. Either individual AMEs or two or more AMEs working together on a task should be able to use the system. The concept is aimed at the future maintenance situation, where the current distinction between line and base maintenance is expected to decrease. Fewer long maintenance visits and more scheduled maintenance executed at night are expected to characterize the future situation.

The core function of the PAMELA system is to assist AMEs in the preparation and execution of the maintenance tasks (see Figure 1). This is achieved by providing technical task information (e.g. Maintenance Manual, parts catalogues, Service Bulletins), as well as a link to onboard and in-flight reported fault information (e.g. to the Built in Test Equipment system or a Digital Electronic Flight Bag). Around this core, additional task support functions have been defined with the aim to improve efficiency and safety (e.g. through error reduction). A set of administrative functions allows the accumulation of digital information related to the execution of the task. This type of information consists for example of sign-off, task handover, non-conformance, damage and incident information which is currently captured on paper forms. A planning function, linking to the planning department, informs the AME directly of the tasks to be executed in the near future (e.g. the shift). Likewise, links to shop and stock database systems are provided to allow checking the availability of tools and parts and ordering them online. Furthermore a notes function is included for digital notes to be made comparable to the described current 'black books' either for personal use or to share across the organisation. These notes are digitally linked to the official maintenance documentation they relate to and stored in a database. As such they are available for review and editing by the organisation (e.g. the Engineering department) and make up a much more controlled information source.

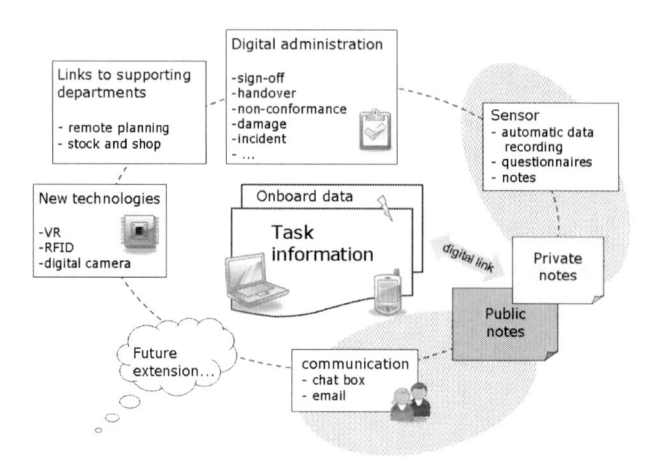

Figure 1. Schematic presentation of the PAMELA concept

Three technologies for performance support are included in the concept: Virtual Reality (VR), Radio Frequency Identification (RFID) and a digital camera. VR can support the visualisation of the task, including human factors information, which enables previewing and rehearsing tasks and to provide on-the-job training. RFID technology can be used for recognition of specific parts or equipment. The RFID information can serve as a quick reference to related information, as such omitting for manual data entry (errors). The digital camera can be used for documentation purposes and to facilitate distant consultation, for example in case of damage.

For a team of AMEs additional functions for information sharing are included in the concept. These allow for consultation of other engineers, comparable to a chat box or email application. Furthermore notes and experiences can be shared, for example by making notes publically available. Functionality to overview the status of different tasks executed in parallel on the same aircraft is expected to improve shared situation awareness.

At the level of the maintenance organisation PAMELA serves as a 'sensor' to collect human factors and performance related information from the work floor. For example, information could be gathered that relates to the workability of the maintenance procedures to recordings on the time spent on a specific repair. This is achieved in three ways. First, data are automatically recorded whilst the tool is being used. For example data could be captured on the information sources accessed and the duration of consultation. Second, information is collected with the use of (short) digital questionnaires to be filled-in by the user, either to voluntarily provide feedback or upon request by the system at pre-defined instances. Third, review of the digital notes made through the notes function will provide the organisation with feedback on the actual way of working. The collected information needs review for validity and relevance and could be used as input for improvement of procedures and processes. The development of the tools for data management and analysis and the organisational system to utilise the data for such purposes is outside the scope of PAMELA. This is however taken on as part of the HILAS project, as described by Ward and McDonald (2007).

As part of the TATEM and HILAS projects the user interface design for a number of the above described functions and different hardware devices was undertaken in a contextual and user-centred manner.

PAMELA developments in the TATEM project

The TATEM maintenance perspective

The TATEM project investigates new technologies and techniques as a means to considerably reduce operators' maintenance related direct operating costs. A health management approach is developed to improve the understanding of the aircraft status. Subsequently fewer costly delays and cancellations are expected that are the result of unscheduled maintenance. In addition TATEM aims to improve the efficiency of both scheduled and unscheduled maintenance through the application of a process-oriented approach. Improvements in relation to personnel and their equipment are based on the introduction of a Portable Support System (PSS) comprising the use of nomadic computer devices and process oriented information application. In short this means the AME has direct access to the information needed at a certain process step, without the need for searching. In Buderath *et al.* (2008) the TATEM ground crew support system, including the technical data, the implementation and the related operational impact are described in more detail.

User interface design in TATEM

The TATEM PSS consists of a set of commercial off-the-shelf nomadic computer equipment, including a laptop, PDA and helmet mounted display. The selection was based on a combined approach of applying defined functional and technical specifications and a selected use case. The use case consisted of a troubleshooting task on the brake system where two AMEs had to work together and execute task steps in a sequential manner; one AME was positioned at the landing gear while the other was in the cockpit. The selected PDA contained a 7 inch colour TFT touch display with 480 x 640 pixels resolution in portrait format. As part of TATEM, NLR focussed on the development of a functional PDA user interface prototype using NLR's rapid prototyping tool Vincent (Verhoeven & De Reus, 2004).

The task-centred user interface design was based on the selected use case, fed by inputs from project partners, previous research on PAMELA and current PDA applications. Three main constraints were taken into account. First, the display size of the PDA is very limited and does not allow for full maintenance procedures to be presented. Second, the system needs to be fed with responses from the user during execution of the task, in order to determine the next branch of information to present. Third, the two AMEs working together on the task need to be informed of each others actions and progress. Due to unavailability of the future technical data during the initial interface design phase, the prototype was based on existing technical data (maintenance procedures), adapted to the prospects of the future system and the constraints of the PDA interface.

A number of workflow aspects were incorporated in the interface design to demonstrate its future application: a main menu containing the foreseen main functions of the PDA, a timeline presenting the planned tasks for an AME (Figure 2), pages containing the instructions for the task to be executed and a summary page of the work executed. The latter gives access to a sign-off page using fingerprint identification when a task is completed.

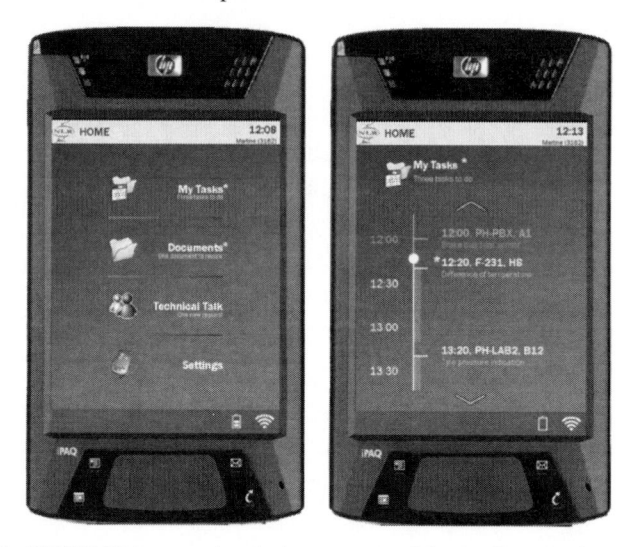

Figure 2. The TATEM PDA presenting the home page and the timeline. A colour image can be found at http://extras.hfes-europe.org

The task instructions were rephrased into concrete and short action statements, containing checkboxes or yes/no radio buttons to be ticked by the user. The latter provides necessary feedback for the system to determine which statement to present next. They were split up in logic sets of statements to fit on a page, thus omitting the need for scrolling. An icon in the shape of a green check-mark or a red cross appears in front of the line to indicate if the inputs provided by the user were correct or not, as expected by the system. This gives a quick overview of the status of the task statements on a page: a list of green check-mark icons indicates that the task statements on the page have been completed correctly. When a red cross appears, something may have been omitted or may not be in the correct state for continuation of the task. A 'continue' button appears at the bottom of the page when all the required ticks are set, as such preventing task steps being omitted. Warnings and cautions are implemented in red and orange coloured textboxes that need to be ticked off in order for the user to be able to continue. When ticked they are minimised to one line and can be expanded again for review. Notes are presented in a green text box that is by default minimised to one line. By ticking it and holding the stylus, the user can expand the box and read the full information. Figure 3 presents a number of the display features described.

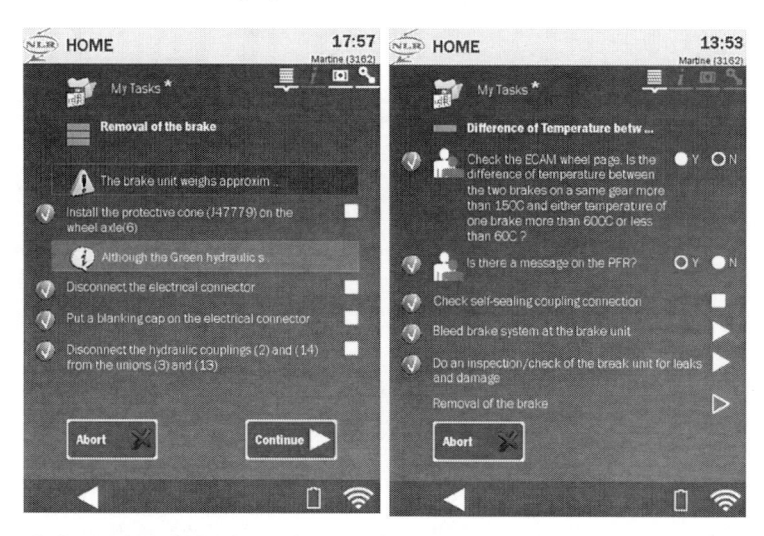

Figure 3. Screenshot of PDA pages during task execution. A colour version is available at http://extras.hfes-europe.org

When two (or in theory more) AMEs work together on a task, they can access the same set of task information together with each others inputs. An icon, resembling two people, is presented in front of a task statement, to indicate that the instruction belongs to the other AME. Both AMEs can make inputs for their co-worker if necessary (e.g. based on radio communication). This option was deliberately not blocked to prevent for unnecessary delays.

PAMELA developments in the HILAS project

Role of PAMELA in the HILAS project

The main goal of the HILAS project is to establish continuous improvements based on the integration and awareness of human factors across the life-cycle of aviation systems. The idea is that if tasks and procedures are designed with the human in mind (human centred design) efficiency gains can be made, and errors can be prevented. The project is divided into four strands of which one focuses on Maintenance (Ward, *et al.* 2008). New methodologies and technologies for monitoring and evaluating organisational system performance have been developed as well as a framework for data collection and exchange; the HILAS Knowledge Management System. As part of HILAS, PAMELA is further developed as one of the tools for the AME which aims to improve performance as well as to capture important human factors information. As such, PAMELA plays a role at both ends of the HILAS scope (Figure 4).

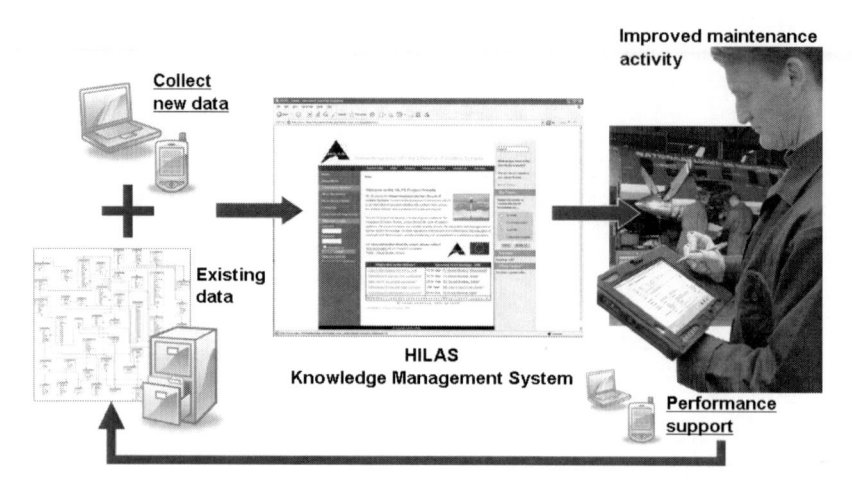

Figure 4. PAMELA as part of the HILAS process

The user interface design in HILAS

The tool that was developed primarily aims to assist the AME during task preparation and execution. This was demonstrated through the use of a portable device which enables the AME to perform activities in an integrated manner. Currently such activities are distributed throughout the workspace (preparation in the office, task execution at the aircraft, documentation printing at another location). A state-of-the-art ruggedized notebook was chosen as a suitable candidate. With a resolution of 1024 x 768 pixels in landscape format, the device strikes a balance between information density and the readability of the screen. The user interface was developed in HTML format based on current web-standards that allow for an easy integration into most standard operating systems.

A design was chosen which incorporated three distinct areas, a status bar on the top, a central content area and a navigation area at the bottom. By placing the navigation bar horizontally at the bottom both left- and right-handed AMEs can navigate the system without obscuring information when buttons are pressed.

The user interface was developed based on previous research on PAMELA, input from consortium partners and field research within the HILAS project. Part of the research focussed on creating a flowchart which structured the task of an AME and served as an input for the user interface development. It was found that the jobcard played a central role in each of the activities of the AME. On the jobcard, tasks are planned, task details are provided, activities are reported, engineers are assigned, tasks are signed off, etcetera.

The digital jobcard (Figure 5) in PAMELA has areas for pre-determined information about the task and the aircraft that can be populated by the planning department. Other areas are available for information entry by the AME. Information that is pre-determined consists of aircraft type, registration, skill requirement, date, location, expected time required for the task, manuals, tools and material required, etcetera. User input consist of actions taken once finished. The layout of the jobcard changes depending on the progress of task execution; at the start of a task it provides the AME with the required information for preparation and provides access to the task procedures, while during task execution it additionally serves as a data entry tool. Switching between the task procedures and the digital jobcard allows the AME to immediately enter specific actions taken into the jobcard. The information collected on the job card provides a quick overview of the work done and as such facilitates sign-off of the task. After the task has been executed a sign-off occurs through the use of a badge or fingerprint identification.

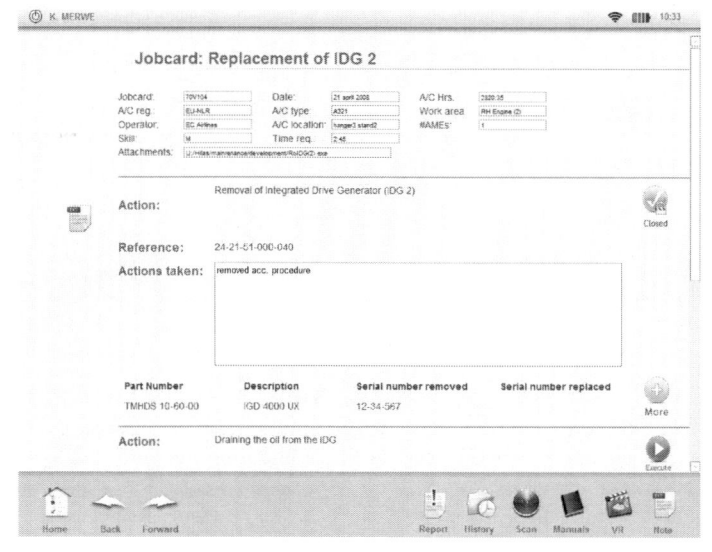

Figure 5. The digital jobcard. A colour version is available at http://extras.hfes-europe.org

During task execution the user is presented with the maintenance procedure belonging to that task. The presentation of the procedure is context-based which means that it is presented as a link in the digital jobcard; the AME does not have to search for the correct procedure (correct aircraft type, effectivity, etc.). The procedure is accompanied by a number of checkmarks that are required to be ticked once the related part of the task is executed. This way the engineer is guided through the procedure and stricter adherence is ensured. In addition the checkmarks provide a means for progress monitoring for both the AME and the system, which can for example be used for information provision in case a task handover takes place. At any point during task execution the user has access to a number of support functions such as VR based task models and RFID technology based support.

An RFID function was included to scan RFID tagged equipment in the vicinity, making use of actively transmitting RFID tags, and to select the relevant equipment for the task being executed. The identified part number and serial number of the specific part at hand are used as reference to access the related documentation and equipment information (e.g. part history, EASA form 1). Additionally, they are used to record information on the removed and installed parts on the digital job card, if applicable.

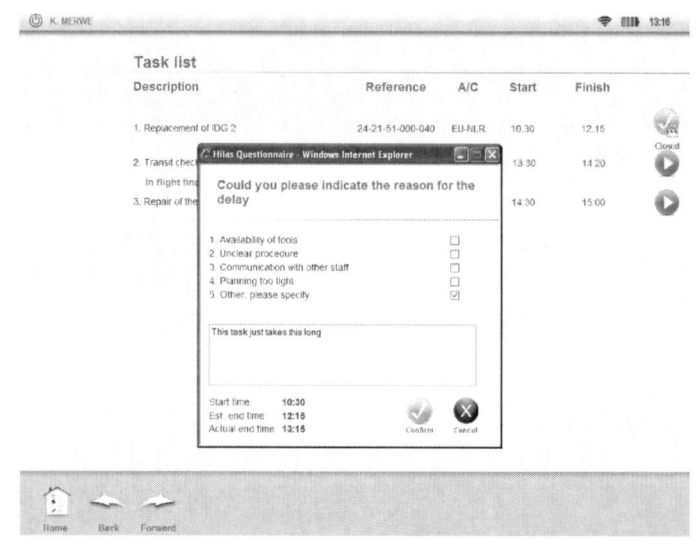

Figure 6. Sensor on the work floor: a digital questionnaire

As part of HILAS, the sensor function of PAMELA was further developed in three ways. First, the information on task progress (timings) as can be collected through the use of the checkmarks is intended to be automatically recorded. Second, during the task execution the user is prompted for information input when specific situations have occurred, for example when a task has taken significantly longer than expected. The causal factors of such overruns are currently not systematically monitored. The AME is presented with a pre-defined list of responses and a free text option for efficient completion of the questionnaire (Figure 6). Third, throughout the procedure the user can add notes to a specific part or the entire procedure. This function allows the user to create personal reminders on how to execute a task as a better controllable digital variant of the current 'black books'. In addition the note system can serve as input for procedure improvements.

Initial evaluation trials in HILAS

Initial evaluation trials with eight Aircraft Maintenance Engineers from two European maintenance organisations showed a very positive general response to the PAMELA concept. Significant improvements related to time savings are expected. An example cited is the reduction of time spent travelling to obtain information, particularly with increased security measures at airports. In the current maintenance situation the PAMELA concept was considered to be more suitable for base than line maintenance. The latter was in general considered too hectic and not structured enough to continuously use a portable support tool for task execution. This is an interesting outcome, given the future situation where more scheduled maintenance is anticipated, for which the PAMELA concept was developed. The user interface design was rated very positively in terms of the amount, relevance, logic and usefulness of the presented information and the workflow structure implemented. Certain elements may need further customisation to cater for differences between organisations. The sensor function was appreciated, provided the organisation would explain its purpose and take privacy considerations into account. Particularly the gathering of data on how long tasks actually take was welcomed as a means to relieve some pressure of unrealistically high expectations off the front line production. AMEs tended to be slightly sceptical about the organisations' willingness to invest in the hardware and worried about the vulnerability of the equipment.

Discussion and outlook

Both the TATEM and HILAS projects are ongoing and the user interface designs as presented are being further progressed towards functional systems for evaluation. In the TATEM project, a more advanced prototype including an active link to the kernel containing the technical data is being developed, whereas in the HILAS project the interface design is being integrated with additional support functions as developed by other project partners.

In the TATEM project, the process oriented approach brings major changes to the way AMEs interact with the maintenance documentation. Two topics that were found to require more attention in this future system are the situational awareness and the user's sense of control. In the current situation, maintenance tasks are organised in a hierarchical manner and often executed as subtasks of tasks at a higher level. Links to subtasks, task names and reference numbers are clearly documented and function as hyperlinks in digital versions. While executing the work AMEs follow this structure and actively look up the correct information, relative to the situation at hand. This allows them to build a mental picture of the situation and to be very much involved in the task process. In the process oriented system however, the task instructions are step by step dictated by the system. Potentially this could lead to an experienced lack of situational awareness and sense of control. In the TATEM PDA prototype a certain amount of hierarchical information was provided as a compromise between the current and a full process oriented system.

In the described HILAS prototype the sensor function to directly collect data from the work floor is new and promising. Only a small number of examples of what data and information could be collected in this way have currently been implemented. The HILAS project aims to further demonstrate the use of this information for procedure and process improvement. Gathering data automatically by recording of user interactions to some extent requires adaptation of the user interface (like the checkmarks in the procedures of the HILAS PAMELA design), and as such has an effect on user interaction with the system. This may take additional interaction time and may as such interfere with the work to be done. Further evaluations will have to determine whether this is acceptable and to what extent, both in terms of task performance and related cost.

The PAMELA concept largely focuses on efficiency and safety improvement in the future maintenance context and contains a number of features to help executing the maintenance tasks in a more guided and controlled way. In the TATEM and HILAS projects, some of these features were implemented in the user interface. Initial responses were promising and potential for time savings and efficiency improvement has been identified. Current trends in aircraft maintenance, like the reduction of aircraft downtime intervals, an increase in outsourcing of maintenance leading to less familiar aircraft to be maintained, and a changing population of technicians even further stress the need for performance support tools.

Acknowledgements

The authors would like to acknowledge the European Commission for co-funding both the TATEM and HILAS projects under the 6th Framework Programme and the TATEM and HILAS project partners for their participation in the current work.

References

Buderath, M., McDonald, N., Grommes, P., & Morrison, R. (2008). The operational impact to the maintainer (Ground Crew Support and Human Factors). In *Proceedings of 2008 IET Seminar on Aircraft Health Management for New Operational and Enterprise Solutions*, London, England: The Institution of Engineering and Technology (IET) (pp. 1-48). Retrieved February 23 2009, from http://ieeexplore.ieee.org

Van Avermaete, J.A.G. & Hakkeling-Mesland, M.Y. (2001). Maintenance human factors from a European research perspective: results from the ADAMS project and related research initiatives. *15th Annual FAA/TC/CAA Maintenance Human Factors Symposium,* London, England Retrieved 23 February 2009, from http://www.hf.faa.gov/docs/508/docs/avermaete15.pdf

Verhoeven, R. & De Reus, A. (2004). Prototyping interactive cockpit applications. In *The 23rd Digital Avionics Systems Conference, 2004 DASC 04 proceedings, Volume 2, 24-28 Oct.* (pp. 9.A.3 - 1-10). Piscataway, USA: IEEE

Ward, M. & McDonald, N. (2007). An European Approach to the Integrated Management of Human Factors in Aircraft Maintenance: Introducing the IMMS. In *12th International Conference, HCI International 2007 proceeding* (pp. 852 – 859). Heidelberg, Germany: Springer.

Ward, M., Gaynor, D., Nugent, T., & Morrison, R. (2008). HILAS maintenance solutions; challenges and potentials for the aircraft maintenance industry. In *Proceedings of 2008 IET Seminar on Aircraft Health Management for New Operational and Enterprise Solutions*, London, England: The Institution of Engineering and Technology (IET) (pp. 1-42). Retrieved February 23 2009, from http://ieeexplore.ieee.org

Human Factors in the display and use of aeronautic information from different sources and of different status

Yvonne Barnard
Institute for Transport Studies, University of Leeds,
Leeds, UK

Abstract

In aeronautic operations and maintenance a large amount of information is provided by documentation and instruments which is needed for the safe operation of an aircraft. However, a development is taking place towards the use of multiple sources, with information being integrated in one display, and the construction of meaningful knowledge in interaction with the user, providing support for decision making and diagnostics. Combining information from different sources means that information could differ in status, age and certainty. The advantage of combining information is in providing the user with a clear picture of the situation, highlighting information that is context-relevant, and ensuring all available information is provided. A wide range of human factors issues is related to perceiving, interpreting and using information from different sources and of different statuses. Cases from different studies are presented in order to address common human factors and design recommendations. These studies deal with the development of demonstrators of information presentation as well as interviews and evaluations with users. Topics include maintenance manuals connected to aircraft systems, electronic flightbags, and cockpit displays. The paper addresses the human factors issues identified, and indicates directions for solutions for information presentation, such as layered information, contextualisation, and integrated information.

Introduction

In flight operations and aircraft maintenance, a large amount of information is provided by documentation. Traditionally, information was provided in paper manuals. In flight operations pilots are also supplied with information from the cockpit instruments and from air traffic control. It is up to the operator to combine information from different manuals and sources to decide which action to take or how to perform a task. Now that documentation is becoming available in electronic formats, new possibilities arise to combine information from different sources, and to provide operators with integrated and complete information (Barnard & Chandra, 2004; Barnard et al., 2004). The advantage of combining information lies in providing the user with a clear picture of the situation, highlighting information that is context-relevant, and ensuring all available information is given. However, there

In D. de Waard, J. Godthelp, F.L. Kooi, and K.A. Brookhuis (Eds.) (2009). *Human Factors, Security and Safety* (pp. 143 - 157). Maastricht, the Netherlands: Shaker Publishing.

may also be a danger attached to having information from different sources, especially when the information is contradictory. For example, information about the weather conditions in a destination area may be based on different predictions by different meteorological services. Combining information from different sources implies that information provided may differ in status, age and certainty. In this paper first different sources of information will be discussed, and examples will be given of information provided by electronic systems for both flight operations and maintenance. Then several human factors are presented, related to the use of information from different sources and of different status: perception and interpretation, situation awareness and workload, dealing with uncertainty, dealing with factual and interpreted information, and biological aspects. In the subsequent section several display solutions are discussed, especially the concepts of layered presentation of information, contextualisation of information, and providing an integrated picture. The paper concludes with a discussion of advantages and risks of providing integrated information, and a short discussion on transfer to other domains.

Different sources and properties of information

Information may come from different sources. Aircraft instruments convey information about the status of the aircraft and its systems, and about the environment, for example information from the weather radar. Manuals, both operational and maintenance ones, provide information coming from the manufacturer. Often different manufacturers are the sources of documentation, because an aircraft may contain a variety of systems from different origins. Other sources also provide information, both in real time and previous to operations, in the form of bulletins, for example meteorological services. Operators may communicate with others, for example with air traffic control, airlines, colleagues, other aircraft, etc.

An example of information coming from different sources is information about the weather. Pilots may receive information about bad weather conditions in the destination area from their aircraft's weather radar, from a cockpit display showing weather information, from a meteorological bulletin received before the flight, warning from air traffic control, information from their own airline, they may have access to internet, and pilots from other aircraft may send out messages. In their flight manuals they also have information on how to deal with certain weather conditions. The pilots need to combine all the information in order to establish a clear picture about the situation at the airfield of their destination, and to decide whether they may safely land there or whether they should start making provisions for diverting to a different airfield. All the information concerned is not of the same nature. The information is of different ages, different scope, and is of differing certitude. The weather radar gives real-time information about the weather ahead of the aircraft; weather bulletins give information about the weather forecast in a larger area.

Next to information coming from different sources, information may also be interactive or not. Older versions of electronic documentation contain pre-composed information: information previously composed into a static composed state (non-interactive). Their displays have consistent, defined and verifiable content, and fixed formats. Newer forms may also contain variable information that can be updated during operations. Software applications allow for selecting and rendering in a number of dynamic ways. In the new generation of information systems, the information displayed will consist of these two kinds of information, providing a mix of interactive and pre-composed information. The question to be asked is whether operators are aware of the nature of the different information elements, and how they integrate and use this information. Users cannot only consult information, but may enter into a dialogue, a conversation, with an information system. Both the system and the user may take the initiative to start a dialogue and determine how the conversation is to be continued. Just as for conversations between humans, both partners have to obey to basic conversation rules in order to be intelligible to each other (Novick & Ward, 2003).

Examples of systems providing information from different sources

In this paper two different applications are discussed that provide information from different sources: electronic flightbags and electronic maintenance manuals. By looking at different aeronautic domains a more general vision may be obtained of underlying human factors problems, and new directions for solutions may be found. Both flight and maintenance manuals are used for understanding given situations and (potential) problems, and for providing support in decision making and performing tasks to address the problems. Although maintenance and flight operations are very different processes, some commonalities may be found in the way in which information can be presented in an efficient and easy to understand manner. A major commonality is the safety-critical aspect; wrong interpretations of the information may lead to major disasters and even loss of human lives. In both operations, some tasks are performed under time-pressure while other tasks allow for more reflection and time to search for information. In flight operations, there are also extremely urgent tasks, to be decided and performed within minutes (or even seconds).

Electronic flightbags

An EFB is an electronic information management device that is used by crew members to obtain information currently provided in paper form. EFB devices can display a variety of aviation data, and perform basic calculations (e.g. performance data, fuel calculations, etc.) The scope of the EFB functionality may also include various other hosted databases and applications. Physical EFB displays may use various technologies, formats and forms of communication (Shamo, 2000; FAA, 2003; Chandra & Yeh, 2006; Yeh & Chandra, 2007). Applications that are currently available in EFBs on the market are: Electronic charts, Electronic checklists, Electronic documents, Flight Performance Calculations, Flight Planning, Surface Moving Map, Video Surveillance, Weather information of all sorts, Logbooks, Electronic mail, Terrain awareness system, Note-taking, and Traffic display. This list is not exhaustive, new applications are currently becoming available (Barnard et al., 2007a). Pilots use the information provided by the EFB in order to make decisions about the flight, or to understand what is going on. This may take place during the flight, but also in preparation, or in debriefing after the flight. The EFB may also be used for learning purposes, either in formal training or in self-learning processes (Barnard et al., 2002).

Maintenance manuals

Maintenance operators use a set of maintenance manuals, such as the AMM (Aircraft Maintenance Manual), IPC (Illustrated Parts Catalogue), and troubleshooting manuals. These manuals are available in an electronic format and use a viewer to give access to different manuals. Next to the manuals, information is available about the particular aircraft to maintain, and its history and configuration. Also statistical data are becoming available about the frequency of maintenance problems. Data for maintenance manuals come from the manufacturer of the aircraft, different aircraft systems often being manufactured by different companies, for example, the engine may come from a different company. As modern aircraft integrate more and more electronic systems, the number of producers of hardware and software may increase. A more radical step in having to deal with information from different sources is the development of intelligent aircraft systems which are able to perform self-diagnosis, and may even be able to do self-repair to some extent. We are moving towards a situation in which maintenance manuals are connected to the aircraft system under maintenance. When a part of the procedure is accomplished this is automatically detected and indicated in the electronic manual. For troubleshooting such options are feasible as well. In this way the logging and reporting process will become automatic. If the system cannot detect a part of the procedure itself, the operator has to give input. The system to be maintained may also take the initiative, and directly open the right procedure necessary to perform the right maintenance action, either because it is time to do so or because a fault has been detected (Barnard et al., 2007b).

Human Factors involved in using data from different sources

When an operator uses information in order to make decisions or to perform a task, he/she has to take the following steps. The operator has to assess the situation and interpret it in order to determine whether he/she should take an action, and if so, what kind of action. Based on the interpretation the operator has to search for information that is relevant. He/she has to filter it from all the information and noise available. When information is found, it needs to be interpreted with regard to the task the operator wants to perform, in other words, the information found should be matched with the operator's interpretation of the system. The usefulness and validity of the information selected needs to be evaluated. Finally the operator uses the information in making decisions and/or performing a task. If, for example, there is a malfunction in a system, the operator has to assess the nature of the malfunction, and search for information in the manual on how to remedy it. The operator has to make sure that the malfunction described in the manual, and the task proposed by it, are indeed what he/she was looking for. Finally the operator uses the information from the manual to remedy the malfunction. This example is rather straightforward, but things may become much more complicated if information comes from different sources and with different statuses. If a pilot has to decide whether to go to the airfield with this malfunction while bad weather has been predicted, he/she will have to put quite some effort into the matching and evaluation step. In all these steps a variety of human factors issues play an important role.

Perception and interpretation of information

Already in the phase of perceiving and interpreting information, errors may be made. Novacek et al. (2001) found that pilots have difficulties with the graphical display of METARs (Meteorological Aeronautical Report), because of the limited information they provide, difficulties in interpreting the display, or their inability to present information timely. Easy perception is a pre-requisite for correct interpretation. However, correctly perceived information does not necessarily lead to a correct interpretation. Potentially dangerous situations may be shown in such a way that it is very easy to perceive them. For example, an area in which a thunderstorm is located may be coloured red. However, there is a danger that in the pilot's interpretation the areas around the danger zone are safe. This may not necessarily be the case, for example because danger has not been detected or because it is relatively smaller, but the situation still contains risks. Absence of a danger alert does not mean that a situation is one hundred percent safe. The weather display in the experimental study of Novacek et al. (2001) in principle provided sufficient weather information, and was easy to use, but many pilots did not fully understand that the weather they had to deal with was indeed bad and dangerous, and thus made incorrect decisions. This is in line with Forman et al. (1999) who found that pilots do not always fully understand weather information.

In order to be able to make decisions in certain situations, both flight and maintenance operators have to build a mental model of the situation their aircraft is in and the events that will take place in the future. So there are two representations of the real situation, the first being the representation made by the aircraft systems, and the second the representation the operator has of the situation, mediated by the representation given by the system. In both representations an abstraction is made out of the richness of detail in the real situation. If the operator does not have sufficient, or correct, information, his/her mental model will be incomplete or even incorrect. If, on the other hand, the systems provide too much information, the operator's mental model will inevitably be a simplification of the real situation (Barnard et al., 2006).

Situation awareness and workload

Having good and easy access to information (such as about routing and runways) may reduce workload considerably. However, if information is displayed in a cluttered way, or if one of the aircrew does not have good visibility, workload may increase (Theunissen et al., 2005). Chandra and Yeh (2003), who perform evaluation studies of commercially available EFBs, identify requirements and recommendations concerning workload. Cf. page 6: "Using an EFB requires effort that may be different from that of using paper. There may be effort involved in locating and orienting the display for use and there is effort in looking at the display, processing the information, and making any necessary entries. Data entry can produce particularly long head-down times and high workload. Visual scanning of the EFB (without data entry) does not require as much effort, but may still be an additional task for the pilot, depending on the function." The additional workload required to use an EFB may distract the pilot from higher priority time-critical tasks, which is particularly detrimental during high workload phases of flight. Next to improving usability, a way of reducing workload is providing only high priority information in high workload phases of flight (Schvaneveldt et al., 2004; Chandrah & Yeh, 2003).

Information is needed to enhance situation awareness, the perception of elements in the environment in a certain frame of time and space, the comprehension of their meaning, and the projection of the status into the future. The time horizon for the projection into the future may differ, both the very near future, for example conflicting trajectories of aircraft, or the more long-term future, for example the weather conditions at the airport of arrival for a long haul flight. The situation awareness of pilots could be enhanced by providing information on potential future aircraft situations related to weather, terrain and traffic, preparing the crew for difficult situations by providing the appropriate procedures or information at an early stage. Lindholm (1999) expresses the need to correlate aircraft situation information (weather and anti-collision functions) with the flight planning. Each phase of flight (taxi, take-off, departure, climb, cruise, descent, approach, landing, and taxi) requires different kinds of information (Nomura et al., 2006). Also, airplanes move rapidly between geographic regions and weather systems. This means that relevant information about the airplane's surroundings can change quickly.

There may be a direct link with the workload issue. If an information system provides too much information, or in a non-optimal way, the operator may be induced to spend too much time and mental effort in regarding the projected future events, instead of on the immediate task, thus losing situational awareness of the current situation.

Dealing with uncertainty

Even the best information systems will not always be able to give complete information about the environment; for example sometimes information about the weather and its evolution is not available or incomplete. This uncertainty may be caused by the system not being able to capture sufficient information, or because the system is not capable of delivering an interpretation of sufficient certainty. Comerford (2004) recommends presenting pilots with "hazard zones", in which different kinds of information are integrated. If there are no data or insufficient data to determine whether a zone is hazardous, she proposes to create and display an "insufficient data" zone. The pilots may, if they wish, access a list of weather variables that are available about that zone, but that are insufficient to define together a hazard zone (or not), and so interpret the situation themselves. In this case the system should also indicate the reason for the uncertainty. Comerford (2004) argues against providing zones with varying degrees of hazards. These kinds of indications may easily lead to misinterpretation. Human beings are not very good at interpreting uncertain data and probabilities. Another option is to indicate go and no-go zones. The problem here is that a "go" zone is not undoubtedly safe, there are always risks, and pilots should not be given false ideas of security.

Information may be of different ages. For example, terrain and traffic information on cockpit displays is usually up-to-date (only a few seconds old) but the age of weather information varies. The weather data captured by cockpit instruments is up-to-date (less than a minute old) but the age of data up-linked from the ground may vary between tens of seconds and several hours. These data should be treated differently. However, dealing with time issues is difficult for most people (Cellier et al., 1996). Weather situations also evolve over time. It is not always easy to form a good representation of this evolution. An important problem in the Novacek et al. study (2001) was the delay in information. The weather information displayed was sometimes 7 to 14 minutes old. This meant that the position of a storm in relation to the aircraft and to the airport was not correctly displayed. The pilots were looking at images that were playing in different time frames. As representation and interpretation of time is in itself already difficult, having to keep in mind (and thus in the mental model) that the position of the aircraft is a few minutes advanced in relation to the position of a displayed weather phenomenon seems rather complicated, especially in a stress situation.

Dealing with factual versus interpreted information

An information system may present factual information or interpreted information. Factual information consists of raw data, without an indication to the user of what to do with it. Interpreted information is information that has already been processed by the system, presenting the user with an interpretation that is related to the task to perform. For example, a weather system may present factual data indicating the location and speed of strong winds, but it may also give interpreted information such as indications of the danger these phenomena present, the position of safe airports, etc.

It is important that the user is aware of whether the information system is providing either factual or interpreted information, in order to know the level of interpretation he/she has to perform him/herself. If the information takes the form of dynamic images or icons, this may become even more important (Curry et al., 1998). If symbolic information behaves in such a way as to give the impression of being realistic, such as, for example, icons representing thunderstorms moving from one area to another, the user may be inclined to think this is a realistic image of the real situation. However, the symbols may just represent areas in which there is a potential risk, or a weather forecast. The pilot may even be inclined to view the information displayed as a map, which can be used to navigate around the storm, even if this use was not intended by the manufacturer (or allowed by the airline and the authorities).

Note that giving detailed factual data may even be impossible. Comerford (2004) cites work on weather types where 200 different types are distinguished. It is impossible to show the pilot all these different types, nor would he/she know how to deal with them. So weather information presented is usually interpreted information. Weather types are often not isolated; weather often consists of multiple types of weather (for example wind, rain and lightning combined). According to Comerford (2004) knowledge is lacking about what the main, combined, weather types are that are most important for aircrews. Also, allocation of different conditional weights to weather phenomena is needed.

Biological aspects

Finally, operators are not just cognitive beings engaged in information processing, but humans who have bodies with biological properties. The biological aspect of humans' reactions to information should not be neglected. Vaa (2005), studying car driving, emphasises this issue. The human body may be seen as a monitoring system, the body is constantly receiving, detecting and interpreting information both from the environment and from the body itself. The biological nature of humans ensures that this mechanism performs automatically, using all the senses available in order to ensure the safety and survival of the organism. However, this monitoring system is not faultless, especially when dealing with highly technological environments which are outside the scope of the evolution of the species. When dealing with information that is not ecological, the organism may not interpret and react correctly. Of course, humans can learn to deal with technology, such as driving a car or flying an aircraft. However, when introducing new systems, one has to be aware that cognitive interpretation, acquired by learning and experience, may conflict with the more biologically based way of interpretation.

Information display solutions

In the previous paragraphs several solutions for displaying information were already discussed. In the literature and in dedicated studies and reports, a wealth of detailed recommendations is to be found on usability aspects of information display. For example, Mejdal et al. (2001) give guidelines on the design of multifunction displays. In this section three general approaches are described that address the display of complex and heterogeneous information: providing information in layers, contextualisation of information, and integrating and interpreting information. Examples are given from studies in the aviation domain, both flight operations and maintenance, performed at EURISCO International.

Layers

Providing information in different layers is a concept that helps to provide the adequate information related to the context of the operator. For example, the concept of having information in different layers was developed for pilots' documentation (Blomberg et al., 2000). Pilots do not need the same amount of information in all circumstances. For example, during a flight the pilot is usually only interested in what to do and how to perform a procedure. If there is more time, for example during a long cruise phase, or in debriefing, the pilot may be interested in the question of why a certain procedure should be performed. In training, in order to understand why the aircraft is behaving as it does, the pilot will want to know more about the workings of subsystems. Having different amounts of information is not only a matter of personal taste, but is closely related to the safe and efficient operation of a system. During operation, especially in a critical situation, only information should be given which is strictly necessary. For these reasons the concept of three information layers was developed:

- Layer 1: information related to the safe operation of the aircraft, providing concise information on what the pilot should do.
- Layer 2: information giving the rationale of the actions described at Layer 1, the philosophy of use, and additional information on operations not directly linked to safety critical issues.
- Layer 3: detailed information on the functioning of the aircraft.

The three layers ensure that the user is provided with information appropriate for his/her goal: during operation only Layer 1 information is needed, if the user wants to know the explanations for these actions and the reactions of the aircraft, Layer 2 information is needed, and if the user wants to understand the working of the aircraft, Layer 3 information is useful. For maintenance work these layers could be defined as:

- Layer 1: information needed to perform the task, such as the steps in a procedure.
- Layer 2: explanations of the reasons for a task, the way in which it has to be performed, and the precautions to be taken, as well as the relations to other tasks.
- Layer 3: explanations about the workings of the systems and all technical details.

A layered approach may present information of different status, certainty and level of detail on different layers. In situations where operators have to act quickly, only one layer may be presented. If they have more time, operators may engage in looking at other layers. In a layered approach, information remains connected, but is filtered out according to the current needs for a certain task.

Another option is to present information on graphically different layers in a 3D image (e.g. Wong et al., 2005). Layers may be transparent so that users can focus on information at a certain level while maintaining awareness of information at other levels. This kind of representations is becoming common in web-based applications dealing with large amounts of data, presented in a graphical 3D format, for example from data mining applications. Such representations might also present new options for aeronautic information.

Contextualisation

Information may be dynamically configured to match the actual situation. This is called contextualisation. In this case information is provided that can be used to characterise the situation of a person, place, or object that is considered relevant to the interaction between a user and its application, including the user and the application itself (Dey, 2001). The context acts as a set of constraints to limit the amount of information (Bazire & Brézillon, 2005).

In a project concerned with innovation of flight documentation (Ramu, 2008; Ramu et al., 2006; Ramu et al., 2004) a categorisation of the constraints was developed related to three direct questions an operator may ask him or herself when coping with a situation:

- What will I do? With this question, the operator wants to anticipate the operational tasks to perform in a given situation. For example: "what about performing a go-around?"
- What do I use? With this question, the operator seeks knowledge about systems and interfaces used in a given situation. For example: "what about the anti-skid system?"
- What if I have? With this question, the operator wants to analyse what will or can happen if a given situation is submitted to certain conditions. For example: "what about a hydraulic fault and low visibility at my arrival airport?"

In this project a demonstrator of pilot documentation in an EFB was developed, in which information is searched for and presented in a contextual way (Ramu et al., 2006; Ramu & Moal, 2006). Information input about the context may be from both the pilot and the aircraft systems. The context is agreed and refined in a dialogue between the pilot and the information system. By selecting environmental conditions, phases of flight, actions and operations, and systems, the pilot formulates a query to the information system. In this way the pilot is presented directly with all the information that is relevant for the current context. Not only information from the manual may be presented, but also other information, such as bulletins. This demonstrator was evaluated with pilots (Ramu, 2008). They found the information system easy to use and helpful. However, it is a different way of dealing with information, and pilots would need to get used to its different logic.

Also in the maintenance area a study was performed on contextualising manuals (Barnard et al., 2007b). Several demonstrators were developed trying out different ways of contextualisation. In this case contextualisation serves as a filter on the manual. The context may be set by the airline or maintenance organisation or by the technician him/herself. Also data from the aircraft systems may provide input for contextualisation, for example about the configuration and the history of the system. During evaluation maintenance technicians indicated to appreciate this approach, but they emphasised that it should at all times be transparent how the filtering was done, and they would want access to all other information if they should wish it. It is also important that safety critical information, such as warnings, should be visible at any time.

Providing an integrated picture

In current EFBs, information is often presented in separate frames or windows. The operator has to select a section, such as information on the terrain or on the weather. Maintenance manuals also have separate sections for different issues.

There are several ways of integrating information. A simple way of doing so is to bring relevant information from different sources into one frame so that the user does not have to navigate from one (part of) a manual to another. In a demonstrator on maintenance manuals the technician was given the possibility to open links to other manuals in the same window as the procedure on which he/she was working (Barnard & Reiss, 2006). More advanced forms are integrating graphical, 3D and animations with textual descriptions. In a study on the use of 3D animations in maintenance manuals (Tapie et al., 2007), several forms of this kind of integration were tried out, such as textual descriptions next to animated images, and starting the animations from a step in the procedure, where the user could follow the animation step by step at his/her own pace. Although technicians were very much in favour of the use of animations and 3D images, they perceived a danger in being too much focussed on the images, and not paying attention to the text, thus running the risk of missing important details and warnings.

Images, and especially animated 3D ones, may grab a user's attention quite strongly. They may be easy to interpret and to use. However, the images are usually a simplification of the reality, or provide a synthesised picture, leaving out details.

Discussion and conclusions

Displaying information from different sources and about different issues, such as information in EFBs about weather, traffic and terrain, in an integrated picture is a solution for enhancing the support of operators in making decisions in potentially risky situations. If the most important information is extrapolated and displayed in a concise way, avoiding cluttered screens, this may bring many advantages to operators. It makes information easier to perceive, and operators are not presented with several different kinds of information at the same time and all competing for attention. As the complexity of information is reduced, operators need to make less effort in obtaining information from the system, and spend less time on interpretation. This means that the workload is reduced and operators can focus on the most important elements. In time-critical situations, decisions may be made faster, enhancing safety.

However, there are also some risks in integrating information. An important issue is trust. Operators may either trust the information too much, and become less critical, not searching for additional information when needed. Or they might not trust the information enough, because they do not understand how it was compiled. This may lead to superfluous search for additional information, for example to look at the details of the information or to inspect the constituting data from which the synthesised information was derived. If operators do not trust the system they might be inclined not to give the information its necessary weight in their decisions, and to use other criteria. Both concerns were expressed in interviews with maintenance technicians.

If the system provides integrated, easy-to-use information, operators may, on the other hand, come to feel too confident and comfortable about the correctness and adequacy of the information. They might be inclined to take more risks. For example, by being provided with hazard zones a pilot might be led to think that zones that are not indicated as dangerous are therefore risk-free. Presenting information relevant for the context in a focussed way, for example on a foreground level, may lead operators to neglect background information that is also relevant.

In order to reduce the risks of integrated information several measures may be taken. Make sure that the operators understand the system and the general way in which information is integrated. This requires training and a careful introduction and implementation of new information systems. Pilots and maintenance technicians are usually trained intensively in using manuals, instruments and other information systems. If they are required to use new systems that present information by using a different philosophy adaptation is needed. In interviews with both pilots and maintenance technicians, who interacted with demonstrators of innovative manuals, these concerns were formulated.

When information is presented on different levels, or is available in both integrated and non-integrated form, operators should have the possibility to switch easily. Transparency is needed on what information is available and may be accessed in order to avoid confusion. Shifting between different kinds of information should be made easy, avoiding confusion. By providing open, transparent information systems, trust will be enhanced.

This paper focuses on information from different sources and of different status in aviation. Other safety critical domains, such as the process industry and road transport, share similar problems. For example, in cars, information systems are becoming available that combine information from different sources. These include, among others, displays with combined traffic and weather information, cooperative systems providing information from intelligent traffic systems and other cars, giving warnings about accidents or traffic jams ahead of the vehicle. Such information may be combined with navigation information, for example advising to take another route. As cars are being equipped more and more with advanced driver assistant systems and nomadic devices providing all kinds of information, the challenge of combining information in such a way that drivers are not distracted from the driving task is pressing. Single information screens, combining information from different sources, are becoming available. As events in road traffic may develop very rapidly, it is of the utmost importance to prioritise information. Warnings, and indicating the behaviour that is immediately required (such as braking) should always have priority over giving information needed for more strategic purposes, such as finding the most efficient route. Some of the concepts presented in this paper may be of interest in other domains, but should then be tailored to the needs related to the specific users' tasks.

Acknowledgements

The work presented in this paper is partly based on projects performed at EURISCO International, Toulouse, France, in collaboration with J-P. Ramu, J. Tapie, M. Moal, and S. Prigent.

References

Barnard, Y.F., Boy, G., Tremaud, M, Payeur, F., & Fauré, X. (2002). Articulation of Operational and Training Materials. In S. Chatty, J. Hansman, and G. Boy (Eds.), *Proceedings of the International Conference on Human-Computer Interaction in Aeronautics* (pp.30-35). Menlo Park, California, USA: AAAI Press.

Barnard, Y., & Chandra, D. (Eds.) (2004). *Proceedings of the Workshop Electronic Documentation: Towards the next generation*, HCI-AERO 2004. Toulouse, France: EURISCO International.

Barnard, Y., Ramu, J-P., & Reiss, M. (2004). Future Use of Electronic Manuals in Aviation. In Y. Barnard, and D. Chandra (Eds). *Proceedings of the HCI-Aero 2004 Workshop on Electronic Documentation: Towards the Next Generation* (pp. 13-16). Toulouse, France: EURISCO International.

Barnard, Y., Reiss, M., & Mazoyer, P. (2006). Mental Models of Users of Aircraft Maintenance Documentation. In F. Reuzeau, K. Korker, K. and G. Boy (Eds.), *Proceedings of the International Conference on Human-Computer Interaction in Aeronautics* (pp. 232-239). Toulouse, France: Cépaduès-Editions.

Barnard, Y., & Reiss, M. (2006). User-centred innovation of electronic documentation for maintenance. In D. de Waard, K.A. Brookhuis, and A. Toffetti (Eds.), *Developments in Human Factors in Transportation, Design and Evaluation* (pp. 129-142). Maastricht, The Netherlands: Shaker Publishing.

Barnard, Y., Moal, M., Tapie, J., & Prigent, S. (2007a). *Human Factors Study on Electronic Flightbags: State of the Art*. (Report T-2007-192). Toulouse, France: EURISCO International.

Barnard, Y., Moal, M., Tapie, J., & Zahiharimalala, H. (2007b). *Contextualisation of Technical Data*. (Report T-2007-208). Toulouse, France: EURISCO International.

Bazire, M., & Brézillon, P. (2005). Understanding Context before Using It. In A. Dey, B. Kokinov, D. Leake, and R. Turner (Eds.), *Modeling and Using Context, Proceedings of the 5th International and Interdisciplinary Conference* (pp. 20-40). Berlin, Germany: Springer-Verlag.

Blomberg, R., Boy, G., & Speyer, J.-J. (2000). Information Needs for Flight Operations: Human-Centered Structuring of Flight Operations Knowledge. In K. Abbott, J-J. Speyer, and G. Boy (Eds.), *Proceedings of the International Conference on Human-Computer Interaction in Aeronautics* (pp. 45-50). Toulouse, France: Cépaduès-Editions.

Cellier, J.M., De Keyser, V., & Valot, C. (1996). *La gestion du temps dans les environnements dynamiques*. Paris, France: PUF, Collection Le Travail Humain.

Chandra, D.C., Yeh, M., Riley, V., & Mangold, S.J. (2003). *Human factors considerations in the design and evaluation of Electronic Flight Bags (EFBs), Version 2*. (Report DOT-VNTSC-FAA-03-07. Cambridge, MA, USA: USDOT Volpe Center.

Chandra, D.C., & Yeh, M. (2006). Evaluating Electronic Flight Bags in the Real World. In F. Reuzeau, K. Korker, K., and G. Boy (Eds.), *Proceedings of the International Conference on Human-Computer Interaction in Aeronautics*. Toulouse, France: Cépaduès-Editions.

Comerford, D. (2004). *Recommendation for a Cockpit Display that Integrates Weather Information with Traffic Information*. (Report NASA/TM-2004-212830). California, USA: National Aeronautics and Space Administration, Ames Research Center.

Dey, A. (2001). Understanding and Using Context. *Personal and Ubiquitous Computing, 5* (1), 4-7.

Federal Aviation Administration (FAA) (2003). *Guideline for the certification, airworthiness, and operational approval for electronic flight bag computing devices*. (Report AC n°120-76A). USA: Federal Aviation Administration.

Forman, B.E., Wolfson, M.M., Hallowell, R.G., & Moore, M.P. (1999). Aviation user needs for convective weather forecasts. In *Proceedings of the American Meteorological Society 79th Annual Conference*, Dallas, TX, USA.

Lindholm, T.A. (1999). Weather Information Presentation. In D.J. Garland, J.A. Wise, and V.D. Hopkins (Eds.), *Handbook of Aviation Human Factors* (pp. 567-590). London, UK: Lawrence Erlbaum Associates.

Mejdal, S., McCauley, M.E., & Beringer, D.B. (2001). *Human Factors Design Guidelines for Multifunction Displays* (Report FAA/AM-01/17). Washington DC, USA: Office of Aerospace Medicine.

Novacek, P.F., Burgess, M.A., Heck, M.L., & Stokes, A.F. (2001). *The Effect of Ownship Information and NexRad Resolution on Pilot Decision Making in the Use of a Cockpit Weather Information Display* (Report NASA/CR-2001-210845). Hampton, VA, USA: NASA Langley Research Center.

Novick, D., & Ward, K. (2003). An interaction initiative model for documentation. In *Proceedings of SIGDOC 2003*, (pp. 80-85). San Francisco, CA, USA.

Ramu, J.-P., Barnard, Y., Payeur, F., & Larroque, P. (2004). Contextualised operational documentation in aviation. In D. de Waard, K.A. Brookhuis, and C.M. Weikert (Eds.), *Human Factors in Design* (pp. 257-269). Maastricht, The Netherlands: Shaker Publishing.

Ramu, J.-Ph., & Moal, M. (2006). Navigation Portal for Flight Operations. In F. Reuzeau, K. Korker, K., and G. Boy (Eds.), *Proceedings of the International Conference on Human-Computer Interaction in Aeronautics* (pp. 250-251). Toulouse, France: Cépaduès-Editions.

Ramu, J-P. (2008). *Efficience d'une Documentation Opérationnelle Contextuelle sur la Performance des Pilotes de Transport Aériens*. PhD thesis. Toulouse, France: Institut Supérieur de l'Aéronautique et de l'Espace.

Ramu, J.-Ph., Barnard, Y., Moal, M., & Boy, G. (2006b). *CRIISTAL Final Synthesis Report*. (Report T-2006-175). Toulouse, France: EURISCO International.

Schvaneveldt, R.W., Beringer, D.B., & Leard, T.M. (2004). *Evaluating aviation information systems: The role of information priorities*. USA: Federal Aviation Administration.

Shamo, M. (2000). What is an Electronic Flight Bag and What is it Doing in My Cockpit? In *Proceedings of the International Conference on Human-Computer Interaction in Aeronautics*. Toulouse, France.

Tapie, J., Barnard, Y., Mazoyer, P., Devun, J., & Moal, M. (2007). The use of three-dimensional animations in aircraft maintenance documentation. In D. de Waard, G.R.J. Hockey, P. Nickel, and K.A. Brookhuis (Eds.), *Human Factors Issues in Complex System Performance* (pp. 253-267). Maastricht, The Netherlands: Shaker Publishing.

Theunissen, E., Koeners, G.J.M., Roefs, F.D., Ahl, P., & Bleeker, O.F. (2005). Evaluation of an electronic flight bag with integrated routing and runway incursion detection functions. *Proceedings of the 24th Digital Avionics Systems Conference, Volume 1* (pp. 4.E.1 - 41-11).

Vaa, T. (2005) Modelling driver behaviour on basis of emotions and feelings: Intelligent Transport Systems and behavioural adaptation. In L.Macchi, C.Re, and P.C.Cacciabue (Eds.), *Proceedings of the International Workshop on Modelling Driver Behaviour in Automotive Environments*. Luxembourg: Office for Official Publication of the European Communities.

Wong, B.L.W., Joyekurun, R., Mansour, H., Amaldi, P., Nees, A., & Villanueva, R. (2005). Depth, Layering and Transparency: Developing Design Techniques. In S. Balbo, and T. Bentley (Eds.), *Proceedings of OzCHI 2005, the Annual Conference of the Australian Computer-Human Interaction Special Interest Group, Human Factors and Ergonomics Society of Australia.* Canberra, Australia.

Yeh, M., & Chandra D.C. (2007). *Electronic Flight Bag (EFB): 2007 Industry Review*. (Report DOT-VNTSC-FAA-07-04). Cambridge, MA, USA: USDOT Volpe Center.

Transportation - ADAS
(Advanced Driver Assistance Systems)

SITUATION AWARENESS IN A FULLY AUTOMATED DRIVING SCENARIO

In D. de Waard, J. Godthelp, F.L. Kooi, and K.A. Brookhuis (Eds.) (2009). *Human Factors, Security and Safety* (p. 159). Maastricht, the Netherlands: Shaker Publishing.

Is drivers' situation awareness influenced by a highly automated driving scenario?

Natasha Merat & A. Hamish Jamson
Institute for Transport Studies
University of Leeds, U.K.

Abstract

This paper presents results from a study conducted for the European FP6 project *CityMobil*. The experiment described here is part of four cross-site experiments designed to study the human factors issues associated with various degrees of automated driving. Thirty-nine drivers were asked to drive a simulated route with two zones in a within-subjects design, with a main factor of automation. Driver behaviour in "manual" driving, where all driving manoeuvres and decisions were made by the drivers, was compared to "highly automated" driving, where lateral and longitudinal control of the driving task was dictated by the "automated system". In this condition, drivers were asked to take their foot off the pedals and their hands off the steering wheel and allow the vehicle to be driven for them. Situation awareness in both driving environments was measured by computing drivers' response time to a series of unexpected/critical traffic events. Results showed that drivers' response to these events was significantly later in the highly automated condition, implying both reduced situation awareness and perhaps an excessive trust in the automated system.

Introduction

The driving task is becoming more and more automated and it is now possible for various aspects of driving to be controlled by a range of automation and assistance systems. Examples of such systems include Adaptive Cruise Control (ACC), Intelligent Speed Adaptation/Assistance (ISA) and Lane Keeping Assistance System (LKAS), as well as various collision warning and avoidance systems, which use radar detection devices. The idea behind the implementation of most such systems is that they will provide assistance and comfort to the driver, reducing the number of road accidents by increasing safety. Indeed, in the case of a highly automated driving scenario, there is no longer a need for the driver to be involved in the driving task, and his/her role moves from one of an operator to a system supervisor, simply monitoring the functioning of the automated vehicle. However, as suitably highlighted by McKnight & McKnight (2003), the task of maintaining a vehicle in the centre of the road and ensuring a steady speed are perhaps not the most difficult aspects of the driving task, and would indeed be relatively easy to achieve on an empty road, even by novice drivers. However, problems arise when the automated vehicle is required to interact with more complex road environments, as well as

In D. de Waard, J. Godthelp, F.L. Kooi, and K.A. Brookhuis (Eds.) (2009). *Human Factors, Security and Safety* (pp. 161 - 171). Maastricht, the Netherlands: Shaker Publishing.

pedestrians and other road users, the (unpredictable) behaviour of which can result in unwanted and unsafe interactions with the automated vehicle. Since the devices currently available in the market are not capable of dealing with all such eventualities, the role of the human as an attentive and capable supervisor is even more important. Unfortunately, however, it is well accepted that *"monitoring is a role for which humans are generally ill-suited"* (Endsley & Kaber, 1999). Despite such obvious and understandable concerns, advances in technology mean that there is now a general move towards the introduction of progressively more automation and assistance systems into vehicles, and the ensuing research issues are now more about how best to design systems to exploit their capability with the driver in mind, including how they might influence the driving task and how to ensure that safety is not comprised.

Whilst a great degree of research has been conducted on examining the effect of automation on performance, much of it has been carried out in aviation. Therefore, many of the concepts around how automation might affect car drivers' performance are based on results from studies on pilots (e.g. Hancock & Parasuraman, 1992). Clearly, creating such associations between the pilot and car driver must be done with some caution, as the task of monitoring the road is a much more continuous one, with the possibility of many more interactions for the car driver. In addition, the car driver has to be vigilant at all times, keeping their eyes almost exclusively on the road. In contrast, the pilot's attention to the outside world is really only required during take off and landing, and during emergency situations (see Harris & Harris, 2004 for further discussion).

To date, a large proportion of the work investigating the effects of automation in car drivers has concentrated on examining the effects of ACC on performance (e.g. Rudin-Brown & Parker, 2004; Seppelt & Lee, 2007; Stanton & Young, 1998). In general, both theoretical and empirical papers suggests that the introduction of automation to tasks traditionally done by human operators can change the role of the driver and result in a new set of human factors issues which need to be addressed, if they are not to compromise safety (see also Bainbridge, 1983). These include unwelcome and unexpected changes in workload (both underload and overload, Parasuraman & Riley, 1997), potential for a loss of skill (Stanton & Marsden, 1996), and reduced situation awareness (SA) (Parasuraman, Malloy & Singh, 1993), to name but a few. For instance, the intention in introducing an automated system which handles the longitudinal and lateral control of the vehicle is that it causes a comfortable reduction in drivers' workload, providing them with more processing resources for other tasks, such as consulting the satellite navigation system, or reading an email. However, any sudden and unexpected faults or limitations in the automated system will require the driver to come back into the loop and could result in unmanageably high levels of workload. Similarly, if such high automation of the vehicle is the default approach to driving, drivers' skill in controlling the vehicle will slowly diminish with time, resulting in problems when they are required to regain control of the driving task, in case of faults or limitations in the system or infrastructure.

There is currently a growing need and desire in Europe to increase the amount of automation in road transport systems, to allow better management of the road network, increase road safety and reduce fuel consumption. This has been one of the aims of the EU funded project CityMobil, which has involved the introduction of advanced urban transport systems on a large scale. Whilst much of the research conducted in this project is concerned with 'driverless' vehicles such as Cybercars and Personal Rapid Transits (PRTs), there is also some effort dedicated to the implementation of 'dual-mode' vehicles, where the driving task can either be controlled completely by the driver (manual driving) or various aspects of the driving task can be automated.

As outlined by Flemisch et al. (2008) dual-mode driving, involving the transition between manual and fully automated driving, is not necessarily a two-stage process, and includes a number of intermediate stages (see Figure 1). For instance, whilst the addition of an ACC can be considered an example of 'assisted' driving, 'highly automated' driving is more likely to involve full longitudinal and lateral control of the vehicle by an automated system, although the driver is still required to monitor the task in this condition and driving is still not 'fully automated', which is when the driver is effectively acting as a passenger and is totally removed from the driving task. Similar models, with varying levels of automation have been offered by Sheridan & Verplanck (1978) and Kaber & Endsley (1997), as well as others.

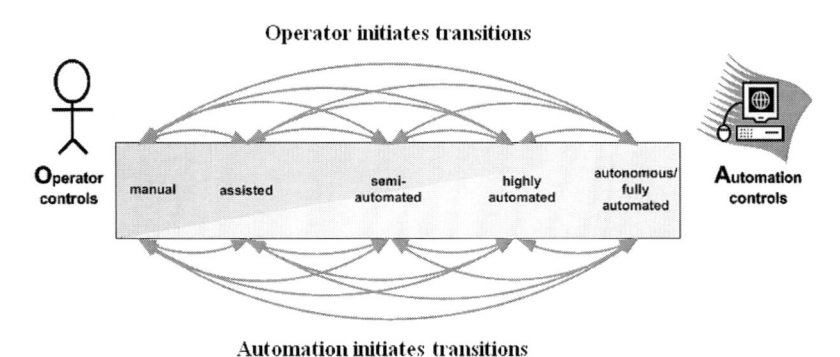

Figure 1. Automation spectrum, regions and transitions as described by Flemisch et al. (2008).

With respect to the human factors of such dual-mode driving, it can be argued that (unwanted) changes in workload, situation awareness and skill can apply to each of the above stages, whilst there are also many research questions about the processes involved during transitions between the driver and the system, from one region of automation to another (and back again, as depicted by the arrows in Figure 1). Whilst many of these issues still remain to be investigated, the study described in this paper attempted to investigate the nature and degree of any changes in drivers' *situation awareness* between manual and highly automated driving.

Situation awareness as outlined by Endsley (2000) refers to a 3 level process of "knowing what is going on". This involves the *perception* of stimuli and cues (Level 1), the *comprehension* of their meaning and relevance to the task at hand (Level 2) and finally *projection* and the ability to anticipate a future status (Level 3). Therefore, in terms of the driving task, good situation awareness (SA) can include an awareness of where the vehicle is in relation to the road and other vehicles, how it can be driven and how its various controls can be used to respond to unfolding events in the road. As the levels of automation increase in the driving task, and more automated and assistance systems are included, drivers' understanding of the capabilities and limitations of these systems, and issues such as whether (and how) the system handles different traffic scenarios are also part of the general concept of their situation awareness.

Method

Participants

A total of thirty nine participants (20 male, 19 female) took part in this study. The participants were all regular drivers, driving an average of 10,000 km per year and aged between 23 and 63 years (mean: 41 years). Average driving experience was 21 years (range: 21 to 42 years).

Design and procedure

The experiments were conducted in the University of Leeds Driving Simulator. A within-subjects design was used, where drivers' situation awareness in manual driving was compared to that of highly automated driving, by assessing driver response to three 'critical' longitudinal events. Automated driving involved control of the car by a lateral and longitudinal *automated system*.

Upon arriving at the driving simulator, all drivers were provided with a detailed written description of the workings of the simulator and its controls, followed by further clarification by the experimenter, if required. They were then given the opportunity to drive the simulator, practicing the operation of the car, and the automated system (lateral and longitudinal controllers). Once drivers were familiar with the simulator and had driven the practice road, they had a short break, followed by the experimental drive which lasted around 40 minutes. This experimental drive consisted of a section of urban road with two zones. Each section zone was approximately 18km in length and 7.3m wide, with a lane width of 3.65m in each direction. The road consisted of straight and curved sections with radius varying between 750m and 1000m. The layout and geometry of the two zones was almost identical, although one zone was driven manually (i.e. with the driver controlling all aspects of the driving task) whilst for the other, drivers were encouraged to hand over the longitudinal and lateral control of driving to the vehicle's automated system by pressing a button on the steering wheel. The order of these 'manual' and 'highly automated' driving tasks was counter balanced across subjects.

situation awareness and a highly automated driving scenario 165

The highly automated driving occurred on 'eLanes', and was supported by an in-vehicle interface, which communicated the workings of the automated system to the drivers (see Figure 2). The automated system's lateral controller kept the vehicle in the centre of the lane, and the longitudinal controller kept the speed at 40 mph and maintained a headway of 2 seconds with a lead car. If drivers failed to transfer control to the vehicle after about one minute of entering the eLane, they were reminded by the experimenter that they should switch the automated system on, although all drivers pressed the button as soon as the automated system was available. Upon transferring control to the car, drivers were asked to take their foot off the accelerator pedal and their hands off the steering wheel, effectively supervising the automated system whilst it drove the vehicle for them. However, drivers were reminded that they should oversee the driving task at all times.

Figure 2. The eLane (left) and in-vehicle interface (right) used in this study

If, for whatever reason, drivers decided to switch the automated system off, they were able to regain control of the driving task using any or a combination of the following methods: by pressing the button on the steering wheel, by moving the steering wheel, or by depressing the brake pedal. If this occurred in the eLane, they would then hear the following message: "you have control", and both lateral and longitudinal management of the vehicle was given back to the driver at the same time. However, in order to maximise data collection in the eLane, drivers were encouraged to pass control back to the car as soon as they were ready.

Participants were reminded that both the lateral and longitudinal controllers were comfort devices that could only manage gentle manoeuvres, but could not respond in a critical situation. In other words, there was a limit to how much the longitudinal controller could decelerate, whilst the lateral controller would always maintain lane centre regardless of any static obstacle(s) in the road. Drivers were told that if the longitudinal controller was not able to respond to a situation, it would warn them to take control of the vehicle. This was signified with an auditory alarm, which the drivers were familiarised with during the practice drive. This alarm was based on a time to collision warning: at 60Hz, the system calculated the deceleration of the simulator required to match the speed of the lead vehicle without collision. If this deceleration exceeded the maximum available system deceleration (0.25 g), then the audible warning was presented.

In order to maximise longitudinal data collection, most of the driving task involved a car following scenario, for both the highly automated and manual drives. To measure drivers' situation awareness to the events taking place in the driving scene, three longitudinal events were implemented for each driving zone (manual and highly automated). All of these critical events occurred at intersections and required a quick and appropriate response from the driver. To reduce any learning effects the order of these critical events was changed between the manual and highly automated drive. Learning was also minimised by interspersing the critical events with at least two non critical longitudinal events where the lead car was forced to decelerate between 0.5 m/s^2 and 2 m/s^2. This lead vehicle deceleration was therefore well within the limits of the longitudinal controller. Whilst there was no data collection for these scenarios, their presence was thought to create a more realistic driving task.

For the three critical longitudinal events, the lead car decelerated at a rate of 6 m/s^2 in response to an upcoming traffic scenario, and the brake lights of the lead vehicle illuminated when its deceleration exceeded 0.1g. In each case, drivers were required to take action in order to avoid a collision with the lead car. The longitudinal events used were as follows:

i. A vehicle emerged from a side road and joined the experimental road, pulling in just in front of the lead car ('emerger from left' event). The emerger was visible some 3s before the lead vehicle actually started to brake, allowing the participant to anticipate the event.

ii. A set of traffic lights changed from green to amber to red as they were approached by the lead car ('traffic lights' event). The lead vehicle only braked as the lights changed to red, allowing the 3s amber signal available for the participant to anticipate the event.

iii. An oncoming vehicle turned across the traffic to enter a side road ('oncomer turns across' event). Again, participants had an opportunity to anticipate the event, as the oncomer began its turn 3s before the lead vehicle began to brake.

Since in each case participants were able to observe the unfolding critical event, in the manual driving condition, participants' situation awareness was a measure of how quickly they anticipated and responded to these events. In the highly automated condition, the longitudinal controller was unable to decelerate sufficiently to these events, which meant that drivers had to regain control of the car. As outlined above, the driver was also warned about an imminent collision, with an auditory alarm. The rate at which drivers regained control of the driving task was used as a measure of their anticipation of the unfolding events and therefore their situation awareness. This was measured from the time at which they started steering again or when they depressed the brake pedal, or both.

Results

Drivers' situation awareness was assessed by comparing minimum time to contact, minimum headway to the lead car and 'anticipation' in the manual and highly automated drives, using a 2 (Drive: manual, automated) x 3 (Event: oncoming turns across, traffic lights, emerger from left) repeated measures ANOVA.

Anticipation was measured as the difference in time between the initiation of the lead car's brake lights and that of the participant applying their brakes. Each critical longitudinal event was choreographed to allow at least 3s for the participants to predict the impending deceleration of the lead vehicle. Therefore, if the driver braked before the lead car, they had a better anticipation of the unfolding events, and a negative value was achieved for anticipation. Similarly, if they braked after the lead car, they had less of an anticipation of the emerging events and the higher the number, the less their anticipation.

Anticipation

There was a significant effect of drive on anticipation (F(1,38) = 212.83, p < .0001), with drivers braking 0.4 seconds after the lead car braked in the manual condition, compared to 1.892 seconds in the highly automated condition. The repeated measures ANOVA also showed a significant effect of 'event' (F (2,76) = 85.21, p < .0001), where anticipation was found to be best for the traffic light event with an average value of 0.043 seconds across the two drives. Finally, the ANOVA revealed a significant interaction between drive and event (F (2,76) = 49.36 p < .0001), with a significantly earlier anticipation of the traffic light event by drivers in the manual condition. As shown in Figure 3, participants always braked earlier in the manual driving condition than the automated condition, regardless of event type.

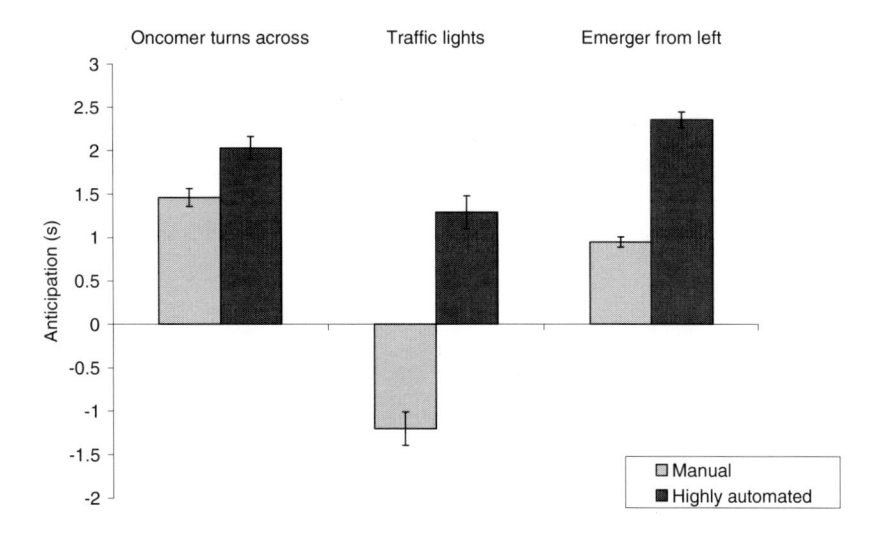

Figure 3. Drivers' anticipation of each oncoming longitudinal event

Minimum Time to Contact

The ANOVA showed a significant effect of drive on minimum time to contact with a significantly longer time to contact during the manual drive (1.82 seconds, versus 1.44 seconds in the highly automated drive, (F (1, 38) = 13.77, p = .001). There was no difference in this value between the three event types, but there was a significant interaction between drive and event (F (2,76) = 5.31, p < .01), where the event involving the emerger from left instigated a particularly short time to contact with the lead car in the highly automated driving condition.

Minimum headway

During the critical longitudinal events, there was a significant difference in minimum headway between the manual and highly automated drive (F (1,38) = 60.47, p < .0001), with a much smaller minimum headway during the highly automated drive (1.60 seconds versus 2.54 seconds). The ANOVA also showed a significant effect of event (F (2,76) = 11.20, p < .0001), with a significantly longer headway during the traffic light event, compared to the other two events. There was a significant interaction between drive and event type, (F (2,76) = 5.02, p < .01), where the longest minimum headway was seen during the traffic light event in the manual drive, whilst the shortest was observed during the highly automated driving when a car joined the road from the left.

Response to the automation alarm

Almost all of the 39 drivers failed to respond quickly enough to the 'emerger from left' event, perhaps having the least awareness of this event. In contrast, the traffic light event was perhaps easier to spot, which may explain why only 20 out of the 39 drivers braked after the alarm (see Figure 4).

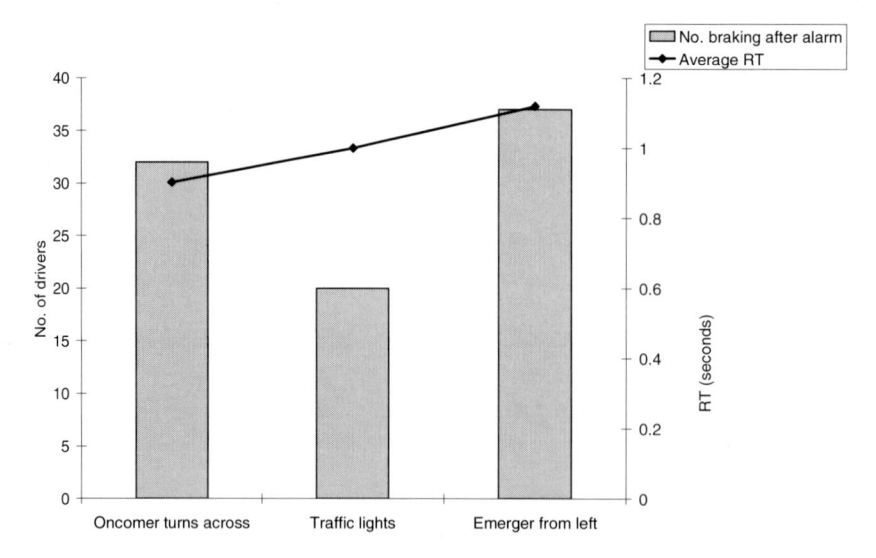

Figure 4. Drivers' response with respect to the auditory alarm

Discussion

The aim of this study was to compare drivers' situation awareness for events which occurred in the driving scene during two different driving conditions: one in which control of the vehicle was managed completely by the driver, and one in which lateral and longitudinal control of the vehicle was dominated by an automated system.

Situation awareness was measured by studying drivers' response to a number of critical events, all of which required an immediate reaction from the driver (braking) to avoid collision with a lead vehicle. In each case, drivers' understanding of the unfolding events and therefore their response (and rate of this response) to the event was used to infer their situation awareness.

Results showed significant differences between manual and highly automated driving, whereby drivers' response to critical events was always much later in the highly automated driving condition. This was confirmed by driver behaviour measures such as time to contact and minimum headway with the lead car, which were both shown to be considerably lower in the highly automated condition. We also used a new measure of 'anticipation' whereby drivers' ability to foresee and understand the unfolding events was used as a measure of their situation awareness. Anticipation was measured as the difference in time between the lead car's brake lights coming into sight and when drivers depressed their brake pedal. Therefore, if drivers braked before the lead vehicle's brake lights were seen, they had better anticipation and were more situation aware. Results showed a much better anticipation of the unfolding events in the manual driving condition than the automated driving condition.

Whilst all three of the above driver behaviour measures showed a better understanding of the significance of each critical event in the manual driving condition, they may also suggest a high (perhaps too high) degree of trust of the automated system by participants. In other words, the late brake response which lead to a low time to contact and small minimum headway in the highly automated condition may have simply been because participants expected the automated system to manage the situation and were then forced to take control when they realised (perhaps later than was safe) that the system was not able to handle the critical situation. Therefore, drivers' situation awareness in the highly automated driving condition was certainly influenced by their trust in the automated system. This is certainly confirmed by results which show that many drivers braked after the alarm was emitted.

Some interesting interactions were seen between the two driving conditions and the three longitudinal events which are worth some discussion. In particular, participants' reaction to the sudden change to red of the traffic lights was quite good in the manual condition, probably because the changing traffic lights were the most visible of the three longitudinal events and therefore allowed the highest anticipation by drivers. In contrast, drivers were worst at predicting the merger of a vehicle from the junction to their left during the highly automated drive, which produced low time to contact and time headway values to the lead car, in this driving condition. This is partly confirmed by looking at drivers' response time with respect to the alarm, which shows the largest number of brakes occurring after the alarm in the 'emerger from left event', whilst the lowest number is seen in the traffic lights event.

To summarise, this study has revealed some interesting results about drivers' attitude to and behaviour with a highly automated vehicle. In particular, the drivers in this study were found to react much later to critical events when driving was controlled by an automated system. Whilst this may have implications about their situation awareness about unfolding critical events, it may also suggest an artificially high degree of trust in the system, which may well be detrimental in a real driving situation.

Acknowledgements

The authors would like to thank the EC for funding this research.

References

Bainbridge, L. (1983). Ironies of automation. *Automatica, 19*, 775-779.
Endsley, M.R. (2000). Theoretical underpinnings of situation awareness: A critical review. In M.R. Endsley & D.J. Gardland (Eds). *Situation Awareness Analysis and Measurement* (pp. 1-23). Mahwa, NJ: Lawrence Erlbaum Associates.
Endsley, M.R. & Kaber, D.B. (1999). Levels of automation effects on performance, situation awareness and workload in a dynamic control task. *Ergonomics, 42*, 462-492.
Flemisch, F., Kelsch, J., Löper, C., Schieben, A., & Schindler, J. (2008). Automation spectrum, inner/outer compatibility and other potentially useful human factors concepts for assistance and automation. In D. de Waard, G.R.J. Hockey, P. Nickel, and K.A. Brookhuis (Eds). *Human Factors Issues in Complex System Performance* (pp. 257-272). Maastricht, The Netherlands: Shaker Publishing.
Hancock, P.A. & Parasuraman, R. (1992). Human Factors and Safety in the Design of Hancock, P.A. & Parasuraman, R. (1992). Human Factors and safety in the design of Intelligent Vehicle-Highway Systems (IVHS). *Journal of Safety Research, 23*, 181-198.
Harris, D. & Harris, F.J. (2004). Evaluating the transfer of technology between application domains: a critical evaluation of the human component in the system. *Technology in Society, 26*, 551–565.
Kaber, D.B. & Endsley, M.R. The combined effect of level of automation and adaptive automation on human performance with complex, dynamic control systems. In *Proceedings of the Human Factors and Ergonomics Society*, 41st Annual Meeting. Santa Monica, CA.
McKnight, A.J. & McKnight, A.S. (2003). Motor vehicle technology: automation of driving tasks. Retrieved from http://www.isd.mel.nist.gov/ research_areas/ research_engineering/Performance_Metrics/PerMIS_2003/Proceedings/McKnight.pdf
Parasuraman, R., Molloy, R. & Singh, I.L. (1993). Performance consequences of automation-induced "complacency". *International Journal of Aviation Psychology, 3*, 1-23.
Parasuraman, R., & Riley, V. (1997). Humans and automation: Use, misuse, disuse, and abuse. *Human Factors, 39*, 230-253.

Rudin-Brown, C.M. & Parker, H. (2004) Behavioural adaptation to adaptive cruise control (ACC): implications for preventive strategies. *Transportation Research Part F: Traffic Psychology and Behaviour, 7,* 59-76.

Seppelt, B.D. & Lee, J.D. (2007). Making adaptive cruise control (ACC) limits visible. *International Journal of Human Computer Studies, 65,* 192-205.

Sheridan, T.B. & Verplanck, W.L. (1978). Human and computer control of undersea teleoperators. *Technical Report.* Man-machine systems laboratory, Department of Mechanical Engineering, MIT, Cambridge, MA.

Stanton, N.A. & Marsden, P. (1996). From fly-by-wire to drive-by-wire: Safety implications of automation in vehicles. *Safety Science, 24,* 35-49.

Stanton, N.A. & Young, M.S. (1998). Vehicle automation and driving performance. *Ergonomics, 41,* 1014-1028.

Evaluation of a generic warning for multiple intersection assistance systems

Stephan Thoma[1], Thomas Lindberg[1], & Gudrun Klinker[2]
[1]BMW Group Research and Technology
[2]TU München
München
Germany

Abstract

Intersections are accident hotspots and thus the development of Advanced Driver Assistance Systems (ADAS) is promoted within the scope of research projects like PReVENT INTERSAFE[1] and Vehicle Infrastructure Integration (VII)[2]. Many of these future assistance systems follow a common warning scheme already implemented in present ADAS like forward and side collision warning systems. The driver's attention is directed to the source of the threat by presenting a visual/acoustical alert that might be combined with a haptic feedback.

Even with only these two functions implemented in a vehicle, it is a challenge to avoid unwanted side effects due to inconsistent or simultaneous warning messages. Additional intersection assistance systems will increase the demand for integration and prioritization of ADAS information even further.

Introducing a generic warning similar to the "master alert" used in aviation might be an approach to simplify warning management within a car, where multiple warning systems share a common warning scheme. This avoids the need to prioritize different warning systems but also imposes the interpretation of the warning within the current driving context on the driver.

In order to verify that the loss of distinguishability of visual alerts when using a generic warning has no negative effect on driving performance, two video scenario experiments were conducted in a driving simulator. In both experiments, videos of a normal urban traffic environment ending with a critical situation at an intersection were presented. In the first experiment (N=60), the comprehension of the cause of the warning was compared under the two conditions "generic warning" and "specific warnings". In the second experiment (N=40), reaction time was measured by allowing the participants to freeze the video using the brake pedal.

---

[1] http://www.prevent-ip.org/en/prevent_subprojects/intersection_safety/intersafe/
[2] http://www.its.dot.gov/vii/

In D. de Waard, J. Godthelp, F.L. Kooi, and K.A. Brookhuis (Eds.) (2009). *Human Factors, Security and Safety* (pp. 173 - 188). Maastricht, the Netherlands: Shaker Publishing.

Introduction

A high percentage of all injury accidents occur at intersections (France 30%, Great Britain 60%, Germany 42%, according to Fuerstenberg, 2007). Passive safety systems in vehicles have helped to reduce fatalities substantially in the past years. However, to continue this positive trend, there is a need for active safety systems that are able to intervene before a potential crash. Thus, a lot of effort is put into the development of intersection assistance systems (e.g. Fuerstenberg, 2006; Ehmanns et al., 2005; Meitinger et al. 2004; Benmimoun, 2005).

A commonly used human machine interface (HMI) is a warning message being presented through one or more sensory channels in case of an imminent critical event. In order to provide a seamless assistance functionality to the driver, integration of multiple warning systems is necessary. Complex traffic situations may require multiple warnings presented simultaneously or in close temporal proximity. Without any further warning management, this could lead to masking of messages or inappropriate or slower driver reactions. Chiang et al. (2006) found such a warning interference effect when presenting a forward and side collision warning in close temporal proximity. The participants were less likely to crash when only the forward collision warning was presented. Apparently the output of the second side collision warning slowed down the participants' reaction to the first warning.

Considering the information processing model by Wickens (1984), simultaneous warnings presented on different sensory channels can be processed more effectively by the driver than messages sharing one modality. Lermer et al. (2007) show that reactions to an auditory forward collision warning are delayed when a vehicle status warning is presented acoustically at the same time. However a simultaneous side collision warning which is presented via a counter torque on the steering wheel does not slow down the reaction to an acoustical forward collision warning.

Additional intersection assistance systems will increase the demand for integration even further. One approach is to prioritise the different warnings according to their relevance in the current driving context. The complexity of the required prioritisation algorithm is a challenge considering the huge amount of scenarios and situations which need to be covered.

Another way of dealing with the increasing number of driver assistance systems is to group multiple warnings so that different warning systems share a common warning message. Figure 1 shows a spectrum of possible grouping strategies. Using one unique warning message for every warning system in the vehicle leads to the least integrated solution (top row in the figure). The other extreme is to group all warnings in the vehicle, which would result in a "master alert". Many other "mixed" approaches lie between these two extremes (only two possibilities are shown in the figure). The grouping strategy has to be considered for every sensor modality separately. E.g. the auditory part of a warning message could be grouped according to the direction of the threat and combined with a visual "master alert".

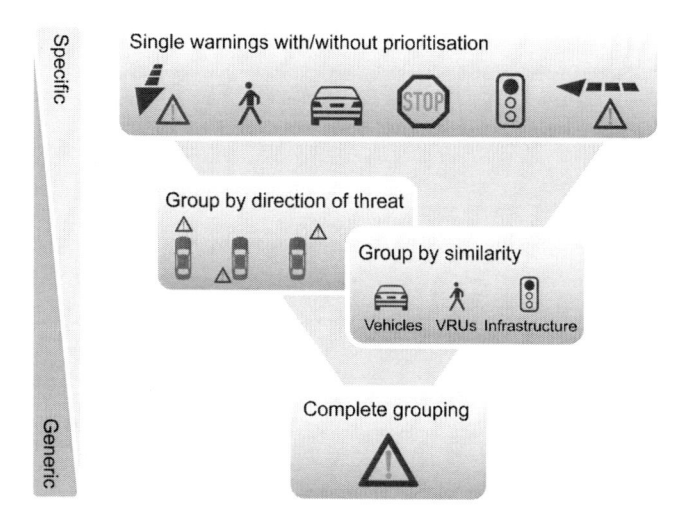

Figure 1. Warnings integration strategies

The use of a generic warning might be an approach to simplify warning management within a car because multiple warning systems share a common warning scheme. This avoids the need to prioritize different warning systems, but also imposes the interpretation of the warning within the current driving context on the driver. Cummings et al. (2007) investigated a master alarm warning scheme for several collision event types: frontal collision warnings, left and right lane departure warnings, and warnings for a fast-approaching following vehicle. The results show that the master alarm did not negatively influence the driving performance or the reaction times. This indicates that the drivers concentrate more on the surroundings of the vehicle in order to identify the cause of the alarm.

On the one hand, replacing multiple warnings by one single generic warning message might be a valid approach to integrate different warning systems within a vehicle. But on the other hand, it could also worsen comprehension and in consequence lead to longer reaction times or inappropriate reactions.

Two experiments were conducted in a driving simulator to evaluate the use of a *visual* master alarm in comparison to specific *visual* warnings with regard to intersection assistance scenarios. Evaluation of grouping approaches for haptic or auditory warning messages or "mixed" integration solutions is object of further research.

Concepts and hypothesis

For this study six representative intersection assistance systems were selected. Table 1 shows the systems, the corresponding specific warning icons and the icon used for the generic warning message.

Table 1. Icons used for the various warning systems

Situation description	Warning icon
Forward collision warning (FCW)	(car icon)
Left/right cross traffic collision warning (CTW)	(arrow + triangle icon)
Oncoming traffic warning (OTW)	(arrow + triangle icon)
Pedestrian warning (PW)	(pedestrian icon)
Red traffic light warning (TLW)	(traffic light icon)
Stop sign warning (SW)	(STOP sign icon)
Generic warning (GW)	(warning triangle icon)

The forward collision warning icon is used in BMW series-production cars. The left/right cross traffic collision warning and oncoming traffic warning icons were originally presented in Fuerstenberg (2007) and further simplified for this experiment. Pedestrian, red traffic light and stop sign warning use standard symbols. The generic master warning does not provide any information as to the cause of the alarm. Also, no action is proposed to the driver (e.g. brake or steering).

The experiments were performed in order to answer the question if replacing the various system specific icons by a generic icon leads to worse driving performance, i.e. reaction. Two hypotheses were formulated and evaluated separately in two similar experiments:

1. "Comprehension"
   The driver needs specific symbols in case of a critical situation to understand the reason for the warning.
2. "Reaction"
   The driver reacts faster and more appropriately when a specific warning is presented.

The visual warning was combined in all cases with a standard auditory warning which is used for forward collision warning in BMW series-production cars.

Method

Evaluating active safety systems or warning strategies in real world tests (Benmimoun et al., 2007; Klanner et al., 2008) is very expensive in terms of time and money. As the drivers must never be put at risk, these experiments are often limited to a few scenarios. Even creating adequate driving simulator scenarios is a complex task. Regarding the number of events relative to the mileage, an accident is fortunately a very rare event. Therefore, reproducing normal urban traffic in a driving simulator will not lead to critical situations in a statistically relevant amount. Unnatural and uncooperative behaviour of simulated vehicles and secondary tasks are used to force participants into hazardous situations. But still, due to different styles of driving, not all designed scenarios will lead to an utilizable warning event. Particularly experiments focusing on the integration of multiple warning systems face the problem of presenting an unnaturally large amount of messages in a very short period of time. Repetitions of a scenario within one single simulator session would cause anticipation and learning effects. As a result, driving performance (e.g. reaction times) would be much better than compared to an unexpected event in a real traffic environment. A complete between-subjects design can be used to avoid these negative side effects. Depending on the number of independent variables under investigation, the number of required participants reaches an impractical level. Even disregarding this fact, a between-subjects design requires accurately comparable scenarios. Otherwise, variations in the style of driving and the resulting large standard deviation of the reaction time distribution make it impossible to verify smaller effects.

To address some of the mentioned disadvantages, two video scenario experiments were conducted in a fixed-base driving simulator to evaluate the different warning strategies. Artificial video sequences of hazardous situations were designed and iteratively improved during the development process. By using video sequences no logic or trigger programming is necessary to deal with different ways of driving. These predefined sequences ensure a high reproducibility which is required for a between-subjects design.

Experiment 1

The first experiment was designed to examine the "comprehension" hypothesis. This hypothesis questions whether a specific warning icon is beneficial for the drivers' understanding of the situation and the warning itself. Therefore, the experiment was designed to determine the understanding of the warning by giving the participants the chance to verbally express their opinion regarding the reason of the warning.

Material and design

For both experiments a static driving simulator was used which was equipped with five plasma displays which offered a field of view of approximately 180° (see Figure 2). The warning icons were presented in a head-up display and in a freely programmable instrument cluster display.

Figure 2. Static driving simulator with plasma screens and mock-up

The first independent variable in the experiment was the type of HMI. The four factor levels (1) "No warning", (2) "Auditory warning only", (3) "Auditory and specific visual warning" and (4) "Auditory and generic visual warning" were used. However, in the analysis, only the last two conditions were included as they represent the specific and generic visual warning of interest.

The auditory warning was equal for factor levels (2), (3) and (4) as the experimental focus was on comparison of different visual warning strategies.

The second independent variable in the experimental design was the type of warning situation. Four different scenario types were used:

- The warning-cause is easily detectable at the point in time when the warning occurs. The driver does not need any additional information but the view through the windscreen to asses the situation. Therefore, the expected influence of a specific warning icon is very low in these situations.
- The warning-cause is difficult to detect (see Figure 3). The reduction of perceivable cues in the vehicle environment might force the driver to take additional information into account which is given by in-vehicle warning systems in order to understand the reason for the warning.
- A secondary event in the scene complicates the detection of the cause of the warning (see Figure 3). There are two potentially hazardous objects in the scene but only one justifies a warning. The in-vehicle warning system might be able to influence the driver's interpretation of the situation.
- An incorrect warning icon is presented in a hazardous situation. A warning is justifiable but the icon does not match the cause of the warning (only relevant under the specific warning condition). This condition was introduced to reveal potential negative effects of specific warning icons. Presenting a wrong specific warning might be worse than presenting an uninformative generic warning.

Table 2 shows representative situation descriptions for the Cross Traffic Warning System.

Figure 3. Simulation screen shot for Situation "Difficult to detect" (1) and "Secondary event" (2)

Table 2. Example scenario descriptions for Cross Traffic Warning system

Warning scenario type	Scenario description
Easy to detect	Driver does not perceive cross traffic on a rural intersection. Slight line-of-sight obstruction.
Difficult to detect	Driver does not give way to cross traffic from right crossroad. Massive line-of-sight obstruction by buildings.
Secondary event	Driver waits at a priority road for an adequate gap in the crossing traffic. After accelerating he does not detect a pedestrian who started crossing the side road a few seconds earlier.
Incorrect warning	Driver violates the right of way to crossing traffic. In the specific icon condition, a stop sign warning icon is presented.

Finally, the third independent variable represented the various warning reasons respectively the warning systems listed in Table 1.

A between-subjects design for the factor "HMI type" was used and a within-subjects design for the factors "scenario type" and "warning system". Therefore twenty-four video sequences for each warning system and scenario type needed to be designed. However, due to technical problems, the secondary event scenario was missing for the pedestrian warning condition.

The twenty-three videos were presented in randomized order to every participant with regard to order effects. Additionally, the HMI type was changed randomly for every video to avoid habituation effects. The HMI type was permutated across every group of four participants in such a way that for every video, every HMI type was presented exactly once (see Figure 4).

The dependent variable was the correctness of the articulated warning reason.

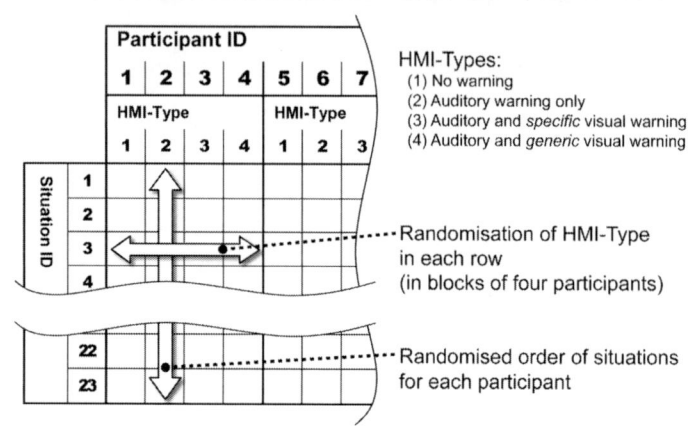

Figure 4. Randomization scheme for first experiment

Procedure

Driving experience, vehicle miles travelled per year, previous simulator experience and standard demographic data was collected using a short questionnaire. After taking seat in the driving simulator, the participants were informed that they would see various recorded video sequences of short drives through an urban environment. The participants were asked to put themselves in the position of a driving instructor (while being seated in the driver's seat), where they are supervising a learner who is likely to make driving mistakes. Additionally, they could benefit from a warning system which is installed in the vehicle (no details were given on the type of warning system). They were told that the warning system may also give nuisance alarms.

As this experiment was designed to test the timely comprehension of a warning, the videos were all faded out approximately one second after the warning was presented. The fade-out was used as the participants should not have had a chance to continue visually analysing the scene. The warning was timed in such a way that the critical event (e.g. impact) would have happened roughly three seconds after the warning occurred. Under the "no warning" condition, the warning message was omitted. The participants were asked to articulate their understanding of the reason for the warning as fast as possible after the warning (after the fade out in the case of the baseline). They were told that there are no wrong answers and that they should answer as spontaneously as possible. Accordingly, no feedback was given on the correctness of the response. All answers were recorded as WAV files. The recording was started simultaneously with the output of the warning message. In the case of the "no warning" condition, the recording was started with the imaginary warning point in time (which exists as all video sequences were combined with all HMI types). To avoid anticipation of the occurrence of the critical events, the duration of the video sequences varied between 10 and 60 seconds.

Participants

In order to have fifteen events for each HMI type/scenario combination, sixty participants, fifteen female and fourty-five male, were invited to participate in the experiment. The average age was 29.4 years (S.D. = 7.7). Thirty-one participants had no prior driving simulator experience, thirteen participants had attended at least one or two prior simulator experiments and the rest had attended tree or more prior experiments. Thirty-three participants were driving more than ten thousand kilometres per year. All participants were full, temporary or external employees of BMW Group and they did not receive an incentive for participation (twenty-six interns/master students engineering, fourteen employees engineering, seven employees finance, thirteen employees other departments). The unequal gender distribution of the sample mainly reflects the composition of the BMW research and development department. Recruitment of external participants was not possible for this experiment. No other inclusion or exclusion criteria were used.

Results

The responses (1380 sound recordings, sixty participants multiplied by twenty-three recordings each) were manually classified as correct or incorrect with respect to the corresponding scenarios. Thirty of them were excluded due to invalid answers (e.g. no answer within given time frame, incomprehensible or meaningless utterances). The thirty missing values were randomly distributed over the whole sample.

Table 3 shows the number of participants who named the correct reason for the warning (respectively the video fade out for the "no warning" HMI type). As an example only the data for the condition "Difficult to detect" is presented here. The maximum possible number of correct responses is fifteen. Generally, the experimental setup led to a very low number of false responses even in the "no warning" condition (e.g. Forward Collision Warning, FCW or Cross Traffic Warning, CTW). Apparently the scenarios were not complex enough and/or the participants' mental and visual workload was too low. Increasing the situation complexity was hardly possible using the BMW driving simulator environment. Therefore, a secondary task was used in the second experiment to artificially increase the distraction from the road scene.

The exact Fisher test was used to compare the number of correct responses for each condition. Table 4 shows the test results for the comparison between the conditions "generic icon" and "specific icon" for all scenario types. None of the "easy to detect"-scenarios showed a significant difference between the number of correct answers. This is not very surprising, as the warning-cause was visible at the point in time when the warning message was displayed. In two of the "difficult to detect" scenarios, however, significantly more correct answers were given under the "specific warning icon" condition. It seems that the participants used the additional visual information given to interpret the situation. This fact is substantiated by the significant difference in the TLW "incorrect warning" situation: participants tended to misinterpret the situation by taking the wrong warning icon into account. In a worst case scenario this incorrect system message could lead to a delayed reaction to the real threat.

Table 3. Number of correct responses for scenario type "Difficult to detect" under the four HMI type conditions

		Warning type			
		No Warning	Auditory only	Auditory and *generic* visual	Auditory and *specific* visual
Warning system	FCW	15	15	15	15
	CTW	14	13	15	15
	OTW	14	12	13	10
	PW	2	1	2	9
	TLW	8	5	6	13
	SW	5	2	2	7

Table 4. p-values for Exact Fisher Test comparing HMI type "generic icon" and "specific icon"

		Scenario type			
		Easy to detect	Difficult to detect	Incorrect warning	Secondary event
Warning system	FCW	1.000	1.000	1.000	1.000
	CTW	1.000	.483	.598	.139
	OTW	1.000	.390	1.000	.102
	PW	1.000	.021*	1.000	(n/a)
	TLW	1.000	.021*	.041**	.025*
	SW	1.000	.109	.700	.272

* Significantly more correct responses under the *specific* icon condition

** Significantly more correct responses under the *generic* icon condition

Experiment 2

The second experiment addressed the reaction time hypothesis. While maintaining the video scenario experimental setup, the task was to find a method of measuring reaction times. The idea was to allow the participants to interact with the simulation by pressing the brake pedal.

Design

The first independent variable, the HMI type, was reduced to (1) "Auditory and *generic* warning icon" and (2) "Auditory and *specific* warning icon".

The main intention behind the reduction of HMI types was to lower the number of required participants by focusing on the most interesting HMI type conditions.

The levels of the independent variable "scenario type" were changed in comparison to the first experiment. Four different scenario types were used:

- Difficult to detect
  The warning is appropriate in the respective situations but the warning-cause is difficult to detect in the scene.
- Correct warning, secondary event
  A hazardous situation occurs and a warning message is displayed. The warning is appropriate to the respective situations. There is a secondary event which requires the drivers' attention.
- False positive ("false alarm")
  A warning message is displayed although there is no critical situation.
- False negative ("miss")
  No warning message is displayed even though there is a reason for doing so.

For each of the six warning systems four video sequences were recorded (in contrast to the first experiment, all 24 scenarios were implemented). Most of the scenes were reused from the first experiment with only slight changes. Introducing the false negative and positive event types should avoid skill based reactions of the participants. The number of correct (brake) responses and the reaction times for the false positive messages are not part of the analysis presented.

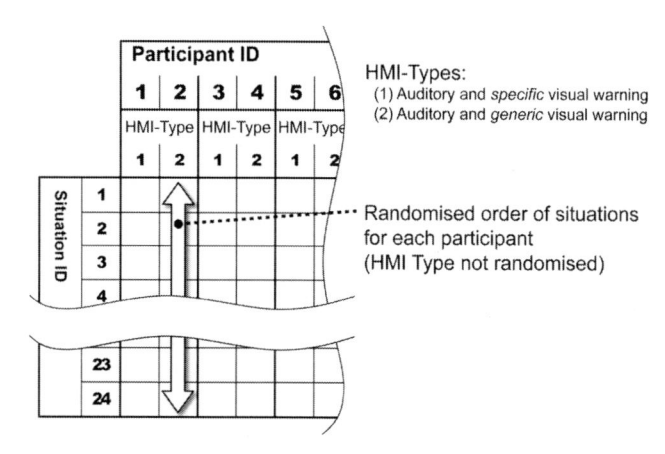

Figure 5. Randomization scheme for second experiment

The dependent variable was reaction time (not applicable to the "false positive" condition). It was determined by measuring the elapsed time between the onset of the warning message and the beginning of a brake reaction. Again, an in-between design was applied using two groups for the two different HMI types. The scenes were presented in randomized order (see Figure 5). In contrast to the first experiment, the HMI type was kept constant for one participant. A baseline reaction time measurement was performed to detect potential predisposed reaction time differences between the two treatment groups.

Procedure

Just like in the first experiment, the participants were asked to put themselves in the position of a driving instructor who should avoid potential driving mistakes of a learner. This time they were instructed to press the brake pedal whenever they judged a situation to be a potentially hazardous. The participants were informed that a warning system would be assisting them. However, the warning system might warn even though no critical event is happening, and it might also miss a critical situation, thus, omitting a warning. A quick but well judged response was demanded from the participants. Whenever the brake pedal was pressed and the brake pressure exceeded a certain threshold, the video was frozen immediately. This represents a slight change with regard to the first experiment, where the videos were faded out. In this experiment the experimenter needed to have the possibility to easily resume the video in case of a false brake reaction. Multiple less hazardous non-warning situations were inserted in the videos to force the participants to decide on pressing the brake pedal or not. The investigator was able to continue the video in case the brake reaction was preformed without a hazardous situation being imminent. In the first experiment, the number of correct answers was generally very high; even under the "no warning" condition. Thus, while viewing the video scenes, the participants were asked to perform a secondary task in the second experiment. The idea behind this measure was to simulate a distraction pattern like it is caused by in-vehicle information systems like navigation systems or entertainment functions. Returning the driver's attention to the road is one major use-case for in-vehicle warning systems. A Critical Tracking Task (Jex, 1988) implementation by Dynamic Research (see Figure 6) was used to simulate a realistic and adjustable distraction pattern. The task was presented in the mock-up's central information display (normally used for BMW iDrive menu). The participants need to keep a horizontal line in the centre of the screen by using the up and down arrow keys on a keypad. A disturbance function is used to move the line away from the centre position. In this experiment a CTT-lambda value of 0.8 was used.

Participants

To achieve reasonable reaction time distributions, the number of participants per treatment group was increased from fifteen to twenty for this experiment. At the same time, the number of factor levels was reduced to a number of two. Therefore, forty participants (three female, thirty-seven male) participated in the second experiment (average age 29.5 years, S.D.=7.6). Twenty-eight participants had no prior driving simulator experience, eight participants had attended at least one or two prior simulator experiments and the rest had attended tree or more prior experiments. Thirty participants were driving more than ten thousand kilometres per year. Only

BMW Group employees were invited, hence the unbalanced gender distribution. Due to the similarity of the used scenarios, only persons who did not participate in the first experiment were invited. No other inclusion or exclusion criteria were applied.

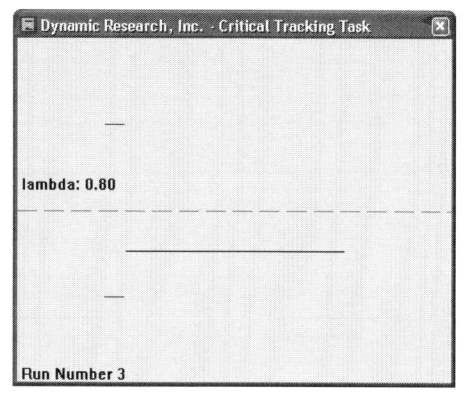

Figure 6. Screenshot of Critical Tracking Task

Results

Event protocols were recorded for each scenario and automatically evaluated after the experiment using a custom software tool. The baseline reaction time measurement showed no significant difference for the two treatment groups (generic vs. specific). The reaction times for the false positive warning scenarios were not used for further analysis (there were only very few reactions to these warning events). The remaining average reaction times are shown in Table 5. The reaction times for the "no warning" scenario type originate from both the generic and specific warning icon condition. Due to the experimental design, comparison of reaction times is only possible within one scenario type as the implemented traffic scenarios are not the same.

In the red traffic light warning scenario (TLW) "difficult to detect" a Kolmogorov-Smirnov test showed significant faster reaction times (p=.037) under the condition "specific warning icon" than under the condition "generic warning icon". The non parametric test was used as the reaction times were not normally distributed. In the respective scenario the traffic light was occluded by a truck until approximately one second after the warning message was displayed. The difference in the mean reaction times indicates that the drivers used the specific traffic light warning icon to interpret the situation and execute a brake reaction.

However, in all other scenarios the display of a specific warning icon did not show a significant positive effect with respect to reaction times. In the cross traffic warning scenario (CTW) "difficult to detect" a t-test (under this condition, the data was normally distributed) even showed significantly slower reaction times (p=.026) under the condition "specific warning icon".

Table 5. Mean of reaction times in seconds and standard deviation

		Scenario type				
		No warning/miss	Difficult to detect		Secondary event	
			Generic warning	Specific warning	Generic warning	Specific warning
Warning system	FCW	0.75 (0.4)	1.5 (0.4)	1.4 (0.4)	1.0 (0.3)	1.1 (0.3)
	CTW	1.3 (0.3)	*0.9 (0.3)***	1.2 (0.5)	1.4 (0.2)	1.4 (0.3)
	OTW	1.4 (0.5)	1.9 (0.3)	2.1 0.3)	1.5 (0.7)	1.9 (0.7)
	PW	1.77 (0.8)	0.9 (0.4)	0.7 (0.3)	1.5 (0.4)	1.4 (0.5)
	TLW	1.6 (0.8)	*1.7 (0.5)*	*1.3 (0.5)**	1.1 (0.4)	1.0 (0.4)
	SW	1.81 (0.3)	(not enough samples)		1.4 (0.3)	1.5 (0.3)

* Significantly shorter reaction times under the *specific* icon condition

** Significantly shorter reaction times under the *generic* icon condition

Discussion

The comprehension test experiment showed a significant influence of the warning icon, especially for the traffic light and pedestrian warning scenarios. Both warning systems use very simple and intuitively understandable icons. This might lead to better recognition performance during a hazardous event and explain the missing influence of the other warning icons. By trend, the number of correct answers under the "difficult to detect" scenarios were lower than under the other conditions, which indicates a coherent influence of the situation complexity. Specific warning icons might be valuable for the driver, especially in situations where the cause of the warning is not directly perceivable. For instance, car-2-car communication systems are likely to detect objects which are not visible to the driver due to line-of-sight obstruction.

Despite these effects, both experiments revealed limited differences between the two visual warning strategies. This could be an artefact of the selected experimental setup. The design of the first experiment led to relatively high numbers of correct answers (see Table 3). Even under the "no warning" condition, almost no participants gave wrong answers. This is a disadvantageous precondition, as there is only little room for improvement of comprehension by adding a warning output. Even though the traffic situations were implemented as complex as technically possible, it seems that the participants' visual and mental workload was not high enough to cause misinterpretations by just watching the videos. Therefore, in the second experiment a secondary task was added, to visually distract the participants and simulate the operation of in-vehicle information systems like entertainment functions. However, the impact of the secondary task is marginal. The reaction times for the "no warning" scenarios in the second experiment were not noticeably higher than for the other scenarios (see Table 4). A significance test is not possible here as the traffic scenarios do not exactly match by experimental design. A higher level of distraction should cause longer reaction times if no warning is presented. Apparently, the participants prioritised the secondary task with respect to the traffic situation. Forcing the participants into a secondary task right before a warning occurs seems to be the only solution.

The video scenario technique used in this experiment might suffer a lack of validity, but conducting a simulator experiment with participants driving on their own seems to be impracticable considering the high number of scenarios required. The participants would change their driving behaviour after a few warnings, thus making it complicated to trigger the traffic situations. Measuring reaction times using video scenarios is an interesting alternative, in terms of time, money and reproducibility. Even though it creates reasonable reaction time distributions with respect to the scenarios, further research is needed to prove the validity of that method. Otherwise no clear statement about the potential advantages or disadvantages of a single generic warning replacing multiple specific warnings is possible.

References

Benmimoun, A., Chen J., & Suzuki, T. (2007). *Design and Practical Evaluation of an Intersection Assistant in Real World Tests.* Proceedings of the 2007 IEEE Intelligent Vehicles Symposium, June 13-15, 2007, Istanbul, Turkey.

Benmimoun, A., Chen, J., Neunzig, D., Zuzuki, T., & Kato, Y. (2005). *Specification and Assessment of Different Intersection Assistance Concepts Based on IVC (Inter-Vehicle-Communication) and RVC (Roadside-Vehicle-Communication).* Paper presented on 12th World Congress on ITS, 6-10 November 2005, San Francisco, USA.

Chiang, D., Llaneras, E., & Foley, J. (2006). *Driving Simulator Investigation of Multiple Collision Alarm Interference Issues.* In Proceedings of Driving Simulator Conference (DSC) Asia/Pacific. Tsukuba, Japan.

Cummings, M.L., Kilgore, R.M., Wang, E., & Kochhar, D.S. (2007). Effects of Single Versus Multiple Warnings on Driver Performance. *Human Factors, 49,* 1097-1106.

Ehmanns, D., Hopstock, M., & Spannheimer, H. (2005). *ConnectedDrive: Advanced Assistance Systems for Intersection Safety.* Paper presented on the 12th World Congress on ITS, 6-10 November 2005 (Paper 2515), San Francisco.

Fuerstenberg, K. (2006). *Intersection Driver Assistance System - The EC project INTERSAFE.* Paper presented on the 13th World Congress & Exhibition on Intelligent Transport Systems and Services. London, United Kingdom.

Fuerstenberg, K., Hopstock, M., Obojski, A., Rössler, B., Chen, J., Deutschle, S., et al., 2007. In INTERSAFE Final Report, SP Deliverable 40.75.

Jex, H.R (1988). The Critical Instability Tracking Task – Its Background, Development and Application. In W. B. Rouse (Ed.) *Advances in Man Machine Systems Research*, vol. 5, Greenwich: Jai Press.

Klanner, F., Thoma, S., & Winner, H. (2008). Driver Behaviour Studies and Human-Machine-Interaction Concepts for Intersection Assistance. In Proceedings of 3rd Conference on Active Safety through Driver Assistance. Garching bei München, Germany.

Meitinger, K.-H., Ehmanns, D., Heißing, B. (2004). *Systematische Top-Down-Entwicklung von Kreuzungsassistenzsystemen*. VDI Berichte No. 1864, Düsseldorf, Germany: VDI Verlag,.

Wickens, C.D. (1984*)*. Processing resources in attention. In R. Parasuraman and D. R. Davies, *Varieties of attention*. New York, USA: Academic Press.

Enhanced information design for high speed train displays: determining goal set operation under a supervisory automated braking system

Anjum Naweed, Bob Hockey, & Sam Clarke
The University of Sheffield
Sheffield, UK

Abstract

As a task traditionally reliant on external sources of information, integration of European railway initiatives will change traditional British train driving by migrating a new signalling standard for exclusive use inside the cab. On-board computer systems and driver-interfaces will instruct throttle changes for speed and braking points under a supervisory automated braking system. However, it is unclear what effects such a fundamental transformation in human-machine expertise and automatising of goal setting activity is likely to have on driver control. The present study used a blues-skies approach and process control analogy to investigate performance under automatic (closed loop) and driver goal set (open loop) operations in ATREIDES (Adaptive Train-Research Enhanced Information Display and Environment Simulator). Eighteen postgraduate participants, all with engineering backgrounds, were trained before performing a short train journey under tracking, predictive or basic driver display modes. The results showed that performance under automatic goal setting was more optimal, required fewer safety interventions, and considered subjectively easier than driver goal set modes. However, modes were equally demanding on objective measures, suggesting supervisory automation and overt tracking did not reduce cognitive effort. The findings are discussed in relation to characteristics of control activity with implications of future changes to the UK train driving task.

Introduction

Motivational background

The proposed integration of the European Railway Traffic Managing System[*] (ERTMS) in the UK will make fundamental changes to the train driver's task. Though the reasons for change are motivated by European initiatives (e.g., international standardisations, interoperability regulations) and appear to weigh heavily on one side of the performance-safety divide (e.g., maximised capacity,

---

[*] Information accessible at http://www.ertms.com/

In D. de Waard, J. Godthelp, F.L. Kooi, and K.A. Brookhuis (Eds.) (2009). *Human Factors, Security and Safety* (pp. 189 - 202). Maastricht, the Netherlands: Shaker Publishing.

service frequency, bridging train gaps, Department for Transport, 2007a), integration of ERTMS will provide domestic trains with an in-cab enhanced display in the form of the European Train Control System (ETCS). However, the planned changes will not only introduce a new in-cab signalling convention, but migrate signals from the track altogether, effectively rendering conventional lineside signage obsolete (Department for Transport, 2007b) and bringing about fundamental changes to the UK driver's task.

The new ETCS signalling display is effectively a speedometer with a movement authority pane that previews upcoming infrastructure. However, the system is also taking a mediatory approach by introducing supervisory automation; a goal set mode of normal operation that displays a braking indicator with an in-built latency, and re-aligns movement authority with a series of supervised braking zones. Speed correction in accordance with supervised braking is monitored, timed, and in the event of delay, subject to an automated brake application, enacting a fail-safe mechanism dropping them to line speeds. In short, the system is expected to effectively instruct and manage throttle changes for both speeds and braking distances (OKTAL ERTMS Simulator, Cambrian ETCS training team, personal communication). Though the system is purported to allow trains to reach maximum speeds whilst maintaining safe braking distances, it will do so by translating the key proponent of train driver expertise (Naweed, Bye & Hockey, 2007a) into the onboard computer, principally reinventing UK train driving as a tracking task not unlike the process control analogy of closed loop control (Hollnagel & Woods, 2005). The change is likely to reduce overall task components and restructure coupling in favour of engineering philosophies such as ecological interface design (Burns & Hajdukiewicz, 2004). However, some may express concern at the new knowledge requirements and impact this level of automation will have on the driver, particularly as their role may change fundamentally to that of a monitor or anomaly handler. (Woods & Hollnagel, 2006).

As a task characterised by the requirement for sustained speeds, train driving is amenable to increased automation. However, the regulatory nature of throttle control actions and interaction with goal setting open loop activities (Moray, 1997) best define cognitive train driving skill. By recognising the skill-based component baselining the task (Edkins & Pollock, 1997), in-cab displays have classically aimed at adopting a decision support or advisory role, and generally displayed environmentally verifiable information, e.g. signal preview, routing details, track gradients, and so on. Research has also explored the effect of enhancing this support by layering displays with predictive and trial manoeuvring components (Askey & Sheridan, 1996; Stjernström, 2001; Einhorn et al, 2005; Cole, Bosomworth, Hayman, McLeod, & Croce, 2006) in an advisory capacity. Whilst some industry-led initiatives have adopted in-cab displays in a purely advisory capacity, (e.g., Freightmiser, Rail Innovation Australia, 2008) goal setting automation appears to be gaining popularity. As a highly contentious topic, automation is typically host to a range of interpersonal human factor issues (i.e., mistrust, complacency, passivity), arising from the perceived loss of operator control, further compounding the effective calibration of shared responsibility.

However, the trend for technology takeover within the rail domain is not entirely unfounded. In principle, train driving has much in common with other analogous domains (e.g. maritime navigation, aviation, process control), where collision avoidance plays a key role. It is the threats and obstacles that arise outside of railway control (vandals, trespassers, environmental debris, level-crossings, etc.) that necessitate external visibility, and have been responsible for some of the most severe UK rail accidents in recent years. Until railways evolve into structures that operate under the parameters of a closed-system, different levels of partial-automation will no doubt continue to take precedent. For future UK ERTMS integration however, operator response to ETCS will no doubt be linked to the resilience of the display's supervisory system, and how consistent their own mental models are with the automated algorithms underpinning its goal setting components.

Objectives

The present study used the principles underlying ETCS and process control analogy to investigate whether train driving with the fail-safe component of supervisory automation elicited better performance under the mode of closed loop driver-tracking (i.e. automatic goal set), or open loop modes of driver-predictive or driver-basic (i.e. driver goal set) operation. The closed loop driver-tracking mode was designed to implement speed changes by guiding throttle-actions along a route optimised-trajectory inside a movement authority envelope. The driver-predictive mode retained the optimised-trajectory but omitted the tracking element and substituted it with the presentation of an enhanced braking feature. The driver-basic mode omitted all such features. Though it depicted the standard advanced track profile and signal preview range, the mode required more overt quasi-mathematical user-input, equating it more closely to the traditional train driving task. The experimental approach taken by this study grounded train driver cognition in traditional task requirements, but adopted a blue skies perspective of a closed-system architecture where (1) collision threats only arose from rail-networked traffic, (2) direct sources of external information were obsolete, and (3) all information requirements were provided by the interface. The simulation platform was an integrated driver-cab display designed exclusively to research such initiatives.

On the basis of previous research (Askey & Sheridan, 1996; Einhorn et al, 2005) used to inform the platform's primary display design, the two operator states containing enhanced features (i.e., driver-tracking and driver-predictive) were expected to perform better than the basic mode. However, the research was purposefully explorative. Though ecologically driven feature enhancements may generally spare mental resources, the loss of strategic goal set behaviour limits the number of tasks available. Whilst closed loop tracking requires little effort, performance would depend on the parameters of coagency, and the operator's contextual perception of supervisory automation. Performance degradation may then occur for any number of reasons, such as unwillingness to accede to an automated authority (Parasuraman, Barnes & Cosenzo, 2007), automation reliance (Dzindolet, Peterson, Pomranky, Pierce & Beck, 2003), or excess trust and complacency (Parasuraman, Molloy and Singh, 1993). Similarly, though the substitution of closed loop tracking with open loop predictive features may reduce workload compared to the driver-basic state, the use of automated braking may have unpredictable implications for safety and performance.

Methods

The ATREIDES platform

The simulation platform used to present the integrated display was known as the Adaptive Train-research Enhanced Information Display & Environment Simulator (ATREIDES, Naweed, 2009). ATREIDES was designed entirely within a MatLab® environment (The MathWorks, Inc. 1984-2006) and developed from a model of train driving information needs (Naweed, Bye & Hockey, 2007b) adapted, structured and hybridised from a series of cognitive task analysis techniques and field research methods (Klein, Calderwood, & Macgregor, 1989; Rasmussen, Pejtersen, & Goodstein, 1994; Vicente, 1999; Annett, 2003). ATREIDES was influenced by display elements already explored in existing research (Askey & Sheridan, 1996; Stjernström, 2001; Einhorn et al, 2005; Dorrian, Roach, Fletcher & Dawson, 2006; Alcatel, 2008), but integrated more enhanced features and predictive components (e.g. enhanced track structuring and profiling, fuel, ETA, routing/schedule information) to sufficiently extricate the role of a 3D world view.

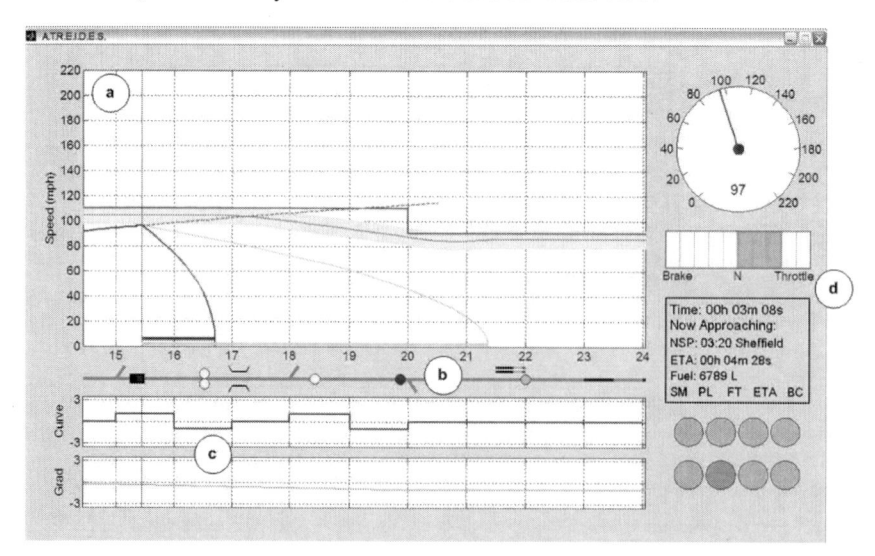

Figure 1. ATREIDES. System interface showing the (a) speed-envelope, (b) track structure/signal preview, (c) track profile core component, and (d) control panel panes. A colour image is available at http://extras.hfes-europe.org

ATREIDES had adequate levels of detail and environmental realism to explore performance related issues, housed several levels of complexity, and an array of customisable display variables for a range of process operation conditions (see Figure 1). It was programmed to simulate realistic train physics, advance or brake appropriately in view of track resistance characteristics, and equipped with a fully customisable design of the rail infrastructure. In view of traditional train driving task and growing trend for technology takeover, ATREIDES was built to incorporate all technologically practicable driver information needs divulged from preliminary studies (Naweed et al, 2007a; 2007b), and featured a variable range of adaptive control settings and intervening safety measures. Primary task performance was ascertained from real-time output of speed, throttle-use and time. Spare mental capacity was captured using momentary load on a 'scan and cancel' secondary reaction time task.

Participants

Eighteen mixed-nationality participants, speaking English as a first or second language (mean age of 27 years, age range from 19-37 years; 15 males and 3 females), and with an engineering background, took part in the experiment. Participation was permitted only after the completion of two 90 min training and testing sessions, where the core interface, optimum driving parameters, general principles of pursuit-tracking, and rudimentary versions of all of the present study's display modes were familiarised. Candidates were chosen following satisfaction of all objective (error-rate, time-keeping, task proficiency) and subjective (task comprehension, task awareness) selection criterion. All had normal hearing, and normal or corrected-to-normal vision. Seventeen were right-handed, and two were left-handed by self-report. The experimental session lasted for approximately 90 min. The participants received cash payment in return for taking part in this study. All experiments were conducted in accordance with the guidelines laid down by the Department of Psychology, University of Sheffield.

Apparatus

Prior to experimentation, a whiteboard was used to illustrate new feature mechanics, and remind participants of previously encountered display components. ATREIDES was displayed and run on a 39.1 cm (15.4 inch) WXGA LCD Acer® Aspire® 5630 laptop (refresh rate of 60 Hz, anti-reflective matt screen) and controlled using a customised X-Keys® Desktop (PI Engineering, Michigan, USA) response-box (see Figure 2a). In-built laptop speakers were used to emit the simulator's secondary task audio (8-bit 11kHz mono, 0.83 s intermittent alert, retrieved from http://www.sounds.beachware.com/2illionzayp3may/ilssz/CRASHBUZ.mp3; downloaded on 25-04-2007) using 75% of the available threshold on the master volume switch. The participants were seated near the middle of a general purpose laboratory (atmospheric conditions amounted to background ambient noise/internal soft-light) and controlled ATREIDES with the response box (see Figure 2b), placed at a handedness-compliant location. The laptop was positioned approximately 70 cm in front of them. An A5-sized card presenting the journey Schedule Diagram (see Figure 2a/b) was placed on the laptop's open keyboard, showing a detailed overview of all journey stations. A supply of workload assessment forms and a pen were placed within easy reach to complete when required.

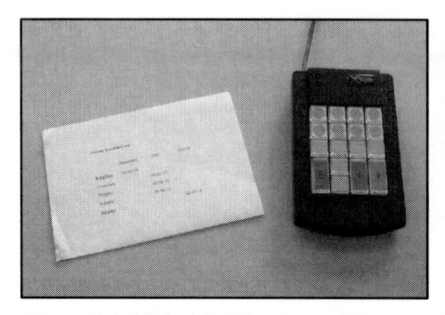

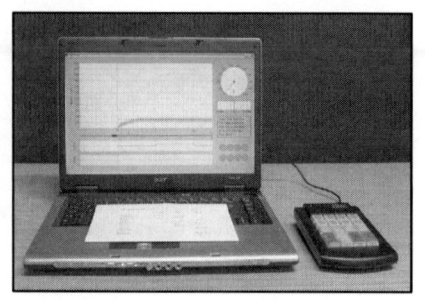

Figure 2. (a) Schedule Diagram and Response box used during experimentation and (b) arrangement with ATREIDES during actual experimentation

Display variables

All task modes operated under an automated braking system that intervened in the event of excessive speed (>4 *mph (> 6 km/h)*) for a sustained (>5 *s*) period. The system assumed control by applying 60% of the maximal braking effort, and returned it only when speed fell below permitted line speeds. Textual speed warnings were shown in control pane. The control pane's advisory section was configured to omit fuel and ETA data, and only show the time and next upcoming station. The speed-envelope pane displayed side-scrolling inertia with a variable resolution (i.e., focus honed at lower speeds and stretched at higher ones). The closed loop driver-tracking (i.e. automatic goal set) mode consisted of a fixed optimum-trajectory curve surrounded by a translucent movement authority zone profiled across the pane. A speed marker mapped to the speedometer and anchored beside the pane's vertical speed axis was used to track the trajectory (see Figure 3a).

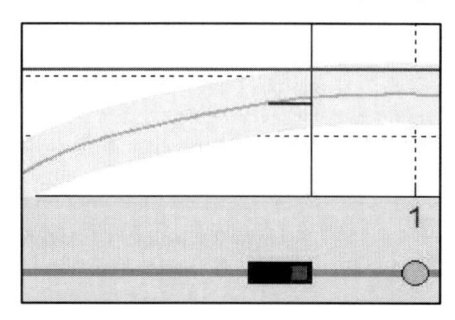

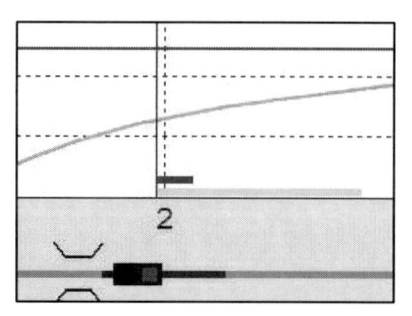

Figure 3. (a) Driver-tracking and (b) driver-predictive display configurations. Colour images are available at http://extras.hfes-europe.org

The zone's threshold above the trajectory allowed for continued user operation (i.e., without triggering timed automated braking invocation) as long as speed changes remained within its boundaries. The pursuit-trajectory was pre-calculated and informed by the schedule, track gradients, and throttle, to deliver a journey profile effectively optimised for fuel-efficiency, punctuality and passenger comfort. The open loop driver-predictive (i.e., driver goal set) mode stripped all tracking components (i.e., speed maker, movement authority zone). The optimal trajectory was retained but paired with real-time braking distances (i.e., full-service, emergency) on the speed-envelope floor, mapped to the train-head on the track structure/signal preview pane (see Figure 3b). Braking information was only previewed once throttle grading crossed from the neutral to the negative. Precise positioning relative to the trajectory was discerned from the speedometer and vertical speed axis. The open loop driver-basic (i.e. driver goal set) mode was the basic ATREIDES configuration, devoid of enhanced features.

Design

The session consisted of one block of three practice trials and one block of five experimental trials. Practice journeys were approximately 3 minutes long, with simple track design. The experimental journey was based on a real route, structured for low-level complexity, and contained a total of five signals and four stations (a departure point, two passing stations and the end destination). The journey was optimised for a duration of 7 min 30 s and contained a moderate speed envelope ranging from 30-90 mph. The journey was repeated five times in an alternating ABACA design that enabled the driver-basic (A) mode to act as a control condition and track baseline improvement. The two enhanced feature conditions (B, C) were counterbalanced, with half the participants receiving the order B-C and half C-B. Short assessment sheets containing retrospective self-reports of task difficulty (3 questions, 1-9 point scale) arranged with 'very easy' and 'very hard' at the two extremes captured subjective workload.

Procedure

At the beginning of the session participants were informed of trial number, session length, and then (with the use of a whiteboard), reminded of all display components presented during previous training sessions. The functions and mechanics of each mode were explained and questions answered. Standard instruction for driving in the closed loop tracking mode was to target and track the pursuit-trajectory as closely as possible within the confines of the overlaid movement authority. Participants were instructed to operate the train to the best of their ability in all modes, and respond to the secondary task immediately. They were also asked to be mindful of safety and performance and consider the three tenets of ride quality (passenger comfort, punctuality, fuel-economy) taught during training as goal setting criterion. Participants carried out the practice journeys, which were informal and explanatory, and with any mistakes or errors commented upon with neutrality. Prior to the start of each experimental trial, participants were told which condition they would be encountering. There was generally no communication once the trial began.

Each trial also required participants to carry out a secondary task, the immediate cancellation of auditory alerts. Delays in onset were randomised and occurred within a minimum (15 s) and maximum (30 s) time frame that approximated to intervals of 30-45 seconds. Alert cancellation was randomised into eight possible response keys arranged in two adjacent rows and mapped to visual indicators in the control pane (top two rows on the response box shown in Figure 2a). The task was timed to reset itself (10 s) in the event of no response. Once the destination was reached, a workload assessment sheet reporting subjective evaluations of task difficulty was immediately completed. New trials were loaded once the current data were saved. Following all trials, participants were paid and signed a payment receipt shortly before departing.

Results

Separate within-participants analyses of variance (ANOVAs) were performed on reaction time (RT), speed error and subjective performance data, from each of the experimental conditions (i.e., driver-tracking, driver-predictive, and driver-basic) to evaluate the effect of task mode on goal set (i.e. closed loop automatic or open loop driver controlled) operation when driving under an automated supervisory braking system. All ANOVAs were tested and corrected for sphericity, where necessary, using Mauchly's test. Post-hoc analyses were all examined using Bonferroni adjustment. The repeated open loop driver-basic (i.e., baseline) trials were averaged only in those instances where preliminary analyses revealed no significant differences. Speed-error data were expressed as a deviation from the optimum second by second speed profile over four progressive journey phases from departure (early, early middle, late middle, and stopping) by computing root mean square (RMS) values. ATREIDES indexed the output of RT data by time of onset. RT values in each minute period were therefore averaged and matched against key route characteristics over journey time to examine any differences in momentary workload.

Speed error

The means for speed error are summarised in Figure 4(a). They indicate a general increase in error over the journey for driver-basic and driver-predictive, with a smaller increase for the driver-tracking condition. ANOVA confirmed the significant main effect of task mode, $F(4,68) = 14.36$, $p < .001$, post-hoc analysis showing that optimal speed management under driver-tracking was better than under either of the driver goal set states; driver-predictive, $p < .001$; driver-basic, $p = .003$. There was also a significant effect of journey phase on speed error, $F(1.25,21.20) = 45.98$, $p < .001$. Subsequent post-hoc analyses revealed significant pairwise comparisons for all phases, $p < .001$, except between the early middle and stopping phases, $p = .069$, suggesting that high speed and station arrival periods were harder to perform optimally. A significant interaction between task mode and phase was also found, $F(3.69,62.80) = 3.32$, $p = .018$, which, given the comparatively flatter driver-tracking profile shown in Figure 4a, suggested automatic goal set operation resulted in near-optimal journey performance, regardless of the phase.

Secondary task

The RT data for response to alerts is shown in Figure 4(b). Consistent with the generally flat pattern of RT over time, ANOVA did not show a significant main effect for either task mode $F(2,34) = 2.44$, $p = .102$, or journey time, $F(2.77,47.07) = 2.39$, $p = .085$. Although there is a hint of an interaction in Figure 4(b), this also was not significant, $F(3.72,63.16) = 1.19$, $p = .322$. These results suggest that any fluctuations in momentary load, across either task mode or journey time, were not large enough to be detectable with this measure.

Speed error

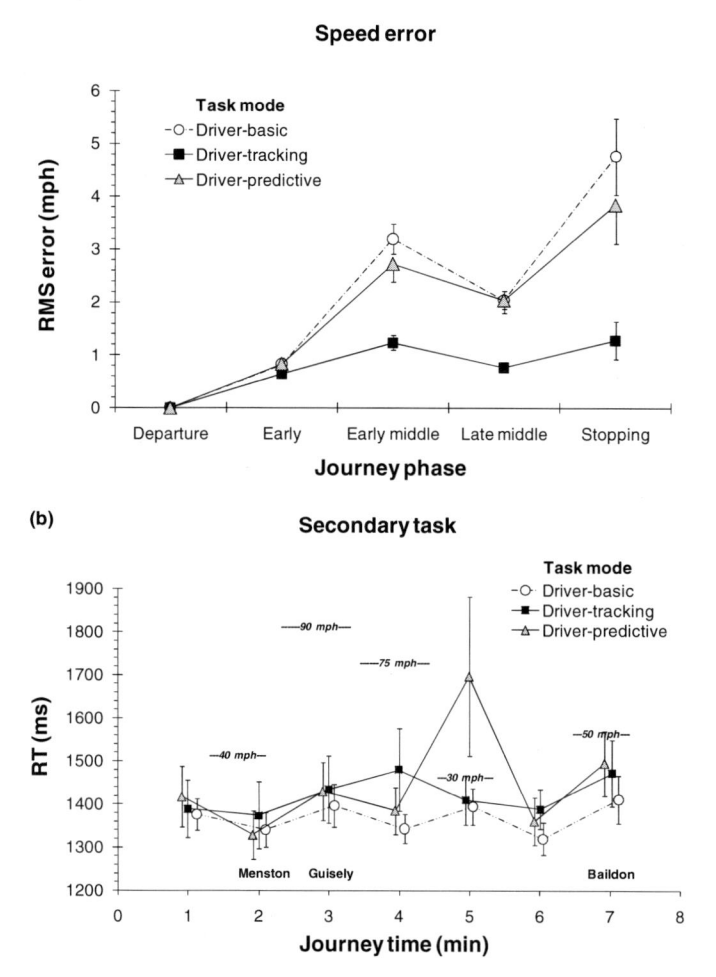

(b) Secondary task

Figure 4. (a) Mean RMS speed error performance compared against the optimum profile and (b) mean RT for the secondary task for all experimental task modes over the journey (driver-basic trials averaged). Error bars indicate standard errors. Station a and key speeds included

Supervisory braking intervention

Data produced by the supervisory braking system were examined descriptively by recording the mean number of interventions made under each mode by each participant. Performance in the driver-predictive (i.e., driver goal set) mode was subject to the occasional intervention ($M = 0.18$, $SD = 0.53$), whilst performance under the repeated driver-basic mode fell gradually over trials (Trial 1, $M = 0.18$, SD

= 0.39; Trial 2, $M = 0.12$, $SD = 0.33$; Trial 3, $M = 0.06$, $SD = 0.24$). No system interventions were recorded for any participant in the driver-tracking mode, suggesting either that there were no physical transgressions outside of the pursuit zone, or that operation under a tracking (i.e., automatic goal set) system encouraged a more timely recovery. Overall, these results suggested that driver goal setting precluded the ability to manage line speeds as effectively as automatic goal set operation, though baseline improvement demonstrated that continued learning and skill development were also important factors, with respect to gaining more proficiency with unaided modes of operation.

Workload self-report

Table 1. Means and standard deviations for ride quality own performance measures

Experimental condition			Subjective measure of task difficulty					
			Passenger comfort		Time keeping		Fuel economy	
Task type	Task mode	Trial	M	SD	M	SD	M	SD
Open loop	Driver-basic (baseline)	1	5.11	1.53	4.28	1.64	4.89	1.37
		3	5.44	1.89	4.50	2.20	4.72	1.60
		5	4.83	2.09	3.83	2.18	4.61	2.03
	Driver-predictive	2/4*	5.28	1.74	4.50	2.07	4.61	1.58
Closed loop	Driver-tracking	2/4*	3.56	2.01	3.39	1.88	3.39	1.72

*Trial order counterbalanced across participants

The means and standard deviations for the subjective reports of workload are summarised in Table 1. They indicate general assessments of moderate ease for driver-tracking, with driver-basic and driver-predictive considered to be a little more difficult. ANOVA showed a significant main effect of task mode on perceptions of difficulty for passenger comfort, $F(3,51.15) = 6.03$, $p < .001$, post-hoc analysis showing that driver-tracking was considered easier than the third driver-basic trial (baseline 2), $p = .008$. Values for every other pairwise comparison were not significant ($p > .1$ for all), indicating largely indifferent appraisals of difficulty between either of the open loop goal set states. Perceptions of difficulty for time-keeping also showed a main effect of task mode, $F(4,68) = 3.15$, $p = .019$, though the post-hoc analysis revealed no clear differences between conditions. Perceptions of difficulty for fuel-economy showed a highly significant main effect for task mode, $F(4,68) = 6.14$, $p = < .001$, post-hoc analyses showing that driver-tracking was rated easier than driver-predictive, $p = .027$; driver-basic trial 1 (baseline 1), $p = .032$, and driver-basic trial 3 (baseline 2), $p = .018$, suggesting that automatic goal set operations were considered better for fuel consumption.

Discussion

The results of the present study show that performance under a supervisory automated braking system is closer to optimal levels under an automatic goal set, closed loop tracking mode than under user goal set modes containing varying levels of open loop modality. However, the results also suggest that driver-tracking under the current closed loop configuration is not an automatic or unattended activity generally characteristic of this type of control. Objective RT performance under the secondary task showed no statistical differences for momentary load, suggesting any mental task demand alleviated by tracking activities were allocated elsewhere. However, though no significant differences in RT were found, subjective data revealed that driver-tracking was generally felt to be easier than either of the open loop modes, suggesting driver goal setting was considered moderately more difficult. The lack of supervisory braking intervention suggests that movement authority layers adequately reinforced pursuit-tracking behaviours, producing safer journeys.

Most surprising was the discovery that presentation of features that advocated driver-prediction (braking, optimal profile) without surrendering goal setting resulted in overall indiscriminate performance both objectively and subjectively. Open loop control was therefore treated in the same manner despite the visual presentation of prescribed optimal routing. Performance issues for the predictive mode may therefore be related to feature conflict or simple distrust. The presentation of an optimum trajectory without the means to track it may have hindered its usability, and simply created a conflict with the presiding speed envelope. Similarly, the mapping of braking features along the floor of the pane instead of in curve form (Askey & Sheridan, 1996; Einhorn et al, 2005) may not have conveyed braking information adequately enough to enable participants to carry out precise manoeuvring around the envelope or trajectory. The slower RTs for alerts observed at the greatest decelerative speed change indicates it was evidently sought by some. However, as a feature that previewed optimal station stopping, it was surprising to find that speed error under the driver-predictive mode was not better than baseline performance. Standard errors at the end phase evidenced clear station-stopping variation in all modes, suggesting a preference to stop well in advance of the optimally designated platform. In the driver-tracking mode, this meant deviations from not only the trajectory but from the boundary of the surrounding movement authority zone. Whilst this may relate to time-keeping and punctuality, its persistence in near-optimal tracking conditions suggested a safety issue. For blue skies purposes, the coupling and effectiveness of open loop enhanced features need re-examining in view of inherent changes brought by an in-cab exclusive environment.

The present study used short journeys and a low level of track complexity in a low fidelity simulation, yet it was enough to render the task equally demanding under different modes of operation, suggesting the in-cab architecture of a closed system is laden with resource requirement. In the absence of a window to the outside world, the mental preoccupation of modelling the moving environment may be the likeliest explanation for the high task demand. However, it may also be an indication of increased attention and effort brought on by the use of a supervisory braking system. Perceptions of an automated authority may have reinforced the desire for modes of operation that minimised any chances of intervention. However, the closed loop tracking analogue used in ATREIDES was such that it may have necessitated a further layer or sub-loop of regulatory control activities (Hollnagel & Woods, 2005). Given the inevitability of deviating away from the optimal trajectory, an averaged performance inside the zone was taught as the desired outcome. Therefore any periods of sub-optimal tracking may have been addressed by carrying out compensatory deviations in the opposite direction. In the case of fuel use, this was most pertinent, whereas, with most trains, a driver-cab fuel indicator is not provided. Given the lack of this information, sustained periods of sub-optimal tracking above the trajectory may have been strategically redressed by driving slower, implying that automatic goal setting thresholds were subject to user overruling, and planned anticipatory control.

ATREIDES was endorsed for this study only after participation in training and performance in preliminary studies showed learning was clearly taking place. However, this did not necessarily confirm advanced skill or ability, particularly as subtle differences between baseline conditions indicated continued learning and task familiarity was taking place. As skill increases with continued operation, the tracking mode may become unattended and automatic, and therefore more characteristic of closed loop tracking tasks that are susceptible to sudden task demands. It was hoped that subjective scoring would help infer a clearer attitude towards the automated braking system, but the patterns of self-reporting indicated an overall insensitivity to the scale. Tracking was more than likely treated preferentially simply because of the reductions in effort for goal related decision making. Implications for Britain's use of ETCS are therefore tentative at best. Based on the driver-tracking mode, it appears that train driving efficiency will be reinterpreted into the ability for tracking maximum speeds under the supervisory mode, without triggering automated safety brakes. Safety and performance will therefore be more overtly in the hands of the automated authority. Intuitive regulation and use of control sub-loops in carrying out the task will depend entirely on the flexibility of algorithms governing the system (e.g., in adverse weather or low adhesion driving states), and any punitive measures set in place. However, whilst it is uncertain how train drivers are likely to react in an open environment, current in-cab automated supervisory practice is still a step closer to a true dynamic allocation of tasks based on the needs of the operator.

References

Alcatel (2008). *ETCS demonstrator*. Retrieved 26 October, 2008, from http://www1.alcatel-lucent.com/tas/etcs/etcs_demonstrator/demon_en.htm

Annett, J. (2003). Hierarchical task analysis. In E. Hollnagel, (Eds.) *Handbook of Cognitive Task Design* (pp. 17-35). Mahwah: NJ, Lawrence Erlbaum Associates.

Askey, S., & Sheridan, T. (1996). *Human Factors Phase II: Design and Evaluation of Decision Aids for Control of High-speed Trains: Experiments and Model Safety of High-speed Ground Transportation Systems*. DOT/FRA/ORD-96-09. Cambridge, MA

Burns, C.M., & Hajdukiewicz, J.R. (2004). *Ecological interface design*. Florida: CRC Press LLC

Cole, C., Bosomworth, C., Hayman, M., McLeod, T., & Croce, J. (2006, April 30). *Field testing of an intelligent train monitor.* Paper presented at the 13th Conference on Railway Engineering, RTSA, Melbourne, AU. Retrieved October 26, 2008, from aCQUIRe on-line digital repository.

Department for Transport (2007a). *Delivering a sustainable railway - White paper* (CM 7176). London, UK. Retrieved October 26, 2008, from http://www.dft.gov.uk/

Department for Transport (2007b). *ERTMS national implementation plan.* London, UK. Retrieved October 30, 2008, from
http://www.dft.gov.uk/pgr/rail/interoperabilityandstandards/interopprogress/

Dorrian, J., Roach, G. D., Fletcher, A., & Dawson, D. (2006). The effects of fatigue on train handling during speed restrictions. *Transportation Research, Part F, 9*, 243-257.

Dzindolet, M.T., Peterson, S.A., Pomranky, R.A., Pierce, L.G., & Beck, H.P. (2003). The role of trust in automation reliance. *International Journal of Human-Computer Studies, 58*, 697-718.

Edkins, G., & Pollock, C. (1997). The influence of sustained attention on railway accidents. *Accident Analysis and Prevention, 29*, 533-539.

Einhorn, J., Sheridan, T., & Multer, J. (2005). *Preview information in cab displays for high-speed locomotives: Human Factors in Railroad Operations.* DOT/FRA/ORD-04/12. Cambridge, MA.

Hollnagel, E., & Woods, D. (2005). *Joint Cognitive Systems: Foundations of Cognitive Systems Engineering.* Boca Raton, FL: Taylor & Francis.

Klein G.A., Calderwood. R. & MacGregor D. (1989). Critical decision method for eliciting knowledge. *IEEE Transactions on Systems, Man and Cybernetics, 19*, 462-472

Moray, N. (1997). Models of models of...mental models. In T.B. Sheridan and A. van Lunteren (Eds.), *Perspectives on the Human Controller* (pp. 271-285). Mahwah, NJ: Lawrence Erlbaum

Naweed, A. (2009). *Enhanced information design for high speed train displays.* Unpublished manuscript, University of Sheffield, England, UK.

Naweed, A., Bye, R., & Hockey, G.R.J. (2007a). Enhanced information design for high speed trains. In D. de Waard, G.R.J. Hockey. P. Nickel, and K.A. Brookhuis (Eds.), *Human Factors Issues in Complex System Performance* (pp. 235-241). Maastricht, the Netherlands: Shaker Publishing.

Naweed, A., Bye, R., & Hockey, G.R.J. (2007b). Enhanced decision support for train drivers: Driving a train by the seat of your pants. In D. de Waard, G.R.J. Hockey. P. Nickel, and K.A. Brookhuis (Eds.), *Human Factors Issues in Complex System Performance* (pp. 131-145). Maastricht, the Netherlands: Shaker Publishing.

Parasuraman, R., Barnes, M., & Cosenzo. (2007). Adaptive automation for human-robot teaming in future command and control systems. *The International C2 Journal, 1*(2), 43-68.

Parasuraman, R., Molloy, R., & Singh, I. L. (1993). Performance consequences of automation-induced complacency. *International Journal of Aviation Technology, 3*, 1-23.

Rail Innovation Australia Pty Ltd. (2008). *FreightMiser.* Retrieved 26 October, 2008, from
http://www.railinnovation.com.au/innovation/technologies/freightmiser.html

Rasmussen, J., Pejtersen, A.M., & Goodstein, L.P. (1994). *Cognitive systems engineering.* New York: Wiley.

Stjernström, R. (2001). *User-centred Design of a Train Driver Display.* Technical Report 2001-016. Department of Information Technology. Uppsala University.

Vicente, K.J. (1999). *Cognitive work analysis. Toward safe, productive, and healthy computer-based work.* Mahwah, NJ, USA: Lawrence Erlbaum Associates.

Woods, D., & Hollnagel, E. (2006). *Joint Cognitive Systems: Patterns in Cognitive Systems Engineering.* Boca Raton, FL: Taylor & Francis.

Warning drivers of approaching hazards: the importance of location clues and multi-sensory cues

Michael G Lenné & Thomas J. Triggs
Monash University Accident Research Centre
Australia

Abstract

Research in experimental psychology has examined in depth the benefits to target detection through the provision of a warning signal, the optimum warning intervals, and has proposed mechanisms such as increased arousal and expectancy that underpin these effects. The rapid acceptance of in-vehicle technologies necessitates examination of these issues to ensure that the anticipated safety benefits are in fact realised. This research set out to measure driver responses to an advisory warning – the context of this examination was detection of emergency vehicles (EV). Using an advanced driving simulator we examined the effects of an advisory warning device (AWD) on driving performance in scenarios known to be high risk for EVs. Each event contained a combination of scenario type (adjacent-lane, turning-across, car-following) and warning condition (control, standard, advisory). For adjacent-lane and turning-across events the AWD was associated primarily with reductions in mean speed. The advisory warning allowed the drivers to react more quickly once the EV was detected, resulting in a lower vehicle speed at the intersection, which is a measure of increased safety. Response priming emerged as a likely mechanism underpinning these benefits and is generalisable to other settings where an advisory warning is presented before the threat is perceived. Recent research shows that performance can be enhanced further by providing information about the target location within the warning signal, and/or by using multi-sensory modalities to present threat information to drivers. These developments provide further avenues for exploring the optimal configuration of warning systems.

Introduction

Technologies are being increasingly used to support road users in achieving safe use of the road transport system. These fall into two broad categories: infrastructure-based and in-vehicle. While the application of both of these is growing, in-vehicle technologies, in particular, are being incorporated within modern vehicles to enhance driver performance and safety. Examples of in-vehicle systems include Intelligent Speed Adaptation Systems, Adaptive Cruise Control, Forward Collision Warning Systems, Rear Collision Warning Systems, Seat Belt Reminders, Lane Departure Warning Systems, Lane Keeping Assistance Systems, and Brake Assist/Forward Collision Mitigation.

In-vehicle systems can be classified as either advisory or imminent collision warning systems and there are underlying differences in these systems and the responses they are designed to elicit. An advisory system is likely to be designed to redirect the driver's attention, rather than to trigger an immediate response. Hence earlier warnings are likely to provide greater benefits for such systems. Collision warning systems however should be designed to trigger an immediate driver response. Early alerts in this situation could undermine the safety benefits of the system, while late alerts are of little utility (Lee, McGehee, Brown, & Reyes, 2002).

This paper is primarily concerned with advisory warning systems. The paper begins by reviewing selected research that has employed a non-specific warning signal. The purpose of this discussion is to help to understand how the theoretical developments of the past might shape the use of warnings in vehicle technologies in the future. Having demonstrated the benefits to task performance observed through the provision of a warning signal, we present a brief discussion of the underlying mechanisms. Following this, a summary of our research into advisory warning systems is presented as it provides interesting data to contribute to the warning signal discussion. While non-specific warning signals do provide performance benefits in some conditions, more recent research shows that performance can be enhanced further by providing information about the target location within the warning signal, and/or by using multi-sensory modalities to present threat information to drivers.

Non-specific warning signals

There has been much research conducted since the 1950's that has explored participant responses to warning signals, the influence of moderating variables such as inter-stimulus interval and modality, and discussed the mechanisms underpinning the performance effects observed. With the continued rapid emergence of technologies into the vehicle fleet, it is instructive to consider the previous research that has examined the benefits associated with the provision of warning cues and what impact this research might have on technology design. The remainder of this paper is not intended as an exhaustive review of the literature but rather a select review of literature to help understand how theoretical developments of the past might assist with applications with the intelligent transport system domain.

Much of early research that explored human performance with warning cues used non-specific warning signals, most commonly an auditory click or tone. Bertleson (1967) showed that response time (RT) was faster, but errors increased, when a non-specific auditory sound was presented to begin the foreperiod. Similar results were also reported by Posner and colleagues amongst others (Posner & Boies, 1971; Posner, Klein, Summers, & Buggie, 1972). The warning intervals, or foreperiods, used in these studies ranged from less than 100 msec to up to 1000 msec, with the maximum benefits observed after 200 msec and observed for up to around 1000 msec. The underlying mechanism here was not related to the rate at which information was acquired by memory systems but rather the rate at which a later system responded to that information.

The classic view is that a non-specific warning signal allows attentional preparation which then facilitates sensory and motor processes during the reaction interval (Hackley & Valle-Inclán, 2003). A neutral (i.e., general) warning signal reduces RT by lowering the threshold of response, or by non-specifically priming all responses. In this respect, the faster RT observed with a warning signal can be related to more rapid response selection. A non-specific warning signal is thought to trigger a phasic burst of arousal. This burst of arousal does not affect the speed at which perceptual evidence accrues, and does not decrease response threshold. Rather, the burst of arousal triggers an increase in response activation that results in a more rapid response from the central processor. In this respect, the benefits to RT observed following a non-specific warning are associated with fast-guess responding, and thus the increased response speed observed comes at a cost of an increased error rate.

In terms of the increases in arousal associated with the presentation of a warning signal, it is thought that there are differences across modalities that translate to performance benefits. For example, using a visual task Ulrich and Mattes (1996) found that an auditory warning signal induced a higher phasic state of arousal than a visual warning signal, and thus faster RTs. Those authors also examined the influences of foreperiod duration on performance. Confirming the research of the 1960's and 1970's, Ulrich and Mattes (1996) found that RT was faster for foreperiods greater than around 200msec. This arises because up to 150-200 msec is needed to obtain maximum response readiness as underscored by Bertleson (1967). Similar to others, Ulrich & Mattes (1996) noted that as immediate arousal is assumed to be a transient effect, the greatest benefits are found at shorter foreperiods less than around 1000 msec.

Many other factors are known to influence the effectiveness of a warning signal, including the reliability of the cue. Cue reliability can take the form of temporal reliability, or reliability in presentation of the appropriate target following the warning. In terms of temporal uncertainty, Posner noted that RT is faster when the foreperiod can be accurately predicted. More recently Vu and colleagues (2006) showed that the reliability of the auditory cue, in terms of whether it was reliably presented, also affected the performance benefit on the visual search task. There is much other research showing that factors that change the consistency and strength of the warning signal-target relationship diminish the benefits associated with the provision of a warning signal.

Summary of non-specific warnings

In summary, the benefits to performance found with the use of non-specific warning signals are believed to be related to a transient increase in arousal that results in more rapid, but more error prone, response selection. Arousal is believed to be transient and so the benefits are thought not to persist beyond foreperiods, or warning intervals, beyond around 1000 - 1500 msec before falling away.

In terms of real-world performance, a recent review article by Spence and Ho (2008) noted the importance of warning timing through the examination of rear-end crash avoidance systems. Like others, those authors noted that the potential for a warning to be interpreted as a false alarm increases as the driver is given more advanced notice, therefore leading to the potential for annoyance, distraction, and desensitisation. In the case of the rear-end crash scenario they suggest that approximately 1500 msec is the maximum warning interval to avoid the signal being classified as annoyance and as a false alarm. Between 700 msec and 1500 msec time-to-collision is seen as the optimum warning interval, while anything lower than approximately 700 msec would be too late to be effective.

In a further examination of these issues, some relevant data from our study involved the examination of driver responses to an in-vehicle system. The prototype system was designed to alert drivers to the presence of an on-call emergency vehicle. The emergency vehicle crash problem, underlying causes, and the detailed study design and results are published elsewhere (Lenné, Mulvihill, Triggs, Regan, & Corben, 2008). Hence only a brief overview of the study design and results are presented here before discussing the implications of the findings for technology use.

Study overview

The major aim of this research was to establish whether an advisory warning had any effect on the safety of driver interactions with emergency vehicles, and if so, whether it had a positive or negative effect on the safety of those interactions. The AWD was designed to provide a safer driving environment for both on-call emergency vehicles and the general motoring public. This was achieved by placing the AWD, comprising both visual and auditory indicators, on the lower left corner of the windscreen in accordance with instructions from the project sponsor. Note that cars in Australia drive on the left-hand side of the road.

The AWD was mounted on the lower passenger side corner of the windscreen. The prototype module was powered by an external power source, and was approximately 50 mm wide by 30 mm high by 3 mm thick. It was a proximity-based warning system that was activated when the EV was approximately 350 m from the participants' vehicle (own-cab). It had both a visual (high intensity LEDS) and an auditory warning. When activated, the three LEDS flashed and the auditory tone pulsed (at around 80 dB) simultaneously at a rate of approximately 2 Hz.

The study was conducted using the MUARC mid-range driving simulator. It consists of a Holden Calais sedan with normal interior features, surrounded by two projection screens. The major projection screen is located at the front of the vehicle, and one is located behind it. From the driver's eye point the screen in front of the car provides a field of view subtending angles of approximately 180 degrees horizontally and 40 degrees vertically. The rear screen provides a field of view subtending angles of approximately 60 degrees horizontally and 40 degrees vertically. Three BARCO 700 HQ projectors convey images of traffic scenes onto the front screen. Special electronics blend these individual images into one quasi-continuous image. The projectors are adjusted so that all scene objects are displayed in their correct geometric locations as viewed from the central eye design point. The images displayed on the screens are generated by a Silicon Graphics Onyx computer that updates the images on all channels at a rate of 30 Hz.

The three most prominent crash types involving emergency vehicles that emerged from an Australian database (the state of Victoria), and that could potentially be resolved using the AWD, were: adjacent, involving a side-impact collision; right-turn, involving a vehicle colliding with another vehicle performing a right-turn across; and car-following, which involves a front-to-rear collision. These types of events were modelled in the simulator and participants drove through these scenarios in the control condition (no emergency vehicle), the standard warning condition (involved an emergency vehicle with lights and sirens active), and the advisory warning condition (where the warning device was activated prior to the lights and sirens being detectable). The warning intervals used in this study were longer than those typically utilised in previous research (between 1.5 and 3.0 sec).

There was evidence of performance benefits for all three scenarios. In all three scenarios the participants exhibited the desired behaviour significantly earlier with the advisory warning device operating than with the standard emergency vehicle lights and sirens. In the case of the adjacent and right-turn across scenarios this was evident as an earlier reduction in mean speed, and in the car following scenario was evident as an earlier lane change manoeuvre to clear a path of the overtaking emergency vehicle. While the simulator data tell us when drivers first responded to the events, the analysis of eye movement data provides an insight as to when drivers first detected the key stimuli. The eye movement data showed increased scanning of the environment in the advisory warning condition. Scanning increased markedly when the AWD device was activated, suggesting that drivers were actively searching the visual environment for the emergency vehicle.

The results from this evaluation yield a number of questions that relate to the implementation of such a system in the vehicle fleet, and implications for other in-vehicle systems more broadly. These are considered in turn.

In terms of challenges associated with implementation, several issues relating to the potential effectiveness of the AWD device in the field should be considered. How drivers' responses to the AWD device might change over time (behavioural adaptation) would be of great interest. The potential for the proximity-based warning to yield an acceptable level of false alarms also needs to be established. It would be interesting to conduct a trial to consider how driver responses to the AWD might change over time and to determine how the system performs in the field. Another issue is whether the changes in behaviour associated with the activation of AWD would in fact be positive and actually result in a reduction in crashes, near misses, and response times involving emergency vehicles. This would likely require several years of exposure to AWD in the community before any meaningful crash analyses could be conducted.

There are some outstanding research issues to consider. The importance of understanding the functional requirements for in-vehicle systems such as this one is widely recognised by researchers and this research leads to a number of questions as we consider the implications for other systems. Much previous work in road safety has explored issues including the influence of warning timing, urgency, and mode of presentation on the behavioural responses of drivers (e.g., Barfield & Dingus, 1997;

Noy, 1997). Clearly, the timing and urgency of a warning should be dependent upon the system aims; that is, whether it is designed to simply advise the driver to redirect attention, or whether in the case of a collision avoidance system it is designed to illicit a more immediate response. For advisory warning systems such as this one, conventional thinking has suggested that warning intervals of less than around 1000 msec are appropriate, a rationale being the potential for driver distraction and the annoyance that might arise from having a warning activated much before the threat can be perceived by the driver. This study however found potential safety benefits using a warning interval of up to around 3 sec (the interval between the AWD activation and when the EV siren was audible). This has implications for other advisory in-vehicle systems that might operate in a similar manner. An example here might be an in-vehicle device that warns a driver to the presence of a train at a level crossing. Here again, the warning might be provided to the driver before he/she perceives the presence of the threat (the train).

In terms of what processes may be underpinning the results found, response priming may be an important underlying mechanism. Response priming has been well researched in experimental psychology and in essence relies upon the strength of an association between a cue and a target to facilitate a more rapid response. Priming is invoked here for two reasons. Firstly, the performance benefits in our study persisted over longer warning intervals than can be explained by the arousal hypothesis alone. Secondly, when presented with the advisory warning system, the participants did not respond until the second signal was detected. That is, in the advisory warning condition, decreases in speed and increases in visual scanning were evident only after the participants detected the emergency vehicle. While possible, the role of priming here requires significantly further examination.

The importance of location clues and multi-sensory cues

The importance of including information or cues about the target location is of critical importance in a system such as this one where the threat can be approaching from any direction. This addition could potentially increase the efficiency of threat detection and therefore minimise any negative effects that might be associated with distraction and uncertainty.

There is some research that has explored location cues in the context of warning signals. By means of example, Posner et al. (1980) noted that knowledge of the region in space where the target would be presented was associated with faster RT in a laboratory task. Posner and colleagues raised the importance of orienting attention and the concept of an attentional spotlight, being where attention is directed, thus leading to faster target detection. The 'Simon effect' also describes the situation where subjects code information about stimulus location, and this affects response selection, and thus faster RTs are observed (Simon, 1990).

Much research has also confirmed the utility of auditory cues in benefiting visual search. For example, as reviewed by others (e.g., Hackley & Valle-Inclán, 2003), Begault (1993) used valid spatial auditory cues and found large benefits in responses to visual stimuli presented in the same spatial location. Perrot (1996) also reported that auditory spatial cues enhanced visual search performance and were more effective than visual cues. Also using a visual search paradigm Vu et al. (2006) reported the fastest search times when the auditory cues were presented from the same location in space on the visual target. Search times were slower when the auditory cue was displaced randomly across horizontal and vertical planes, and when the cue was always presented from the centre of the visual field.

While the importance of auditory cues in enhancing performance in other domains is well recognised (e.g., Parker, Smith, Stephan, Martin, & McAnally, 2004; Stephan, Smith, Martin, Parker, & McAnally, 2006), this is becoming increasingly explored in the road context. The more recent research in this area discusses the concept of cross-modal shifts in visual attention and the potential influence on interface design and modality choice for users. A number of papers have found faster RT to a visual cue when it is preceded by an auditory or virbrotactile cue presented from the same direction (Ho, Reed, & Spence, 2006; Ho & Spence, 2005; Ho, Tan, & Spence, 2005, 2006). Those authors state that multi-sensory interfaces therefore have great potential to increase the efficiency of HMI design in spatial application domains such as driving, particularly in situations of high visual demand. Those authors also provide a comprehensive review of their research and future directions (Spence & Ho, 2008).

Conclusion

This paper aimed to present an overview of research examining the benefits to performance seen with the use of a warning signal, to present an overview of some research we conducted, and finally to note the most recent thinking concerning the optimal use of warning cues to enhance the use of in-vehicle technology to support driver performance and safety.

Research over several decades suggests that the benefits to performance found with the use of non-specific warning signals are related to a transient increase in arousal that results in more rapid response selection. Further, these bursts in arousal are believed to be transient and so the benefits are thought not to persist beyond foreperiods around 1000 - 1500 msec before falling away. Our research showed performance benefits associated with slightly longer warning intervals, and we invoked response priming as a potential mechanism to account for the results obtained. Further work, however, is needed to resolve these theoretical contributions.

In extending the utility of the device we evaluated, the latest research clearly confirms the necessity to provide cues to the location of the target event, and ideally to provide that cue from the same location in space as the target event. The benefits of non-visual modalities, including auditory and vibrotactile cues, to visual target detection and response have been demonstrated. These findings are important to consider for technology design, particularly as we consider the theoretical implications that are associated with the integration of multiple warning systems within the vehicle and the anticipated driver responses.

References

Barfield, W., & Dingus, T.A. (1997). *Human Factors in Intelligent Transport Systems*. Mahwah, NJ: Lawrence Erlbaum.

Bertleson, P. (1967). The time course of preparation. *Quarterly Journal of Experimental Psychology, 19*, 272-279.

Hackley, S.A., & Valle-Inclán, F. (2003). Which stages of processing are speeded by a warning signal? *Biological Psychology, 64*, 27-45.

Ho, C., Reed, N., & Spence, C. (2006). Assessing the effectiveness of "intuitive" vibrotactile warning signals in preventing front-to-rear-end collisions in a driving simulator. *Accident Analysis and Prevention, 38*, 988-996.

Ho, C., & Spence, C. (2005). Assessing the effectiveness of various auditory cues in capturing a driver's visual attention. *Journal of Experimental Psychology: Applied, 11*, 157-174.

Ho, C., Tan, H.Z., & Spence, C. (2005). Using spatial virbrotactile cues to direct visual attention in driving scenes. *Transportation Research Part F, 8*, 397-412.

Ho, C., Tan, H.Z., & Spence, C. (2006). The differential effect of vibrotactile and auditory cues on visual spatial attention. *Ergonomics, 7*, 724-738.

Lee, J.D., McGehee, D.V., Brown, T.L., & Reyes, M. (2002). Collision warning timing, driver distraction, and driver response to imminent rear-end collisions in a high fidelity driving simulator. *Human Factors, 44*, 314-334.

Lenné, M.G., Mulvihill, C., Triggs, T., Regan, M., & Corben, B. (2008). Detection of emergency vehicles: Driver responses to advanced warning in a driving simulator. *Human Factors, 50*, 135-144.

Noy, Y.I. (1997). *Ergonomics and Safety of Intelligent Driver Interfaces*. Mahwah, NJ: Lawrence Erlbaum.

Parker, S.P.A., Smith, S.E., Stephan, K.L., Martin, R.L., & McAnally, K.I. (2004). Effects of Supplementing Head-Down Displays With 3-D Audio During Visual Target Acquisition. *International Journal of Aviation Psychology, 14*, 277-295.

Posner, M.I., & Boies, S.J. (1971). Components of attention. *Psychological Review, 78*, 391-408.

Posner, M.I., Klein, R., Summers, J., & Buggie, S. (1972). On the selection of signals. *Memory and Cognition, 1*, 2-12.

Posner, M.I., Snyder, C.R.R., & Davidson, B.J. (1980). Attention and the detection of signals. *Journal of Experimental Psychology: General, 2*, 160-174.

Simon, J.R. (1990). The effects of an irrelevant directional cue on human information processing. In R.W. Proctor and T.G. Reeve (Eds.), *Stimulus-response compatibility: An integrated perspective* (pp. 31-86). Amsterdam: Elsevier.

Spence, C., & Ho, C. (2008). Multisensory warning signals for event perception and safe driving. *Theoretical Issues in Ergonomics Science, 9*, 523-554.

Stephan, K.L., Smith, S.E., Martin, R.L., Parker, S.P.A., & McAnally, K.I. (2006). Learning and Retention of Associations Between Auditory Icons and Denotative Referents: Implications for the Design of Auditory Warnings. *Human Factors, 48*, 279-287.

Ulrich, R., & Mattes, S. (1996). Does immediate arousal enhance response force in simple reaction time? *The Quarterly Journal of Experimental Psychology, 49A*, 972-990.

Vu, K.-P.L., Styrbel, T.Z., & Proctor, R.W. (2006). Effects of displacement magnitude and direction of auditory cues on auditory spatial facilitation of visual search. *Human Factors, 48*, 587-599.

Different alarm timings for a forward collision warning system and their influence on braking behaviour and drivers' trust in the system

Genya Abe[1], Makoto Itoh[2], & Tomohiro Yamamura[3]
[1]Japan Automobile Research Institute
Ibaraki
[2]University of Tsukuba
[3]Nissan Motor Co., Ltd.
Japan

Abstract

Alarm timing for a forward collision warning system plays an important role in system effectiveness. It is necessary to determine the appropriate alarm timing by considering a driver's response to not only true alarms but also how drivers respond when an alarm should have been issued but was not (missed alarms) to evaluate both the increased system effectiveness and decreased over-reliance on the system. By using a driving simulator, two different alarm timings were compared to investigate the braking behaviour toward alarms and the driver's response to missed alarms according to two different alarm timings: (1) adaptive alarm timing, in which an alarm is given based on the ordinary braking behaviour of the individual; and (2) non-adaptive alarm timing, in which an alarm is given by using a particular alarm trigger logic (e.g., Stopping Distance Algorithm) as a common timing for all drivers. As a result, the timing of non-adaptive alarms was earlier than that of adaptive alarms in this study. The results show that, in comparison with adaptive alarm timing, non-adaptive alarm timing induced early braking behaviour independent of the degree to which a collision was imminent. However, adaptive alarm timing did not impair the driver's subjective ratings of trust. In addition, adaptive alarm timing contributed to consistent trust, even if missed alarms occurred.

Introduction

The reduction of traffic accidents is an important research goal, and rear-end collisions are one of the most common types of accidents. The Forward Collision Warning Systems (FCWS) may be of great potential benefit to drivers who do not pay sufficient attention to driving, because these systems may reduce the number of traffic accidents (Alm & Nilsson, 2000; Ben-Yaacov et al., 2002). However, rear-end collisions tend to occur in time-critical situations; therefore, the expected results depend on when the alarm is triggered (Janssen et al., 1993). When determining the alarm timing, the Stopping Distance Algorithm (SDA) is often used (ISO, 1999). SDA is so termed because a warning distance is defined based on the difference between the stopping distances of the lead and following vehicles. It is known that the braking response time may be improved when several parameters of SDA are manipulated to provide drivers with early alarms, compared the response times in non-assisted driving conditions (Lee et al., 2002).

However, it is not obvious that early alarms achieve safe driving. Drivers need to repeatedly implement situational recognition, decision-making, and action implementation to ensure safe driving. If drivers hear alarms before they recognise risky situations, then decide to implement actions to avoiding critical situations, then drivers decisions might be replaced by alarms. Moreover, if drivers are excessively adapted to such early alarms, they might assume that it should not be necessary to immediately take appropriate actions to avoid collisions as long as a alarm is not triggered. . As a result, there is the possibility that the driver's braking behaviour might be delayed if "missed alarms" which should have been triggered but were not (including alarms which drivers did not notice for some reason also) occur. However, it is known that the degree to which braking behaviour is impaired as a result of missed alarms may vary depending on the alarm timing (Abe & Richardson, 2006).

There also may be individual differences in the timing of braking behaviour before imminent collision situations (Abe & Itoh, 2008). Thus, if alarms consider the driving characteristics for an individual, which is called adaptive alarm timing, then it would be possible to minimize differences in the timing between the driver's collision avoidance behaviour and alarm presentations. Consequently, it is possible to prevent a driver's excessive reliance on alarms, thus resulting in drivers' robust behaviour toward missed alarms.

In the present study, two experiments were conducted. In the first experiment (pilot study), the variation of braking behaviours of individual drivers in response to imminent collision situations in which a lead vehicle suddenly decelerated was explored. In the second experiment, two issues were investigated: the differences in alarm effectiveness and the differences of braking behaviour due to missed alarms with difference timings. Here, two kinds of alarm timings were considered: adaptive alarm timing, based on the ordinary braking behaviour of the individual, and non-adaptive alarm timing, which uses trigger logic (Stopping Distance Algorithm) as a common timing for all drivers.

Pilot study

Apparatus

This experiment was conducted with a driving simulator owned by the Japan Automobile Research Institute. The simulator has six-degrees-of-freedom motion and uses complex computer graphics to provide a highly realistic driving environment. The simulated horizontal forward field of view was 50 degrees and the vertical field of view was 35 degrees.

Participants and experimental tasks

Twenty-four participants (Mean age = 25.5 years, SD = 8.0 years) took part in the experiment. All of the participants were licensed drivers and each participant was given two tasks in the experiment. One was to maintain the speed of the vehicle at a target speed of 80 km/h, and the other task was to follow a lead vehicle while avoiding a rear-end collision. .

Experimental design and procedure for the pilot study

The participants were divided into two groups according to time headway conditions and each participant experienced two different lead vehicle decelerations, namely 5.88 m/s^2 (high time criticality) and 3.92 m/s^2 (low time criticality). Twelve participants were assigned in to each group. Two values, 2.0 s and 1.4 s, were considered in the experiment as the time headway conditions while following a vehicle. For one lead vehicle deceleration, nine potential collision events, in which the lead vehicle decelerated at a constant rate with illuminated brake lights but with irregular time intervals, could occur. The order of the experimental conditions was controlled to reduce the potential effects of the experimental conditions on driver behaviour.

All participants gave informed consent and were instructed regarding the task requirements. Each participant was allowed a 10-min practice session to familiarize themselves with the simulator and experience the lead vehicle decelerations that would be involved in the experiment so that they could learn how imminent collision situations would occur. The drivers were directed to use the brakes instead of changing lanes to avoid a collision with the lead vehicle. The potential collision events were repeatedly presented at irregular time intervals to prevent drivers from predicting their occurrence. Each participant had a 3-min break after finishing nine events for one deceleration condition.

Dependent variables

Two dependent variables describing the collision avoidance behaviour of the driver were recorded:

- Accelerator release time: This is the time period between the braking event and the release of the accelerator by the following vehicle.
- Braking response time: This is the time period between the braking event of the lead vehicle and application of the brakes by the following vehicle.

Results

Figure 1 and Figure 2 show that the mean values of the accelerator release time and braking response time for all drivers for the 1.4 s time headway and 3.92 m/s^2 deceleration of the lead vehicle and for the 2.0 s time headway and 5.88 m/s^2 deceleration of the lead vehicle, respectively. The braking behaviour represented by both variables varies according to individual drivers. Also, it seems that there were differences in braking behaviour depending on experimental conditions. A similar tendency was observed for both variables regardless of the driving conditions.

In other words, it is worth determining the alarm timing based on driving characteristics for individual drivers by considering that the braking behaviour may vary according to the driver.

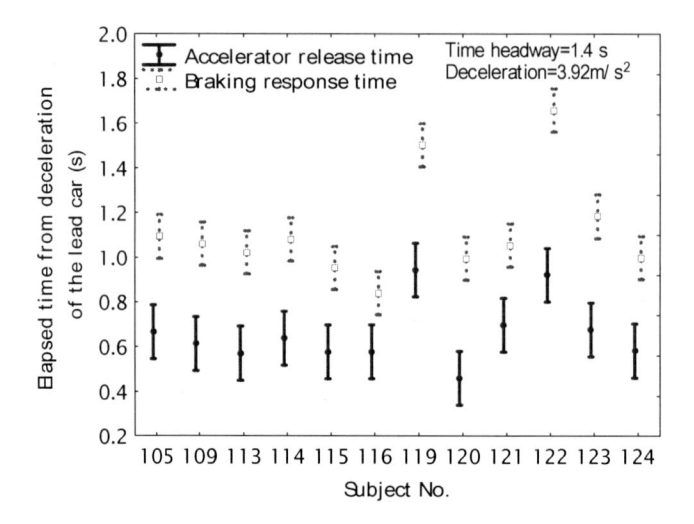

Figure 1. Individual differences in braking behaviour (time headway=1.4 s, Deceleration=3.92 m/s²

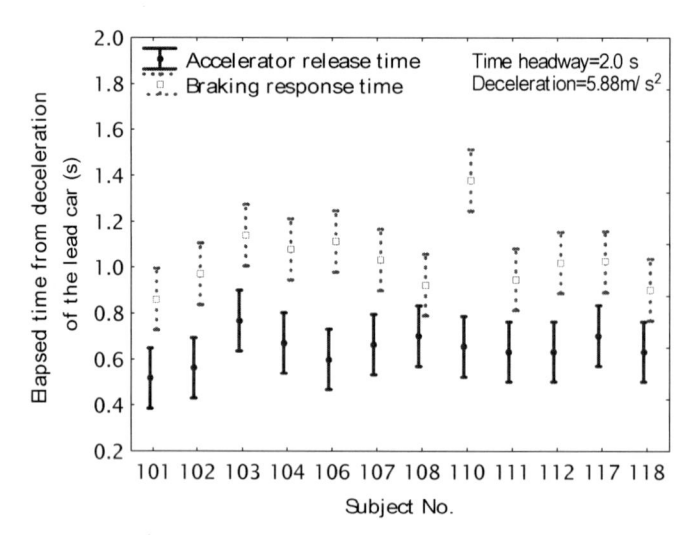

Figure 2. Individual differences in braking behaviour (time headway=2.0 s, Deceleration=5.88 m/s²

Adaptive alarm timing

Adaptive alarm timing was determined for each participant based on the data obtained in the pilot study:

Adaptive alarm timing = (MART + MBRT)/2

MART and MBRT are the median values of the accelerator release time and the braking response time, respectively, for nine trials for each driving condition obtained in the pilot study (Fig. 3). Table 1 contains a summary of the values of adaptive alarm timing which were calculated by using the adaptive alarm timings for all drivers according to the driving conditions.

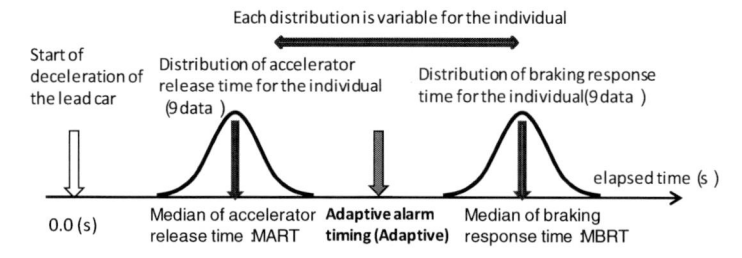

Figure 3. The concept of adaptive alarm timing based on braking behaviour

Table 1. Summary of values of adaptive alarm timing in response to driving conditions

Initial time headway(s)	deceleration/s^2	Alarm timing (s) Max	Min	Average
2	5.88	0.96	0.67	0.78
	3.92	1.21	0.65	0.88
1.4	5.88	0.95	0.64	0.75
	3.92	1.28	0.71	0.87

Non-adaptive alarm timing

The non-adaptive alarm was issued at the same time for all participants based on the Stopping Distance Algorithm, which is one of the preventative alarm trigger logics for FCWS. The SDA is so termed because a warning distance is defined based on the difference between the stopping distances of the leading and following vehicles. This algorithm has three parameters: reaction time (*RT*), deceleration of the leading vehicle (D_l) and deceleration of the following vehicle (D_f). *RT* is the assumed reaction time of the driver of the following vehicle. The warning distance (D_W) is found on the basis of these parameters along with the velocities of the leading (V_l) and following (V_f) vehicle, as follows.

$$D_w = V_f RT + \frac{V_f^2}{2D_f} - \frac{V_l^2}{2D_l} \tag{1}$$

An alarm is trigged when the current headway distance is less than the warning distance (D_W). In this experiment, the values of the three parameters were determined by considering that alarms would be triggered at the timing of the minimum accelerator release time for all participants in response to the driving conditions. Table 2 contains a summary of the values of each parameter and the average alarm timings in response to the driving conditions.

Table 2. Summary of values of alarm parameters and non-adaptive alarm timing in response to driving conditions

Initial time headway (s)	deceleration D_f (m/s^2)	D_l (m/s^2)	(m/s^2)	T (s)	Alarm timing (s)
2	0.6	5	6	1.2	0.44
	0.4	5	6.5	1.2	0.46
1.4	0.6	5	4.8	1.2	0.41
	0.4	5	5.35	1.2	0.34

Experiment on alarm effectiveness and influences of missed alarms (Experiment 2)

The purpose of this experiment was to investigate two issues: how differences in alarm timings influence the driver's response to alarms, and how missed alarms influence the braking behaviour in response to alarm timings.

Apparatus

All materials used in this experiment were the same as those used in the pilot study. In this experiment, FCWS was introduced with a simple auditory beep. All alarms were presented based on the strategies described earlier (see Section 3) in response to experimental conditions.

Participants and experimental tasks

In this experiment, the participants were the same as those in the pilot study. All drivers were required to perform the same tasks as in the pilot study.

Experimental design

The participants were divided into two groups in the same manner as in the pilot study, namely, according to 1.4 s time headway or 2.0 s time headway. That is, each participant experienced the same time headway condition as in the pilot study. The participants who were assigned to the 1.4 s time headway group experienced a deceleration of the lead vehicle of 5.88 m/s^2 at a driving speed of 80 km/h. The participants who were assigned to the 2.0 s time headway group experienced a deceleration of the lead vehicle of 3.92 m/s^2 at a driving speed of 80 km/h.

Each participant of each group was subjected to 14 potential collision events in which the lead vehicle decelerated at 5.88 m/s^2 or 3.92 m/s^2. An alarm was correctly given for each potential collision event, except for the fifth event and the 13th event

(Figure 4). In these two events, missed alarms occurred; that is, alarms were not triggered even when a potential collision event occurred.

As for alarm timings, each participant experienced adaptive alarm timing or non-adaptive alarm timing. Moreover, half of the participants who were assigned into each group experienced adaptive alarm timing and the remaining participants experienced non-adaptive alarm timing. For each group, the order of the alarm timings was counterbalanced according to the number of participants in order to reduce the effects of the experimental conditions on driver behaviour.

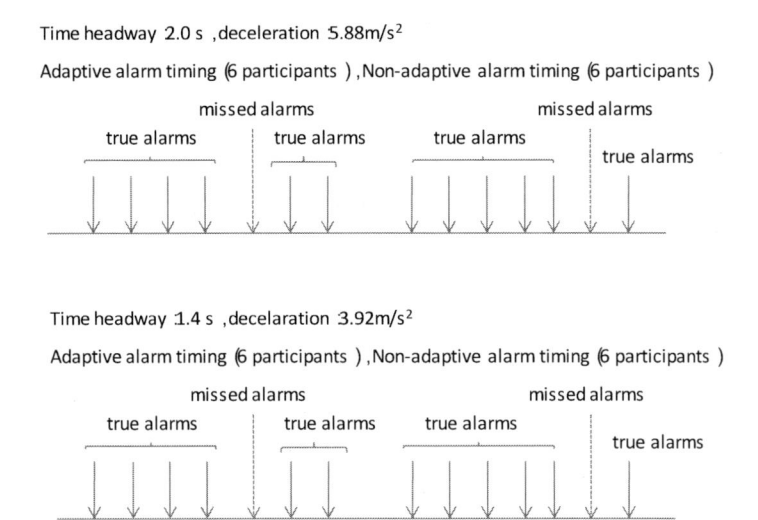

Figure 4. Experimental conditions trial by trial

Procedure

All participants gave informed consent and were instructed regarding the task requirements. Each participant was allowed a 10-min practice session in order to familiarize themselves with the simulator and to experience the lead vehicle decelerations. The participants were directed to use only the brakes instead of changing lanes to avoid a collision with the lead vehicle. Moreover, they were instructed that FCWS provided alarms when potential collision events were happening, although FCWS did not always work correctly, that is, there was a possibility that alarms would not be triggered due to system failures. In the practice session, the participants did not experience missed alarms.

The potential collision events repeatedly occurred at irregular time intervals to prevent drivers from predicting their occurrence.

Dependent variables

The following variable were collected

- Braking response time: This is the time period between the braking event of the lead vehicle and application of the brakes by the following vehicle.
- Trust in the system: The subjective estimation of the driver's trust in the systems was obtained using an 11-point rating scale, in which 0 indicated "none at all" and 10 indicated "completely." All subjective measurements were recorded immediately after the braking event, in which the lead car suddenly decreased its speed.

Results of braking response time

Figure 5 shows the mean values of braking response times for each potential collision event in response to the alarm timings. As can be seen in this figure, the non-adaptive alarm timing induced a swift response of braking, compared to that of adaptive alarm timing when true alarms occur. The result suggests that non-adaptive alarms are presented earlier than adaptive alarms, resulting in timely implementation of the brakes.

With respect to the trial in which a missed alarm happened for the first time (trial 5), independent of the type of alarm timing, the braking response time was dramatically delayed compared to the braking response times obtained in other trials where true alarms occurred. Some interesting phenomena appeared for the trial in which there was a second missed alarm (trial 13). Specifically, for non-adaptive alarm timing, a missed alarm still induced a longer response time for braking, compared to the braking response times for the true alarm conditions. However, for adaptive alarm timing, the braking response time for the second missed alarm trial was almost the same as the braking response time for the true alarms conditions, indicating that drivers managed to apply the brakes in the same manner as they did for true alarms, even if a missed alarm happened. A two-way ANOVA of trials and alarm timings with repeated measures on the factor of trials showed a significant interaction between factors, $F(13, 273)=2.17$, $p<0.05$ and a Tukey's HSD post-hoc test showed that there was a significant difference in braking response time for non-adaptive alarm timing between trial 12 and trial 13 ($p<0.01$) but there was not a significant difference in braking response time for adaptive alarm timing between trial 12 and trial 13.

These results suggest that, by experiencing a missed alarm once, drivers become cautious about malfunctions, including missed alarms. In other words, there is a possibility that awareness features such as alarms must always be triggered when imminent collision situations happen, is diminished. As a results, drivers who experienced adaptive alarm timings exhibited robust behaviour toward subsequent missed alarms. This was not the case with drivers who experienced non-adaptive alarm timing. This can be explained from the braking response times which were obtained in the trials, in which true alarms occurred for non-adaptive alarm timing. That is, true alarms for non-adaptive alarm timing have the potential to lead to early braking response time since a delayed response of braking may become obvious when missed alarms happen compared to the response of adaptive alarm timing in the same situation.

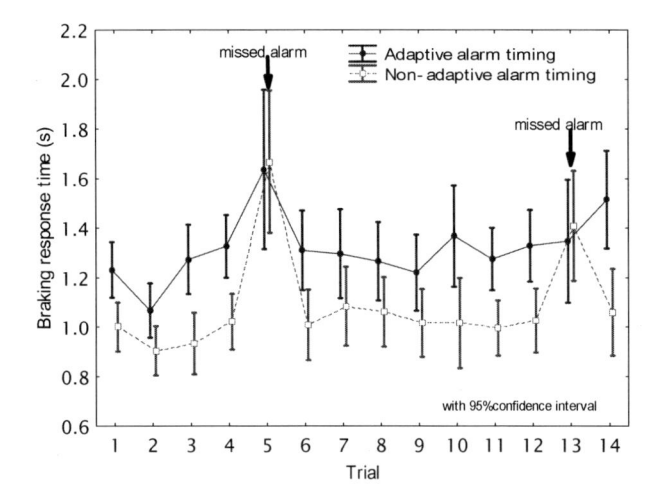

Figure 5. Braking response time for each trial in response to alarm timings

Effects of alarm timings on trust in the systems

Figure 6 shows the mean values of trust ratings for each trial in response to the alarm timings. As can be seen in this figure, trust ratings (an 11-point rating scale) for both alarm timings for the first four trials were relatively high (above 5), indicating that adaptive and non-adaptive alarm timings are reasonably trustworthy as collision warnings. For the 5th trial, in which drivers experienced a missed alarm for the first time, trust ratings dramatically decreased for both alarm timing conditions. Next, seven trials between the 6th and 12th trials were considered to assess how trust ratings changed after experiencing the first missed alarm. It can be said that the values of trust ratings for adaptive and non-adaptive alarm timings reached at the similar levels obtained before the first missed alarm by the 12th trial.

One interesting result was obtained for trial 13, in which a second missed alarm occurred. That is, there were differences in the impacts of a missed alarm on the driver's trust in the system according to the alarm timings. Without a doubt, the second missed alarm decreased the trust ratings for both alarm timings. However, it seems that for adaptive alarm timing, the degree to which trust ratings decreased due to the second missed alarm improved compared to the degree to which the trust ratings decreased was due to the first missed alarm. In contrast, for non-adaptive alarm timing, decreased trust ratings for the first and second missed alarms seemed to be at similar levels. In order to confirm the impact of missed alarms on driver's trust in the FCWS in response to alarm timings, a variable relevant to the rates of decreased trust for missed alarms was introduced, as follows.

Rate of decreased trust for missed alarms = (trust rating for immediately before a missed alarm - trust rating for a missed alarm) / (trust rating for immediately before a missed alarm)

Table 3 contains a summary of the rates of deceased trust for missed alarms in response to alarm timings for the first missed alarm (trial 5) and the second missed alarm (trial 13). For non-adaptive alarm timing, there were no significant differences in trust ratings between the first and the second missed alarm. For the adaptive alarm timing, however, compared to decreased trust caused by the first missed alarm, the second missed alarm produced gentle reduction of trust ratings, $F(1,11) = 6.11$, $p < 0.03$.

The results indicate that there is a possibility that deceased trust caused by missed alarms may vary in response to alarm timings.

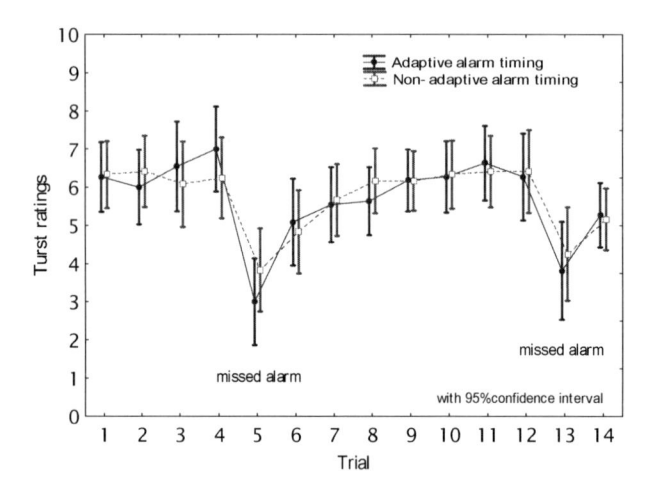

Figure 6. Trust ratings for each trial in response to alarm timings

Table 3. Effect of alarm timing on decreased trust for missed alarms

	Experiences of missed alarms	
Alarm timing	The first missed alarm	The second missed alarm
Adaptive alarm timing	Mean=0.58 (SD=0.28)	0.33 (0.44)
Non-adaptive alarm timing	0.45 (0.31)	0.34 (0.23)

Discussion

The present study focused on the alarm timing for FCWS and investigated driver response to two kinds of alarm timing. One was individually determined based on braking behaviour for the individual (adaptive alarm timing) and the other was for one common trigger logic (SDA) for all drivers (non-adaptive alarm timing). Also, the impact of missed alarms on braking behaviour on the response to alarm timings was investigated.

When drivers experienced a missed alarm for the first time, compared to the presentation of true alarms, the response time to the brakes was impaired by the missed alarm for both alarm timings. With respect to trust ratings, moreover, the first missed alarm resulted in dramatically decreased trust ratings independent of alarm timings. These results suggest that the driver's trust in the system may be impaired when the braking response time is delayed in imminent collision situations, when an alarm is expected.

As for a missed alarm happening for the second time, drivers who experienced adaptive alarm timing managed to apply the brakes in the same manner as when true alarms were presented. From this result, it can be said that adaptive alarm timing has the potential to diminish the impairing influences of missed alarms on braking behaviour earlier than this would occur with non-adaptive alarm timing. Consequently, it is possible for adaptive alarm timing to restrain the decreased trust ratings for missed alarms when the impact of missed alarms on driver behaviour is relatively small.

Needless to say, FCWS should provide drivers with sufficient time to avoid imminent collisions. However, it is necessary to consider how alarm timing should be determined in order to minimize the effects of missed alarms on braking behaviour. For this reason, it is reasonable to conclude that adaptive alarm timing may be useful.

Conclusions

The results support the following conclusions.

1. Both adaptive alarm timing and non-adaptive alarm timing maintain reasonable driver's trust in FCWS.
2. Adaptive alarm timing mitigates the impacts of missed alarms on driver behaviour compared to the impact for non-adaptive alarm timing.
3. It is likely that the effects of missed alarms on driver's trust in FCWS vary in response to not only differences in alarm timing but also to the number of experiences of missed alarms.

References

Abe, G., & Itoh, M. (2008). Exploring appropriate alarm timing for a driver-adaptive forward collision warning system. In D. De Waard, F.O. Flemisch, B. Lorenz, H. Oberheid, and K.A. Brookhuis (Eds.), *Human Factors for assistance and automation* (pp. 103-115). Maastricht, The Netherlands: Shaker Publishing

Abe, G., & Richardson, J. (2006). The influence of alarm timing on driver response to collision warning systems following system failure. *Behaviour and Information Technology, 25*, 443-452.

Alm, H., & Nilsson, L. (2000). Incident warning systems and traffic safety. A comparison between the PORTICO and MELYSSA test site systems. *Transportation Human Factors, 2*, 77-93.

Ben-Yaacov, A., Maltz, M., & Shinar, D. (2002). Effects of an in-vehicle collision avoidance warning system on short- and long-term driving performance. *Human Factors, 44*, 335-342.

ISO/TC 204/WG 14. (1994). Road Vehicles-Forward Vehicle Collision Warning Systems-Performance Requirements and tests Procedures, SO/DIS 15623.

Janssen, W.H., Alm, H., Michon, J.A., & Smiley, A. (1993). Driver support. In Michon, J.A. (Ed.), *Generic Intelligent Diver Support* (pp. 53-66). London: Taylor and Francis.

Lee, J.D., McGehee, D.V., Brown, T.L., & Reyes, M.L. (2002). Collision warning timing, driver distraction, and driver response to imminent rear-end collisions in a high-fidelity driving simulator. *Human Factors, 44*, 314-334.

Human machine interface for a dual mode vehicle

Amon Rambaldini, Antonella Toffetti, & Claudio Arduino
Centro Ricerche Fiat CRF
Orbassano, Torino, Italy

Abstract

The European Project CityMobil is about automation in vehicles. CityMobil cars can have different levels of automation, from highly automated vehicles (e.g. Dual-Mode Vehicles in E-lane scenario) to innovative city cars with some kind of transition of driving controls. In particular, CRF simulated, in the Virtual Reality simulator, a multi-lane road mixed scenario with E-lane infrastructure with a fully automated driving and normal road tracks with complete manual driving. Fully automatic driving means that the drivers neither have to bother about the steering wheel nor the pedals. The aims of the CRF experiment were:
- To design a couple of user interface solutions through an iterative design process (typical for a user-centred design approach) with the purpose to define content, logic and layout to communicate to the drivers the shift of control between driver and automated vehicle;
- To conduct usability tests, involving drivers to find out the best interface solution. In particular, investigate users' attitudes and behaviours during the interaction with a Dual-Mode Vehicle, evaluate the perceived usability of a multi-modal interaction interface and highlight users' judgments and performance with the different interfaces.
This paper presents a description of the experiments (e.g. the virtual scenario, the different user interfaces, the primary and secondary tasks, the experimental design) and some of the obtained results.

Introduction

The purpose of this driving simulator experiment was to design and test a vocal versus a simple acoustic user interface for a dual-mode vehicle driving on a so-called eLane. In the acoustic condition, the human-machine interface was constituted by simple sounds ("beeps") and visual messages displayed at the right of the steering wheel. In the vocal condition an acoustic and visual interface were supported by vocal messages (i.e. spoken words). The main purpose of the interface was to support the transition of control between driver and automated mode of the vehicle, and vice versa, both initiated by the driver and by the system. The users' attitudes and behaviours during the interaction with the dual-mode vehicle and the interface were assessed with questionnaires and observational grids in order to analyze the driver performance.

In D. de Waard, J. Godthelp, F.L. Kooi, and K.A. Brookhuis (Eds.) (2009). *Human Factors, Security and Safety* (pp. 225 - 235). Maastricht, the Netherlands: Shaker Publishing.

Method

Participants

Twenty four volunteers (5 female, 19 male drivers) participated in the experiment, aged between 21 and 46 years. On average they had their driving license for 10 years, with an annual mileage of 15.000 km/year. All participants had experience with the CRF Virtual Reality Simulator (i.e. >3 drives).

Driving simulator scenario

All participants drove the same simulated scenario with different situations (e.g. automatic and manual overtaking) which forced participants to do various driving manoeuvres (see Figure 1). The road consisted of 3 lanes, 4 m wide, plus an emergency lane 2.5 m wide. The multi-lane road was uniformly populated by cars and trucks statistically guaranteeing the presence of 4-5 vehicles visible in front of the participant. Some vehicles had the function to generate dangerous situations in a controlled way.

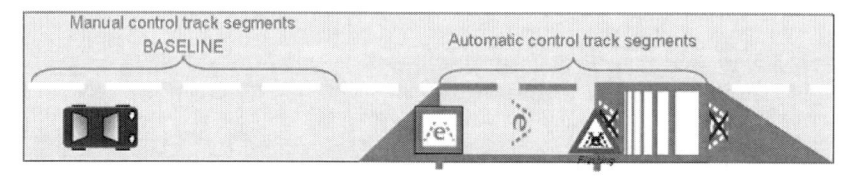

Figure 1. Example of the multi-lane one-way track. The right lane was divided in automatic and manual control segments. The automatic control version (eLane) was identified by specific blue lines and icons on the pavement and by specific road signals

Driving task

Participants were instructed to perform the driving task as their primary task. They were instructed to maintain an average speed of 70 km/h. The participants were not instructed to overtake slower vehicles, but their spontaneous overtaking behaviour was monitored. This allowed, in the case of automatic driving, for example the driver voluntarily taking control of the vehicle while it was in the automatic state, to study the behaviour. There were two scenarios; 1) manual driving, on road segments where the users had to drive themselves, thus maintaining the longitudinal and the lateral control of the car, using pedals and steering wheel; 2) automatic driving on the eLane, on road segments where the longitudinal and lateral control was controlled automatically by the system. The automatic driving was achieved using the Wizard of Oz technique, i.e. the experimenter ("the Wizard") sits in a separate room and acts as the automated system, making all steering wheel actions and longitudinal control. The participants were unaware that the system was not 'really automatic'.

Secondary task

The participants were also asked to perform a secondary task. This task was introduced to distract the users and to avoid that they focused their attention too much outside of the vehicle or to the dual-mode vehicle interface. So, it was possible to test the designed interfaces in a situation of divided attention. The secondary task consisted of the interaction with an in-vehicle information system (IVIS), with different difficulty levels. A recorded voice asked to press specific buttons or a sequence of buttons: i.e. "press TEL button", "press NAV button", "dial 25 47 85" etc. The strings of numbers and the presentation time between the different secondary tasks varied. The task was randomly presented in order to minimize expectation or learning effects.

Dual-mode vehicle interfaces

Two different interfaces were used for communicating the eLane functions to the driver:

- *Acoustic interface* (visual + acoustic messages by means of beeps). The visual information was compensating for the generic level of acoustic information. In fact, a simple "beep" could not be able to explain the exact meaning of the warning.
- *Vocal interface* (i.e. visual + acoustic + vocal messages). The visual display was simpler because the vocal message provided the necessary information. Each vocal message was preceded by an acoustic signal.

A coloured icon was always present on the display, specifying the system status: the grey icon indicated 'off' (but eLane was available) and the green icon indicated 'on'. In case of danger the amber icon indicated a warning, while the red icon indicated the automatic deactivation status. The exact timing of the interface messages is described in the result section.

Experimental design

The design was within-subjects, with all participants testing the two interfaces in three trials (baseline, manual and automatic drive) that were randomized to minimize order and sequence effects. A number of events happened during the drives; 1) entering and exiting an eLane; 2) infrastructure out of order; 3) system breakdown 4) emergency manoeuvre.

Procedure

After being instructed about primary and secondary tasks, participants were trained for about 10 minutes per driving condition. Then, the actual experiment started with a baseline drive, followed by a manual and automatic drive (on the eLane) each lasting 10 minutes. To notify the pre-warning, the same feedbacks (icons and acoustic signals) for vocal and acoustic modality were used (see Figure 2). After each drive an adapted version of the AIDE-HMI questionnaire (2008) was filled in.

The variables investigated were actions on steering wheel and pedals, the users' responses to the system errors, entrance/exit to an eLane, and overtaking. Also, a subjective evaluation of the system (perceived usability, perceived evaluation of the driving performance, willingness to buy the system and perceived image of the system) was provided. Student-t Test and Chi-square Test were used to compare means or percentage values.

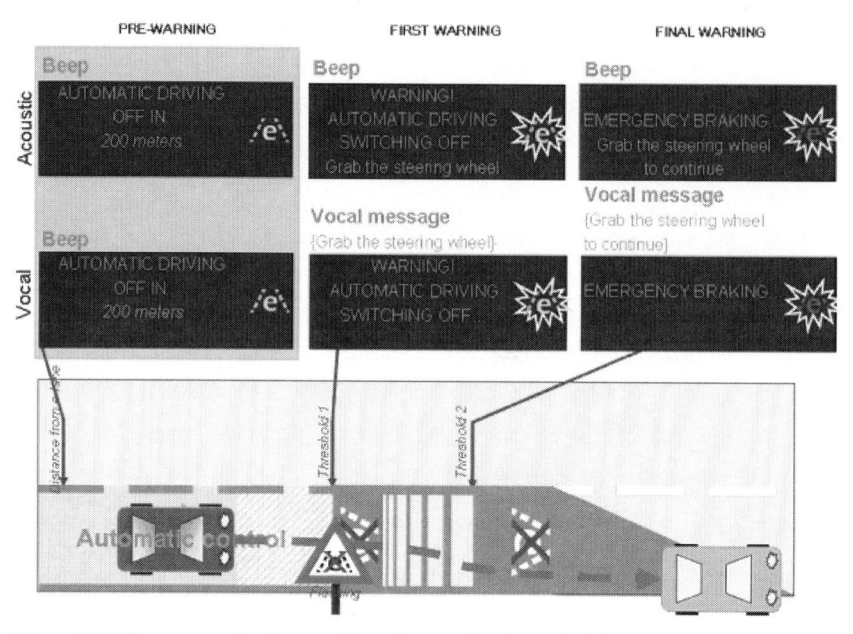

Figure 2. HMI sequence for the transition of control at the end of the eLane

Results

Only the main results will be reported in this paper.

Driving performance

Average speed during baseline
In the vocal modality the participants maintained, in the baseline manual driving segments, a significantly (t (44) =-2.234, p<.05) lower average speed (55 km/h) in comparison to the same situation with the acoustic modality (61 km/h).

Transition of control at the beginning of the eLane
Before entering the eLane, an acoustic pre-warning (Figure 2) advised the driver that the system would be active in 200 m (for both conditions). When the car was in the eLane area a second warning message advised the user that the system was ready and could be activated by pushing the On/Off button, releasing the steering wheel and pedals. As for the vocal interface, the voice only said the last part of the second warning "Push the button, then release steering wheel and pedals". More than 93% (95% CI=±2.53) of users activated the system when the system became ready for the automatic driving. Almost 85% (95% CI=±3.71) of users activated the system in maximum 5s (RT Mean: Mean=3.1s; SD=1.6) after the availability warning (Figure 3). There were no statistically significant differences between the interaction with the

vocal interface prototype and the acoustic interface prototype ($\chi^2(1)=.085$, NS). Considering that it was not a critical situation, the response to both the acoustic and the vocal display can be considered as adequate responses.

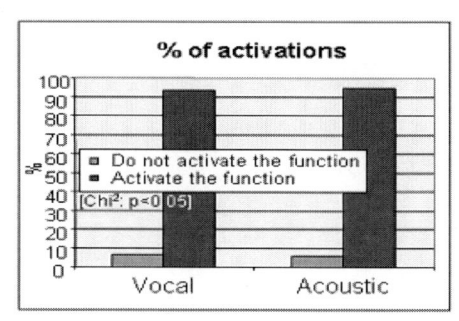

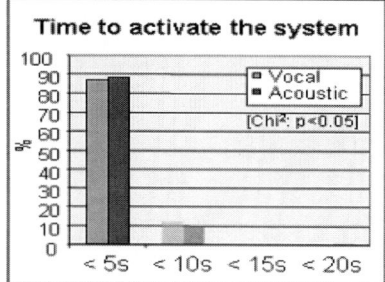

Figure 3. Users' responses to activations of the automatic driving and their reaction time in bins

Transition of control at the end of the eLane

Specific vocal messages were used for the first and the final warning. All drivers took control of the car after final warning at the end of the eLane in all conditions. There was a significant difference in response to the deactivation warning between the two interfaces. The acoustic modality recorded the start of answer at the FIRST WARNING step (Acoustic 36% of occurrence (95% CI=±7.6); Vocal 50% of occurrence (95% CI=±6.7)) whereas in the vocal modality the response was recorded at the start of the PRE-WARNING step (40% Acoustic; 55% Vocal) (Acoustic 55% of occurrence (95% CI=±6.4); Vocal 40% of occurrence (95% CI=±7.5)). When keeping the type of situation in mind, both HMIs seem to result in equally effective responses of the drivers.

Automatic overtaking

In this scenario, the leading vehicle braked unexpectedly, activating an avoidance emergency manoeuvre by the automated system. In this situation, the label and the acoustical signal were the same for both modalities. No vocal messages were used (Figure 4).

The majority of the users (about 85% (95% CI=±3.71)) trusted the system and did not resume control over the car during the automatic overtaking manoeuvre (Acoustic: $\chi^2(1)=36.6$, p<.01; Vocal: $\chi^2(1)=49.02$, p<.01).

The group of users that resumed control over the system was very small (Figure 5). A statistically significant difference in the number of users that took control was observed, between:

- acoustic first repetition and vocal first repetition (first time that the overtaking event was presented) ($\chi^2(1)=4.5$, p<.05);
- acoustic first repetition and acoustic third repetition (last time that the overtaking event was presented) ($\chi^2(1)=4.5$, p<.05).

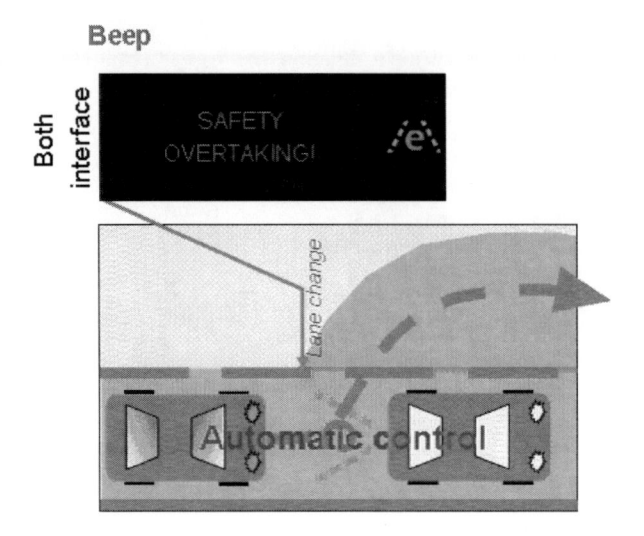

Figure 4. Avoiding an obstacle by the automated system

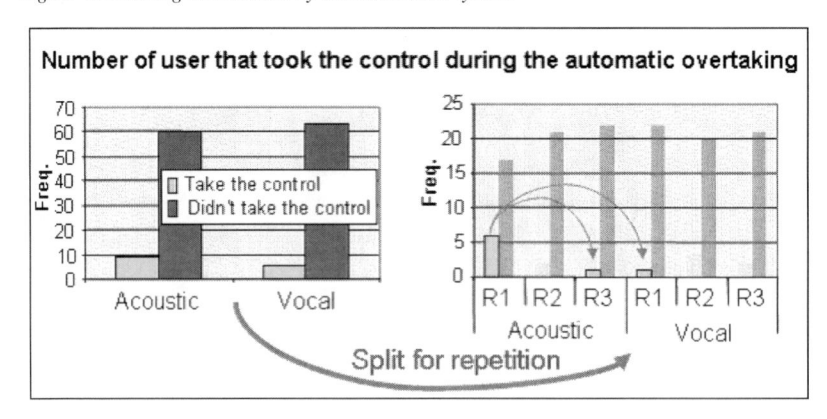

Figure 5. Effectiveness and efficiency of the automated system

In general, users trusted the system and avoided reaction in a proper way. The acoustic modality showed few and only little problems in the first impact interaction, that is to say that users not completely trusted the system, taking over manual control. This behaviour disappeared after a very short experience with the system (see Figure 5). The vocal interface could be considered more adequate even during the first impact trials (the vocal interface is just faster to learn).

Moving to the left lane
To ensure a controlled study, an "experimental-voice" was used to request the performance of specific actions in specific moments during the driving session. During the task a pre-recorded voice ordered drivers to move to the left lane, in the following manner:

- In manual control segments: the voice asked to move to the left lane during the manual driving;
- In automatic control segments: the voice asked to move to the left lane during the automated driving, this meant that the user had to take voluntary control of the car.

In 98% (95% CI=±1.3) of the cases, drivers moved to the left lane, when required; 90% (95% CI=±3.03) of them moved in the first 10s after the voice message. There were no significant differences between acoustic and vocal modality. There were no significant differences between Manual/Automatic control track segments either (Figure 6).

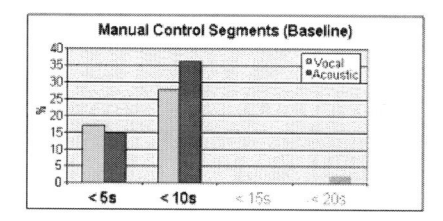

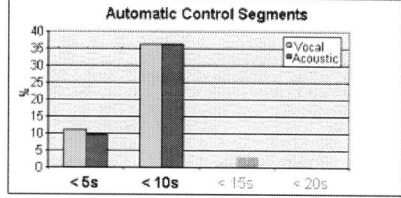

Figure 6. Effectiveness and efficiency

In general, users trusted the system but when a clear command was given they were not afraid to resume the control over the car voluntarily (see also Paragraphs: "*Automatic overtaking*", "*Avoiding an unexpected obstacle*" and "*Manual overtaking of a slow vehicle*").

Manual overtaking of a slow vehicle
In this scenario the leading vehicle slowed down reducing the cruise speed (to less than 40 Km/h) but without activating the avoidance emergency manoeuvre by the automated system (Figure 7). In this situation, the user had to voluntarily overtake the leading vehicle in manual modality (the automated system was not switched off automatically). The voluntary overtaking was monitored both during the eLane and the experimental baseline.

During the automated control only 30% (95% CI=±8) of drivers resumed control over the vehicle voluntarily, to overtake the slow lead vehicle. During the baseline 91% (95% CI=±2.8) of drivers spontaneously overtook slow vehicles. There were no statistically significant differences between vocal and acoustic modality in taking the voluntary control of the vehicle ($\chi^2(1)=.258$, $p>.05$).

It seems that in specific situations the drivers trust the system too much. In this case, the users reacted too conservatively reducing the driving effectiveness, and avoiding overtaking the slow vehicle when it was possible.

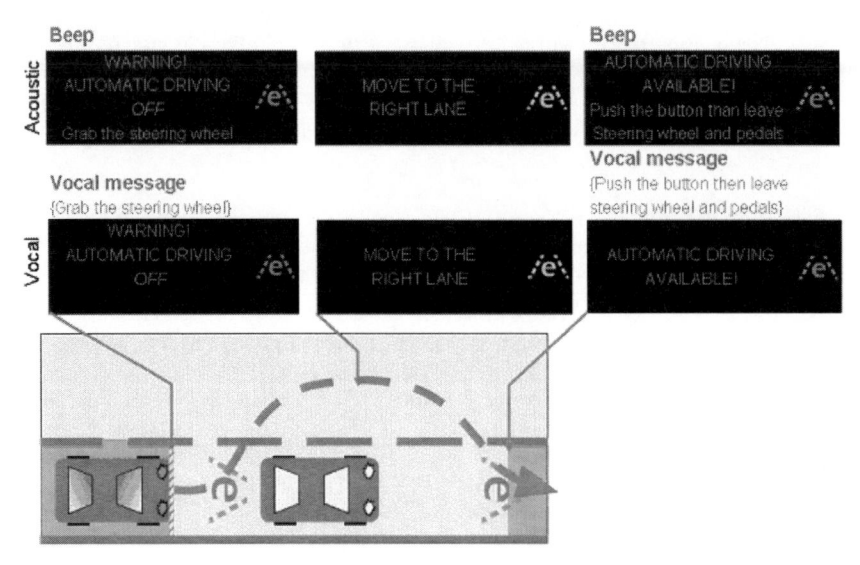

Figure 7. Manually overtaking a slow vehicle

System failure
In case of a system failure, the system gave a pre-warning to inform that the system would be deactivated in a few minutes. A second warning informed that the system had been deactivated and that the emergency parking manoeuvre was beginning. 85% (95% CI=±3.71) of drivers took control over the car in case of failure. For the other 15% of drivers, the Wizard simulated an emergency parking. There are no significant differences between vocal and acoustic modality ($\chi^2(1)=.026$, NS) in taking control over the vehicle. The vocal modality showed significantly shorter response times ($t(37)=2.168$, $p<.05$) compared to the acoustic one.

Infrastructure out of service
When the infrastructure was out of service, at the beginning of the eLane, a flashing road-signal informed the driver about this problem. If the driver pushed the ON/OFF button anyway, the display visualized "WARNING! System out of service" (same for vocal and acoustic modality combined with an acoustic signal). In case of acoustic modality, there were no differences ($\chi^2(1)=.36$, $p>.05$) between the percentage of drivers that recognized the infrastructure out of service and vice versa. On the contrary, in the vocal modality, a significant greater number of drivers ($\chi^2(1)=4.167$, $p<.05$) did not recognize the out of service status of the infrastructure (Figure 8). There were no significant differences between vocal and acoustic modality in the On/Off push button reaction time ($t(26)=-1.896$, $p=.069$).

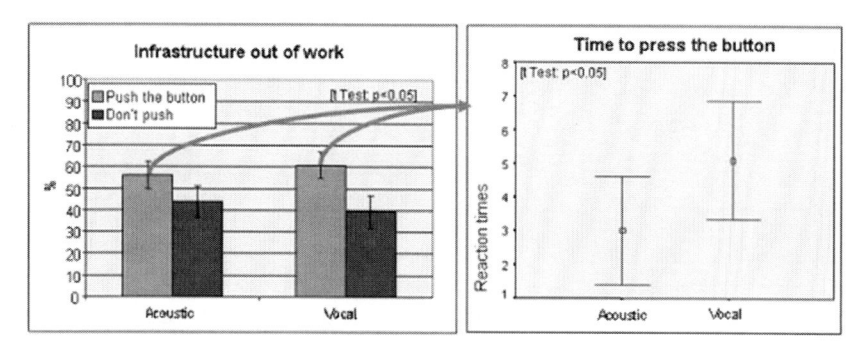

Figure 8. Users' pushing On/Off button and time to do this operation (left). Users' reaction times to press the button (right)

Subjective evaluations

Interface evaluation Questionnaire Responses
The visual information next to the acoustic interface was considered more necessary than the visual information next to the vocal interface (acoustic: t(22)=-4.75, p<.01; vocal: t(22)=-4.5, p<.01). The visual information of both interfaces was considered *"Necessary, Adequate, Pleasant, Comprehensible* and *Legible"*. The acoustic interface is considered neither *"Soothing* nor *Frightening"*. In general, the acoustic/vocal information of the vocal modality is considered more *"Comprehensible"* (acoustic: t(22)=-3.21, p<.01; vocal: t(22)=-25.07, p<.01), *"Adequate"* (acoustic: t(22)=-2.13, p<.05; vocal: t(22)=-5.89, p<.01) and *"Soothing"* (acoustic: t(22)=-3.44, p<.01; vocal: t(22)=-4.89, p<.01) in comparison with the acoustic modality. More than 70% (95% CI=±4.9) of the users preferred the vocal modality. Users reported that the vocal modality reduces distraction during driving; it is considered more comprehensible and could be very useful during the learning phase. The vocal message was also considered a bit annoying but it is possible to design a vocal interface that could be deactivated by expert users (except for the fault event).

Driving performance evaluation questionnaire
Users reported that all conditions were *"Safe"* (acoustic: t(20)=-2.84, p<.01; vocal: t(20)=-3.88, p<.01; baseline: t(21)=-9.21, p<.01) but the vocal modality was considered as *"Safe"* as the manual driving (baseline), while the acoustic modality was not (t(18)=-2.91, p<.01) (see Figure 9). The users trusted the system, and, for this dimension, there were no differences between the acoustic, vocal modality and baseline. 80% (95% CI=±4.29) of drivers preferred to have a mixed automatic-manual system, to be used in the automatic modality in specific situations only: 1) the manual modality; urban area, traffic jam, overtaking situations, mountain and curves streets; 2) the automatic modality; traffic jam, straight and high velocity streets and highways.

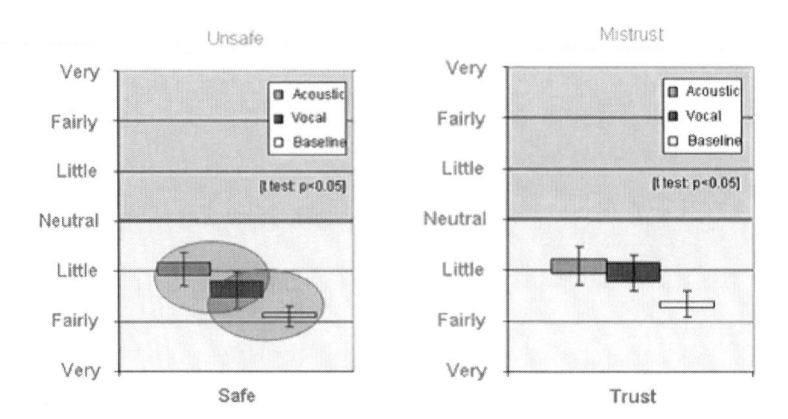

Figure 9. Answers to the questions: "How did you feel during the driving?" and "Your attitudes towards the automatic driving was?"

Conclusion

In general, the vocal modality induced a significant reduction of the average speed during the manual control and induced the drivers to anticipate the right responses when there was only visual information. The vocal modality could be considered as more adequate in the first impact and is very important to highlight the automatic system deactivations. In some cases the vocal modality reduced reaction times, even though response times in this experiment were not time critical.

In general, drivers trusted the system and reacted properly during the automated manoeuvre (automatic overtaking and unexpected obstacle). Sometimes the drivers trusted the system too much, responding too conservatively and reducing the driving effectiveness avoiding overtaking the slow vehicles (manual overtaking). It was also observed that, when an explicit command was given, participants were not afraid to resume control over the car.

The infrastructure was sufficiently informative. It was found that a rather high number of participants did not recognize that the infrastructure was out of order. In this case it could be useful to inform the users not only through road-signals but also with an on-board message especially with a vocal interface. As a side effect, it seems that the vocal message increased the expectations of drivers to be driven automatically by the on-board system.

Regarding the Perceived Usability of the Automatic System, the system evaluation was positive. In fact, it was considered satisfying, amusing, pleasant and interesting, and also easy to use, easy to remember, easy to learn. Driving in automatic mode with the vocal modality was considered as safe as driving in manual mode. The vocal modality was perceived safer than the acoustic one; the vocal modality was preferred. Drivers judged that this modality can reduce the distraction, is more comprehensible and useful during the learning phase. At the same time, vocal messages can be a bit annoying, but this aspect can be solved with a dedicated design that allows the deactivation to expert users. In case of failure the vocal modality seems to be the better one, because it shortens drivers' reaction time.

The visual information of both interfaces is evaluated positively. In general it is considered: "necessary", "adequate", "pleasant", "comprehensible" and "legible". The visual information for the acoustic modality is considered more "necessary" than for the vocal modality; for both interfaces the most problematic labels are: "Automatic Overtaking", "Fault", because these messages are not sufficiently salient and probably require a pre-warning. The vocal information is more comprehensible, adequate and soothing than the acoustic one. Anyway, both interfaces are evaluated necessary, pleasant and easy to hear. A mixed automatic - manual modality is the preferred one, with automatic driving in case of medium and low traffic, straight and high velocity streets and manual driving for medium and high traffic, in the urban context, overtaking and for mountain streets.

Summarizing, both designed interfaces are reasonably usable and satisfactory, although the design could be optimized for specific traffic situations. Also the interfaces should result in the right response in case of unexpected events in open eLanes and mixed-traffic. It is extremely important to be prepared for all types of situations before these systems are actually introduced in traffic.

Acknowledgements

These experiments of CityMobil were funded by the European Union as an Integrated Project in the 6th Framework Programme.

References

AIDE HMI questionnaire (2008). http://www.aide-eu.org/contact.html, accessed July 31 2008.
City Mobil: Deliverable 3.2.2: Test Results of HMI in Use on Cars and with Simulators. (in progress) www.citymobil-project.eu
Norman, D.A. (1990). The 'Problem' with Automation: Inappropriate Feedback and Interaction, no 'Over-automation'. In D.E. Broadbent, J. Reason, and A. Baddeley (Eds.), *Human Factors in Hazardous Situations*. Oxford, UK: Oxford Science Publications.
Stanton, N.A., & M.S. Young. (1998). Vehicle Automation and Driving Performance. *Ergonomics, 41*, 1014-1028.

Transportation - Evaluation

EFFECTS OF ALCOHOL ON SIMULATED DRIVING PERFORMANCE

In D. de Waard, J. Godthelp, F.L. Kooi, and K.A. Brookhuis (Eds.) (2009). *Human Factors, Security and Safety* (p. 237). Maastricht, the Netherlands: Shaker Publishing.

Developing a virtual driving environment to test dose related effects of alcohol and drugs on simulated driving performance

Janet L. Veldstra[1], Karel A. Brookhuis[1,2], & Dick de Waard[1]
[1]University of Groningen, Groningen
[2]Delft University of Technology, Delft
The Netherlands

Abstract

Drug use and especially multiple drug use and drug-alcohol combinations among drivers is an important risk factor for traffic accidents. The incidence of drivers who drive under the influence of psychoactive drugs in actual traffic is considerable (5-17%). Since drug- and medicine use is proportionally increasing over the years, special efforts have to be directed towards gaining better knowledge of the various aspects of this problem before developing appropriate solutions. The objective of the EU-project DRUID (Driving under the Influence of Drugs, Alcohol and Medicines) is to give scientific support to the EU transport policy (White Paper, 2001) by establishing guidelines and measures to combat impaired driving. In the framework of DRUID, a series of experiments will be carried out in driving simulators and on the road studies, to assess the effects of alcohol and drugs on driving performance. In order to standardize the experiments across laboratories a standard virtual world, including relevant scenarios for testing effects of alcohol and drugs has been developed.

Introduction

Traffic safety is compromised whenever people participate in traffic under the influence of psychoactive substances, i.e. alcohol and prescription- and non-prescription drugs (Movig et al., 2004, Ramaekers, 1998). According to Christophersen and Morland (1997) the incidence of drivers who drive under the influence of psychoactive drugs in actual traffic is considerable (5-17%). Since drug- and medicine use is proportionally increasing over the years, special efforts have to be directed towards gaining better knowledge of the various aspects of this problem before developing appropriate solutions.

The objective of the EU-project DRUID is to give scientific support to the EU transport policy, which states that by the year 2010 the number of traffic accidents should be reduced to half (White Paper, 2001). DRUID aims to do so by establishing guidelines and developing measures to combat impaired driving under influence. In the framework of DRUID, a series of experiments will be carried out in driving simulators and on the road studies, to assess the effects of alcohol, prescription and non-prescription drugs on driving performance. For the purpose of standardisation of the experiments across laboratories, a standard virtual world including relevant scenarios for testing effects of alcohol and drugs on driving was needed.

This standard virtual environment has to cover the majority of the circumstances in which hazards occur. One of the aims of the DRUID studies is to determine thresholds for driver impairment as a consequence of drugs or combined drugs and alcohol use. Therefore it has to be possible to predict from the results of the experiments, if and when the probability of an accident significantly increases. To this end three hierarchical levels of skill and control in the driving task are distinguished according to the driver behaviour model by Michon (1985; Janssen 1979) i.e.; the strategic (planning), tactical (manoeuvring), and operational (control) level, respectively. On the strategic level, route planning and speed choice are defined. On the tactical level the driver determines when and which manoeuvres are appropriate in a specific environment. On this level driving tasks include negotiating with other traffic, handling traffic signs, changing traffic lanes and so forth. At the lowest, the operational level, skill-based activities prevail. The carrying out of manoeuvres that are well learned and automated such as handling the steering wheel and brake pedal fall into this category.

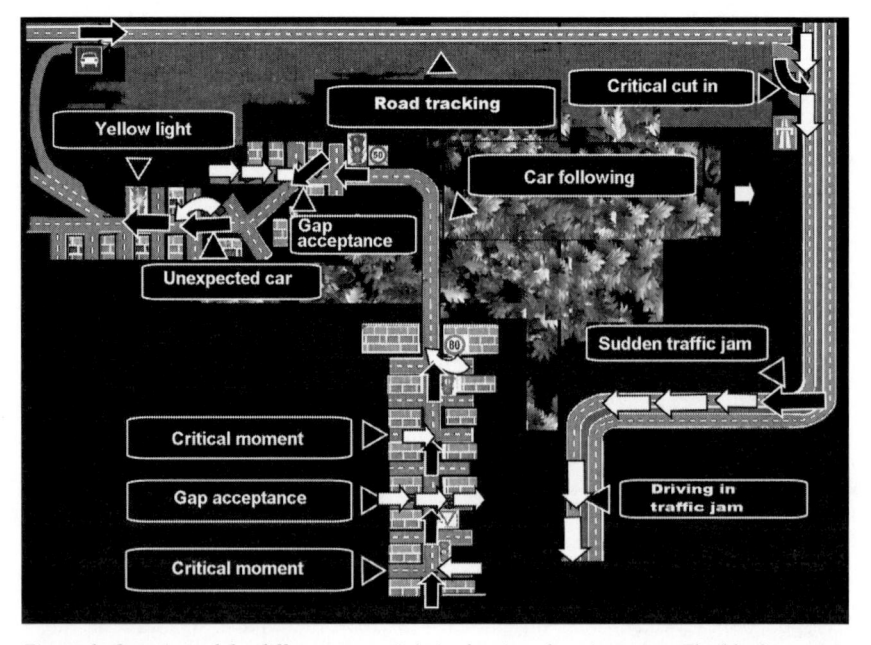

Figure 1. Overview of the different scenario's in the virtual environment. The black arrow is the participant's car; the white arrows are other traffic. Driving the entire route takes approximately 50 minutes and the ride consists of 5 km of city driving, 35 km of rural driving and 30 km of motorway driving

A description of the different scenarios implemented in the virtual driving environment to assess driver impairment due to drug driving is depicted in Figure 1. These scenarios have been described according to their primary measurements, whereby each scenario has been approached from one of the three levels of Michon's model. The classification of the scenarios in a strategic, tactical and operational level has however been rather challenging, as this division not always proves to be that clear cut. For example, a low level manoeuvre such as braking can be thought of as a skill at the operational level whereas the decision to initiate this skill is mediated at the tactical level or even on the strategic level. In other words, Michon's model (1985) recognizes not only different processes but also the related exchange between them. Because the main purpose of this paper is to describe the virtual environment these related exchanges are not taken into account here.

Driving simulator

All scenarios run on an ST Software Driving simulator (Van Wolffelaar & Van Winsum, 1992). This is a (fixed-base) driving simulator consisting of a mock-up car with original controls (three pedals, clutch, steering wheel, safety belt, indicator and handbrake) linked to a dedicated graphics computer, registering driver behaviour while the road environment and dynamic traffic is computed at 30Hz+. Participants have a 180° view of the road environment. The virtual environment was designed in ST software as well. This designing software consists of an interface to design roads and environmental features and of an additional script interface to create traffic situations. Furthermore, vehicles in the virtual driving environment interact with the participant's simulator car autonomously and behave according to hierarchically structured decision rules that are based on human driving behaviour (Van Wolffelaar & Van Winsum, 1992). Therefore, besides measuring the driver's behaviour, the influence of his or her behaviour on the other traffic participants can be assessed as well.

The road tracking task: an assessment tool for driving skills at an operational level

The road tracking task is designed to measure unconscious response errors, or tracking errors, measured as the standard deviation of the lateral position (SDLP; O'Hanlon, 1982). The task has been used in numerous studies to assess effects of drugs and alcohol on driving performance and has shown to be sensitive to drug induced impairments (Louwerens et al., 1987; Kuypers et al. 2006; Leufkens et al., 2007). Although first designed for on the road testing, the test is also widely used to assess drug driving in the driving simulator (Brookhuis et al., 2004; Rapoport & Banina, 2007).

The road tracking task in the standard virtual environment, designed to measure drug induced driving impairment, consists of two straight roads of approximately 10 km. In this part of the virtual ride, the participant drives trough a rural area with moderate density traffic flow and a posted speed of 100 km/h. How the participant manages this part of the route is assessed by measuring the SDLP.

In the road tracking task participants are usually instructed to maintain a lateral position in the middle of the traffic lane and to drive at a constant speed. But, as Brookhuis and de Waard (1994) already pointed out, driving like this is normally not necessary and therefore the instruction is unnatural. In the current scenario the participants will therefore get no other instructions but to drive as they would normally do.

As said the road tracking test has proved to be very sensitive to drug induced impairments at the operational level of driving but it can however be questioned if these effects are indicative of overall impaired driving performance. For example studies in which the direct effect of Methylenedioxi-methamphetamine (MDMA) on driving performance was investigated indicate that MDMA improves or leaves automated (control level) driving performance unaffected, but can cause impairments on other levels of driving by e.g. increased risk taking (Brookhuis et al., 2004; Kuypers et al., 2006; Ramaekers et al., 2006). Therefore driving behaviour at the tactical and strategic level has to be assessed as well.

Assessments of manoeuvring on a tactical level of driving

Car following

Perceptual and attention errors are the most frequently cited errors leading to drug and or alcohol related driving accidents in epidemiological data (Moskowitz, 1984). Decreased attention and or perception may impact the driver's ability to react properly to manoeuvres of other drivers. Such a situation can be when a driver has to respond to changes in speed of a car in front of him or her. To measure this so called car following performance Brookhuis et al. (1994) developed the car following test.

In this scenario the following (i.e. the participants') car is instructed to follow a lead car at a short but safe distance. The lead car is programmed to accelerate and decelerate in a cycle of between 20 and 40 seconds (e.g. between 0.02 and 0.04 Hz). Within this frequency band this frequency is varied randomly. The response of the participant to the speed changes of the lead car is measured by assessing the coherence (the extent to which the pattern of speed change of the lead and follow car correspond), the modulus (an amplification factor between the two signals in which the extent of the reaction is established, when there is an over-reaction the mode is larger than 1 and when there is an under-reaction the modulus is smaller than 1), and most important, the delay (the time the following car needs to respond to the lead car; see Brookhuis et al., 1994, De Waard & Brookhuis, 2000).

Traversing unsignalised crossroads

The ability to estimate the relative speed of other vehicles and accurate awareness of gaps when crossing a road is essential for safe driving. In a split second an achievable manoeuvre can turn out to be a looming accident because the driver judged the time available for safe crossing to be longer than the actual available time. Safe road crossing is a complex task that requires not only the mentioned perception of time and speed but also a fine co-ordination to synchronize this perception with the onset of movement. These are all factors that can be affected by drugs and alcohol.

In the virtual driving environment there are several scenarios used to measure the driver's ability to safely traverse a crossing; the gap acceptance scenario and the scenarios described under violations of traffic laws. The latter are scenarios that measure both the ability to safely traverse a crossing and the ability to comply with traffic regulations.

Gap acceptance
Brookhuis et al. (2004) found that the judgment of safety is deteriorated in drug affected drivers and therefore the perception of risk is reduced. The gap acceptance task is widely used to asses risk taking in traffic. In the gap acceptance scenario the participant has to cross a junction and is faced with traffic coming from both the left and right hand side, or from the opposite side of the road in the case of a y-junction. The gaps in between cars increase with 1 second for each new car. The driver has to choose the appropriate gap to traverse the crossing. A driver's perception of an appropriate gap is dependent on the expected waiting time for a gap involving negligible risk (Adams, 1995). The driver weighs the waiting time versus the risk of causing an accident and comes to a decision to either choose a small risky gap but short waiting time or a larger saver gap but longer waiting time.

The parameters included to assess the drivers risk taking are the size of the chosen gap in seconds, the distance to the car approaching the driver while traversing the crossing and the time required for the two vehicles to collide if they were to continue to drive at the same speed (time to collision, TTC; Van der Horst, 1990). As said the driver's behaviour regarding the other traffic participants can be assessed as well. It can be that the driver chooses a fairly safe gap but because of poor initiation of movement, for example, is still causing nuisance to other traffic participants. This can be assessed by measuring the deceleration of the car approaching the driver while traversing the crossing.

Violating traffic regulations

Analyses of driver records of patients admitted to substance abuse showed that drug users had significantly more traffic violations than a non drug control group (MacDonald et al., 2004). This is an important finding since there seems to be a clear relation between violations of traffic laws and the risk of being involved in a traffic accident. Parker et al. (1995) investigated the relationship between self reported violations, and accident involvement in a study of 1,600 drivers. They found that accident involvement was predicted by self-reported tendency to commit violations. Moreover Rothengatter (1991) studied a database of traffic accidents and found that 92 % of all traffic accidents were preceded by violating at least one traffic law.

In the virtual driving environment four types of violations are assessed; violating the posted speed limit, failing to give right-of-way, running a red light and failing to stop for a stop sign. The first, speeding, is described as a skill at the strategical level and shall not be discussed further here.

Giving right-of-way
In this scenario the participants are driving on a road towards a normal junction at a posted speed of 50 km/h. The driver coming from the right has priority over the participant. The participants are now faced with a situation in which they have to judge if they are able to cross before the priority driver enters the crossing or not. To make this assessment it is critical for participants to have a good judgement of speed of the other vehicle and the time they need to traverse the crossing, and synchronize this perception with the onset of movement. How participants handle this situation is assessed by whether or not they choose to give way or not by measuring deceleration. If participants came to a standstill before the crossing they had no other choice but give way to the other driver. If participants slowed down but did not come to a standstill this means they violated the priority rule over the driver coming from the right hand side. In this case the minimal time to collision (TTC) and deceleration of the other driver can be measured.

Running a red light: the amber traffic signal scenario
Red-light running is a frequent and highly dangerous driving act. Research in the United States showed that running red lights and failing to stop for stop and yield signs are the most frequent causes of urban crashes (Retting et al., 1995). There are several factors that are of importance to be able to react to a traffic light. The driver has to pay attention to the traffic light which is usually on the side of the road or above it. Furthermore the driver has to be able to react quickly and has to be willing to comply with traffic regulations. These are all factors that can be affected by drugs and or alcohol.

For the assessment of drug and alcohol impaired driving a scenario developed by De Waard et al. (1999) was used. In this scenario the participant approaches a crossing with green traffic light at a posted speed of 50 km/h. Two seconds before the participant would pass the traffic light the light turns amber. To equalize the situation for all speeds the traffic light is adjusted to the speed of the participant in a linear fashion in such a way that when the participant drives faster the traffic light will turn amber earlier. In this manner the situation remains critical even if participant does not keep to the posted speed limit. When participants keep the same speed they will drive through a red traffic light, when they speed up they will drive through an amber light and if they want to stop they have to brake hard. The braking reaction can be measured by assessing the brake reation time (BRT).

The choice of the participant can be determined by assessing the colour of the traffic light at the moment of crossing (De Waard, Van der Hulst, & Brookhuis, 1999). Running a red light at the crossing could have three reasons. The participant was not paying attention to the traffic light and did not see it or saw it too late. It could also be that he or she did see the traffic light but was not able to react on it any more. Or the participant could be taking the risk of violating the rules of stopping. To distinguish between the three possible reasons for running the red light, deceleration of the car is measured. If there was a case of inattention one might expect the participant to drive along at a normal pace not decelerating or accelerating. If the participant noticed the traffic light but was not able to react properly any more we would expect to see a deceleration not coming to a standstill. If the participant saw the traffic light but was deliberately violating the stopping rule we would expect to see an acceleration to either run the amber or red light.

Stop before stop sign
A stop signs requires a driver to stop regardless of whether conflicting traffic is present or approaching the junction. In this scenario it is assessed if participants exercise appropriate caution at stop signs. That means come to a total stop before driving further. If participants do not come to a total stand still they are violating the rule to stop at a stop sign.

Reaction to unexpected events

According to Quimby (1986), drivers who take longer to detect and react to potential hazards have more accidents. Furthermore they are more prone to react impulsively and inattentively. Alcohol intoxication and sedation by (medicinal) drugs has been associated with significant deterioration in attention and reaction time (Kelly et al., 2004). This means that alcohol and or drug affected drivers are expected to detect a potential hazard more slowly and moreover have a deteriorated ability to react on it quickly. In the virtual driving environment three scenarios are used to measure the driver's reaction to unexpected events; a car failing to give way, a car suddenly pulling out of a car park, and cars suddenly coming to a standstill on the motorway.

Car failing to give way
In this scenario the participant is driving on a priority road towards a normal junction with a posted speed limit of 50 km/h. A driver coming from the left fails to give way. The participant is faced with a critical situation in which he or she has to react quickly to avoid a collision. In this scenario the participant has the choice of taking priority and risking a collision or give way and be safe. In order to make this assessment it is critical for the participant to make a good judgement of speed of the other vehicle and the time that car needs to traverse the crossing. Furthermore the participant has to be attentive to a car that is taking priority from a direction that is supposed to give way.

How the participant handles this situation is assessed by whether or not the participant chooses to give way or not, and by measuring deceleration. If the participant came to a standstill before the crossing he or she had no other choice but to give priority to the other driver. If the participant slowed down but did not come to a standstill this means he or she took priority over the driver coming from the right side. In this case the minimal (TTC) and deceleration of the car failing to give way can be measured.

Car pulling out of a parking
In this scenario the participant is driving on a straight road and is passing a lay-by with parked cars when suddenly a car pulls out of a parking place. The participant is faced with a critical situation in which he or she has to react quickly to avoid a

collision. How the driver handles this situation is assessed by measuring the BRT and the minimal TTC.

Motorway traffic suddenly coming to a standstill
In the motorway driving task the participant is driving on the motorway with a posted speed limit of 120 km/h in normal traffic density. The traffic unexpectedly comes to a standstill and driver is faced with a critical situation in which he or she has to brake hard to avoid a collision. The reaction of the participant is measured by assessing the BRT and the minimal TTC.

Driving on the motorway

There are two general indicators of driving style that have been linked to accident risk when driving on a motorway with traffic, namely: speed and headway (Ward & Buismans, 1998). Average speed is used a general measurement of skills at the strategic level and will not be discussed here any further. With Headway choice the tendency to leave short headway distances to the vehicle in front is measured as an indication of risky driving.

In the virtual driving environment the participants are faced with several scenarios: first the participants have to filter into traffic to get onto the motorway, second they are driving on the motorway with normal density traffic for 15 km when suddenly the traffic comes to a standstill, from this point on the participants are driving in congested traffic for 10 km, the last 5 km is normal driving again until the participants finally leave the motorway. The scenario in which the traffic suddenly comes to a standstill was described above.

Filtering into traffic
Before coming onto the motorway the participant has to filter into the motorway traffic between two lorries. In this scenario the participants' relative location of filtering in (before, between or after the lorries) is measured as an assessment of risk taking. Moreover the amount of nuisance the participant causes to the other traffic participants and location of lane change from the acceleration lane is registered.

Normal motorway driving
The participant has to drive in normal traffic density on a two-lane motorway with a 120 km/h speed limit. Some of the other traffic participants are programmed to keep to the exact posted speed limit, some drive below or above the posted speed to simulate the natural situation. How the participants deal with this situation is assessed by measuring the number of times they change lanes. The motivation for making a lane change is primarily due to speed inconveniences caused by a slower vehicle. As the inconvenience increases the drivers willingness to change lanes also increases. Thus lane changing is for some part dependent on the drivers speed and for another part on the level of annoyance the driver is experiencing. Furthermore a risk assessment can be made by measuring the mean and minimal time headway (THW) the participant holds to other traffic participants.

Driving in traffic congestion
The situation in which the driver is driving in traffic congestion is a somewhat like the car following task which is described earlier. However since the lead cars are not controlled in frequency of speed changes such as the cars in the car following situation only minimal THW and minimal TTC are measured here.

Speed management as a measurement of driving performance at a strategic level

Since the route has to be fixed for all the participants to encounter the same scenarios route choice cannot be investigated. However, choice of speed can be assessed. Participants under influence of sedating drugs might change their average speed while under the influence to compensate for sedating effects. On the other hand stimulating drugs that reduce hazard perception and increase risk taking might make the participant want to speed up (Brookhuis et al., 2004). Another compensating mechanism in speed management is deliberately varying speed. In the case of severe sedation a participant might temporarily increase speed to increase feelings of arousal as can be seen in an increase in the standard deviation of speed (Brookhuis, 1998). Violation of the posted speed limit, standard deviation of speed and overall duration of the ride are therefore important measurements of drug effects on driving at the strategic level and can be assessed throughout the entire route.

Future research

To determine thresholds for driver impairment as a consequence of drug or combined drug and alcohol use, an "alcohol calibration" study will be performed. In this calibration study the influence of three levels of alcohol (0.3 ‰, 0.5‰, and 0.8 ‰) and a placebo on the measurements within the developed scenarios will be tested. Analyses of these results have to point out if the proposed scenarios are suitable for investigating the effects of MDMA and/or alcohol on driving performance.

References

Adams, J., (1995). *Risk*. London: University College London Press.
Brookhuis, K.A. (1998). How to measure driving ability under influence of alcohol and drugs and why. *Human Psychopharmacology, 13*, S64-S69.
Brookhuis, K.A., De Waard, D., & Mulder, L.J.M. (1994). Measuring driving performance by car-following in traffic. *Ergonomics, 37*, 427-434.
Brookhuis, K.A., De Waard, D., & Samyn, N. (2004). Effects of MDMA (ecstacy), and multiple drug use on (simulated) driving performance and traffic safety. *Psychopharmacology, 173*, 440-445.
Christophersen, A.S., & Morland, J. (1997). Drugged driving, a review based on the experience in Norway. *Drug and Alcohol Dependence, 47*, 125-135.
De Waard, D., Van der Hulst, M., & Brookhuis, K.A. (1999). Elderly and young drivers' reaction to an in-car enforcement and tutoring system. *Applied Ergonomics, 30*, 147-157.

De Waard, D., & Brookhuis, K.A. (2000). Drug effects on driving performance, letter to the editor. *Annals of Internal Medicine, 133*, 656.

Kelly, E., Darke, S., & Ross, J. (2004). A review of drug use and driving: epidemiology, impairment, risk factors and risk perceptions. *Drug and Alcohol Review, 23*, 319 – 344.

Kuypers, K.P.C., Samyn, N., & Ramaekers, J.G. (2006). MDMA and alcohol effects, combined and alone, on objective and subjective measures of actual driving performance and psychomotor function. *Psychopharmacology, 187*, 467-475.

Louwerens, J.W., Gloerich, A.B.M., De Vries, G., Brookhuis, K.A., & O'Hanlon, J.F. (1987). The relationship between drivers' blood alcohol concentration (BAC) and actual driving performance during high speed travel. In P.C. Noordzij & R. Roszbach (Eds.), *Alcohol, Drugs and Traffic Safety-T86* (pp. 183-186). Amsterdam: Excerpta Medica.

Leufkens, T.R. M., Vermeeren A., Smink B.E., Ruitenbeek, P., & Ramaekers, J.G. (2007). Cognitive, psychomotor and actual driving performance in healthy volunteers after immediate and extended release formulations of alprazolam 1 mg. *Psychopharmacology, 191*, 951-959.

Macdonald S., Mann, R.E., Chipman, M., & Anglin-Bodrug K. (2004). Collisions and traffic violations of alcohol, cannabis and cocaine abuse clients before and after treatment. *Accident Analysis and Prevention, 36*, 795–800.

Michon J.A. (1985). A critical review of driver behavior models. What do we know, what should we do? In L. Evans and R. Schwing (Eds.), *Human behavior and traffic safety*. New York: Plenum Press.

Moskowitz, H. (1984). Attention tasks as skills performance measures of drug effects. British. Journal of clininical Pharmacology, 18, 51S-61S.

Movig, K.L.L., Mathijssen, M.P.M., Nagel, P.H.A., Van Egmond, T., De Gier, J.J., Leufkens, H.G.M., & Egberts, A.C.G. (2004). Psychoactive substance use and the risk of motor vehicle accidents. *Accident Analysis & Prevention, 36*, 631-636.

O'Hanlon, J.F., Haak, T.W., Blaauw, G.J., Riemersma, J.B.J. (1982) Diazepam impairs lateral position control in highway driving. *Science, 217*, 79-80.

Parker D., Reason J.T., Manstead A.S.R., & Stradling S.G. (1995). Driving errors, driving violations and accident involvement. *Ergonomics, 38*, 1036-1048.

Quimby, A.R., Maycock, G., Carter, I.D., Dixon, R., & Wall, J.G. (1986). *Perceptual abilities of accident involved drivers* (Report 27). Crowthorne, England: Transport Research Laboratory.

Ramaekers, J.G., Kuypers, K.P., & Samyn, N. (2006). Stimulant effects of 3,4 methylenedioxymethamphetamine (MDMA) 75 mg and methylphenidate 20 mg on actual driving during intoxication and withdrawal. *Addiction Abingdon, England, 101*, 1614-1621.

Rapoport M.J. &. Banina M.C. (2007). Impact of Psychotropic Medications on Simulated Driving A Critical Review. *CNS Drugs, 21*, 503-519.

Retting, R.A., Wiliams, A.F., Preusser, D.F., Weinstein, H.B. (1995). Classifying urban crashes for countermeasure development. *Accident Analyses and Prevention, 27*, 283-294.

Rothengatter, J.A. (1992). The feasibility of driver information and offence detection systems. *Proceedings of the 24th ASATA Intern. Symp. On automotive Technology and Automation* (pp. 135-142). Croydon, England: Automotive Automation Limited.

Van der Horst, A.R.A. (1990). *A time based analyses of road user behaviour in normal and critical enncounters*. PhD Thesis. Delft, The Netherlands: Delft University of Technology.

Van Wolffelaar, P.C. & Van Winsum, W. (1992). A new driving simulator including an interactive intelligent traffic environment. Proceedings of the third international conference on vehicle navigation & information systems (pp. 499-506). Piscataway NJ: IEEE Service Center.

Ward, N.J. & Buismans, J., (1998). Simulation of accident risk displays in motorway driving with traffic. *Ergonomics, 41*, 1478 – 1499.

CarUSE – Developing a framework for IVIS evaluation

Thomas Vöhringer-Kuhnt
Technische Universität Berlin
Germany

Abstract

Usability and driving safety are key aspects when evaluating human-machine-interfaces (HMI) of advanced in-vehicle information systems (IVIS). A central aspect of IVIS evaluation is the potential to distract the driver from the task of driving. Inattentiveness and distraction can have a critical influence on driving performance (stabilization, manoeuvring) especially when the primary task (driving) and the secondary task (operating the IVIS) use the same cognitive resources. Using existing methods to evaluate human-machine-interaction in an automobile in combination with accepted human performance models, the Technische Universität Berlin is developing a methodological procedure that can be used to test advanced driver information systems in early periods of the development cycle. This paper will describe the first steps in the development of that procedure and report the primary results of a feasibility study used to estimate the objectivity and utility of existing methods of IVIS evaluation. Moreover, it was tested to what extent certain assumptions of the "15-second rule" (Green, 2000) are applicable for such a procedure.

Introduction

Advanced in-vehicle information systems have become essential products of the automobile industry. Consequently, specific requirements for the design of the automotive human-machine interaction (HMI) have to be met in order to combine all single systems and functions into a holistic concept of interaction. The ergonomic principles of human-machine interaction, the psychological and physiological characteristics of the driver, and the definition of the driving task and traffic situation mark the external parameters, which need to be considered when designing human-machine interaction as well as advanced driver assistance systems (Timpe, et al., 2002).

CarUSE: On the way to an evaluation toolbox

Beyond the pure functionality of in-vehicle information systems, usability and safety requirements have to be addressed. Guidelines for the design of cockpits and automotive systems have already been developed by industry as well as legal authorities. So far, no standardized method has been established to describe and evaluate the HMI of a vehicle. Nevertheless, certain significant efforts have led to a variety of elaborated and complex evaluation methodologies, e.g., projects such as AIDE[1], HASTE[2], INVENT[3], and RESPONSE[4]. The goal of the 'CarUSE' project, which is run by Harman International, Volkswagen AG and Technische Universität Berlin, is to develop a tool that describes and evaluates the human machine interface of a vehicle. Such a tool should be easy to handle, should be usable not only by HMI-experts, should employ the folding rule character and should be modular, so that isolated parts of the HMI can be evaluated.

Comparing different methods of IVIS evaluation

In the course of an explorative study, a comparison of evaluation techniques such as questionnaires/checklists, subjective ratings about the degree of distraction, operating time, and objective data obtained with the 'Lane Change Test' (LCT) (Mattes, 2003; ISO/DIS 26022, 2007) has been completed. Rather than evaluating in-vehicle information systems themselves, an assessment of established methods of evaluation was the focus of this study. Moreover, it was tested to what extent the "15- second rule" (Green, 2002; Reed-Jones et al., 2007), the assumption that task-duration predicts perceived distraction, is an appropriate starting point for the evaluation tool. This was done by calculating correlations between the time spent operating the system and objective and subjective judgments of distraction and mental demand.

Material

Two different navigation systems were provided in order to conduct the study. One system was a portable navigation device; the other system was a radio-navigation system, which was installed in the centre console. Volkswagen AG provided a seating buck to execute the study. The driving scene was presented on a white plain by a LCD-projector (figure 1). The participants were asked to operate the IVIS in a static (single task without driving) and a dynamic (dual task while driving) setting.

The induced mental demand carrying out the secondary task was estimated using the SEA-scale ("subjektiv erlebte Anstrengung" – subjectively perceived effort[5]) (Eilers et al., 1986). In comparison to other tools capturing the experienced mental demand such as the NASA Task Load Index (Hart & Staveland, 1988), which estimates the effort according to six defined dimension of strain, the SEA scale is an easy to complete, one-dimensional assessment tool. Nevertheless, it is highly correlated to the NASA-TLX overall score ($r = .78$) and its subscale of effort ($r = .81$) (Seifert, 2002).

---

[1] http://www.aide-eu.org/res_sp2.html
[2] http://ec.europa.eu/transport/roadsafety/publications/projectfiles/haste_en.htm
[3] www.invent-online.de
[4] http://www.prevent-ip.org/en/prevent_subprojects/horizontal_activities/response_3/
[5] According to Sanders (1983), effort is the essential criterion to determine the experienced demand.

Figure 1. Seating buck with projection of LCT driving environment (TU Berlin)

Methods

Participants

Twenty-two people between the ages of 20 and 30 years participated in this experiment. Because many different studies show age as a relevant factor (e.g. Jahn et al., 2004; Totzke et al., 2005), we concentrated on a certain age group to eliminate these influences. Thirteen subjects were male, the rest were female. The time holding a driver's license varied between six months and twelve years. The average annual distances driven ranged from zero kilometres to 30 000 kilometres.

Procedure

After completing a demographic questionnaire, the participants drove without operating the IVIS in order to establish a baseline of the Lane-Change-Test. After that, the participants operated the system at different levels of complexity in either a static or a dynamic setting. One task was to change display settings from day to night mode. The other task was to type in a sequence of address information (Berlin, Turmstrasse 10) and to start the navigation module. The sequence of the experimental conditions (system A vs. system B, static vs. dynamic use, simple vs. complex task) was randomized to eliminate effects of learning. After every operation of the system, the subjects were asked to assess the experienced demand using the SEA-scale. Results of the performance of the LCT, especially the standard deviation from the ideal track, were used to determine the driving behaviour. In order to evaluate the IVIS, a modified TRL checklist (Stevens et al., 1999) and a modified version of CarE checklist (Dubrowsky et al., 2001; Schmitz, 2003) were used (detailed results of the two latter methods will not be reported in this paper). Task duration was measured manually by the experimenter with an electronic stopwatch.

After completing all experimental conditions, participants completed a second baseline run.

Data analysis

The evaluation of the 2 x 2 x 2 factorial experimental design was done using analysis of variance, t- tests and regression analysis. The factors were:

- system (A vs. B)
- task (changing settings vs. entering address)
- condition (single vs. dual task)

Driving behaviour while accomplishing the 'Lane Change Task', mental demand when operating the navigation system, and task duration served as dependent variables. The design was a within-subjects post-test-only design for the three main factors. A between-participant design was selected for the factors "gender" and "driving experience" (years since holding a driver's license and average annual driving in km). The level of significance was set to α= 0.05. The normal distribution of the sample was tested using the Kolmogoroff- Smirnov Test. SPSS 15.0 for Windows was used to evaluate the data.

Results

Manipulation check

First, a manipulation check was conducted to determine whether the systems differed in respect to task duration in the static setting (operating the systems without driving). It must be determined that such differences exist before further evaluation of the results can occur. Only if operating the systems without driving leads to differences in task duration, any effects on the dependent variables can be expected and ascribed to the systems themselves, and the two tasks, respectively. The main effect in terms of static time of operation of the factor "system" ($F(1,21) = 5.43$; $p = .03$; $\eta^2 = .21$) and "task" ($F(1,21) = 3497$; $p < .001$; $\eta^2 = .994$) turned out to be statistical significant. Hence, it could be proven that the static operation time for system A differed from that for system B. Furthermore, the static operation time for the simple task is significantly shorter than that for the complex task (table 1). The experimental manipulation was conducted successfully.

Table 1. Static task duration (seconds)

		mean (s)	std
Change Settings	System A	4.83	1.04
	System B	2.31	0.58
Enter Address	System A	28.99	2.49
	System B	33.61	3.72

Results of the performance of the LCT

A standardized lane-change test (LCT, see Mattes, 2003) was used to examine the quality of lane keeping during the single-task condition (only driving) and during the dual-task condition (driving and operating the IVIS). The software "LCT-Analysis", which is included in the LCT software package, was used to quantify the participant's performance. In order to test for any potential effects of learning, the average deviation from the defined ideal line of the first and second baseline ride were compared administering a t-test. No significant difference was found (T = 1.07; df = 21; p = .29); therefore, an effect of learning across the experimental conditions could be eliminated. Another hypothesis relates to the decline in lane keeping when executing a secondary task. The difference of mean deviation from the ideal line between the baseline and the dual task drives could be proven for all experimental conditions (Experimental condition tested against baseline: System A, change settings T = 4.13; df = 21; p < .001; System A, enter address T =10.98; df = 21; p < .001; System B, change settings T = 2,15; df = 21; p = .02; System B, enter address T = 12.82; df = 21; p < .001). However, the hypothesis that using two *different* systems leads to *different* levels of lane keeping degradation could not be confirmed. No significant main effect of the factor "system" ($F(1,21) = 0.86$; p = .36; $\eta^2 = .04$) could be found. Assuming a certain reliability of the LCT, the (more or less diverse) operating concepts of the two systems, which lead to different static operation times, did not evoke differences in lane keeping during task completion.

Differences in task duration

First, it was tested to what extent execution time differed between the two tasks and the two systems (overall in the single and dual task condition). Furthermore, it was examined to what extent the implicit assumptions about the "15- second rule", the idea that longer *static* times of operation (single task) lead to longer *dynamic* times of operation (dual task while driving), is applicable. Moreover, it was tested to what extent longer static and dynamic times of operation lead to a decline in the LCT lane keeping. The main focus here is this: do *longer* times of operation lead to a *higher* distraction while driving? It was also examined whether longer times of operation can be related to a higher mental effort completing the secondary task scenario. Besides an overall main effect of task duration for the factor "system" ($F(1,27) = 5.84$; p = .023; $\eta^2 = .18$), which confirmed the reported differences in static operation times between the two systems, an interaction-effect regarding the single/dual task operation and the two tasks could be found ($F(1,27) = 34.03$; p < .001; $\eta^2 = .56$): The times of operation while executing the more complex task increased significantly more under the dynamic condition, compared to the easy task (figure 2). This interaction effect needs to be evaluated critically regarding the generalization of the static times of operations as a criterion of the degree of distraction.

The hypothesis that systems with longer static times of operations lead to longer dynamic times of operation could not be confirmed. In contrast to the implicit assumption of the "15- second rule", the static and dynamic times of operation of system A and system B showed an ambiguous effect and were inversely related (figure 3). Under the static condition, system A could be operated faster, but under the dynamic condition, system B could be operated faster. (Even though the absolute difference between times is small, it turned out to be statistically significant for both tasks: easy task: T = 11.09; df: 21; p < .001, difficult task: T = 5.43; df: 21; p < .001).

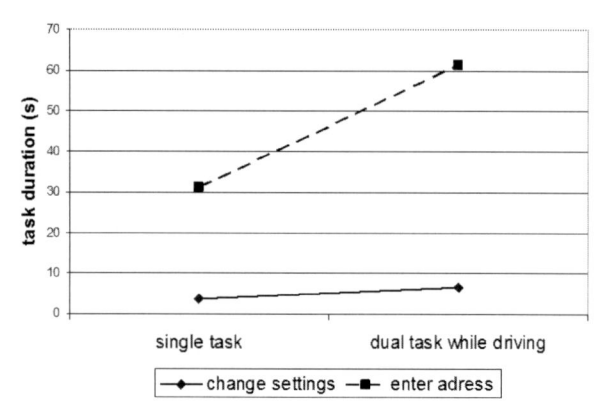

Figure 2. Interaction effect of task duration between the two task and the conditions of task completion

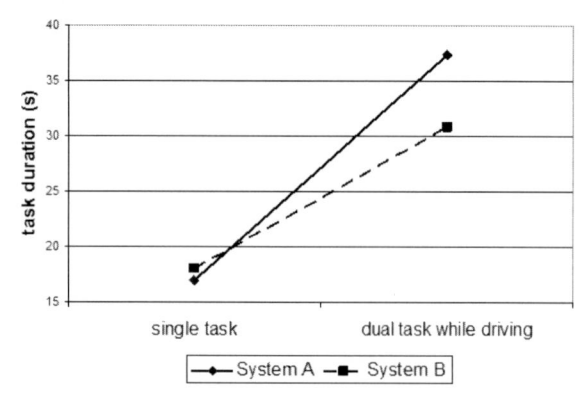

Figure 3. Interaction effect between the two systems and the conditions of task completion

Examination of mental effort

Analyzing the mental effort, a significant difference between the tasks and the conditions of task completion was found (change settings vs. enter address: $F(1,19) = 97.11$; $p < .001$; $\eta^2 = .836$; static vs. dynamic condition, $F(1,19) = 98.40$; $p < .001$; $\eta^2 = .839$). The difficult task was evaluated as more mentally stressful than the easy task. The same applies to the different experimental conditions: The dynamic condition was experienced as more mentally stressful than the static condition. The main effect of the factor "system" ($F(1,19) = 0.28$; $p = .63$; $\eta^2 = .015$) did not turn out to be significant. The operation of the different systems was judged as equally stressful. Significant interaction effects could be found for the interaction "task vs.

condition of operation" (F(1,19) = 79.95; p < .001; η2 = .808): The increase of mental effort during the dual task condition was larger for the complex task than for the easy task (figure 4).

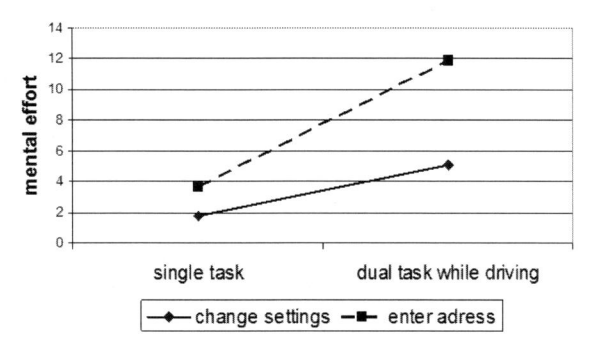

Figure 4. Interaction effect of mental effort between the two tasks and the conditions of task completion

Discussion

The results partly confirm results of similar experiments investigating the degree of distraction induced by a secondary driving task (see Harbluk et al., 2007). Comparing the performance of the single-task condition (driving without operating the navigation system) with the dual-task condition (executing the LCT and operating the navigation systems) revealed a significant greater deviation from the ideal line under the dual-task condition. However, the average deviation from the ideal line was not sensitive for the different system. Given an acceptably high reliability of the LCT, the operating concepts of the two systems (both are touch-screen based) can account for that missing effect. An analysis of the mental effort data reveals a significant difference between the experimental conditions. The baseline ride was experienced as less mentally stressful than the dual-task conditions. The missing effect of the mental effort in comparison of both systems reflects the missing effect in the lateral lane keeping. Consequently, it can be assumed that assessing mental effort with the SEA-scale was a valid measure to determine mental workload.

Considering all factors, the times of operation were significantly different at all experimental levels. The times of operation differed for system A and B, for the easy and difficult task, and for static and dynamic conditions. At this point, it is also important to mention that the increase of dynamic times of operation for the complex task was significantly higher than for the easy task. In reference to the "15-second rule", the times of operations for the easy task are still below 15 seconds (in both conditions), and therefore, executing the easy task while driving could be assumed as safe. However, the performance of the Lane Change Task turned out to be significantly poorer when driving and operating the IVIS and the mental effort significantly increased executing the easy task under dynamic conditions. Therefore, it can be assumed that executing that kind of task while driving could result in impaired lane keeping, even if the static time of operation definitely falls below 15 seconds.

To assure interpretability of LCT measures, an ISO standard is under development (ISO/DIS 26022, 2007), which includes certain ratios as a decision criterion about performance degradations under secondary task conditions. For the "LCT mean deviation performance ratio", the mean deviation (of the lateral position from a normative model) under the dual task condition is related to the mean deviation in the baseline drive. The same is done for the mean task completion time. When either ratio is larger than 1.0, a performance degradation has occurred. These ratios can be used to compare different tasks or systems. So far, a cut off value of barely acceptable performance degradation ratios is missing. The idea to develop alcohol related performance standards for the LCT (Huemer & Vollrath, 2008) also didn't lead to an absolutely interpretable degradation value (such as "using system x while driving is as dangerous as 0.5 ‰ blood alcohol level"). Broadly accepted secondary tasks while driving (such as tuning the radio) could serve as a reference for acceptable lane keeping performance levels. Nevertheless, results of the LCT have to be related to other methodologies of IVIS-evaluation.

To further quantify minutiae of driver distraction, especially in respect to cognitive aspects, certain efforts have been undertaken to detach perceptive and motor aspects of the driving task from its cognitive components in sophisticated experimental settings, but results of such research are ambiguous. Consequently, recent work has used cognitive models to predict distraction (Urbas et al., 2007; Salvucci, 2005, 2006, in press; Wu & Liu, 2007). Cognitive modelling can account for distraction related effects both from perceptual-motor and cognitive aspects of operating technical artefacts under a secondary task condition. In combination with a standardized driving task such as the LCT, an evaluation tool for IVIS could be provided, which accounts for distraction related performance degradation on all levels of task performance.

Acknowledgement

I thank Dr. Peter Rößger and Dr. Miklos Kiss for their support preparing this study and giving advice for the experimental design and interpretation of the results. I'm grateful to Bo Höge, Shengguang Lei, Dan Finnegan and two anonymous reviewers for their helpful comments. The project CarUSE is conducted in cooperation with Harman/Becker Automotive Systems GmbH, Volkswagen AG, and Technische Universität Berlin and sponsored by them.

References

Dubrowsky, A., Hüttner, J., Warning, J., Wandke, H., & Küting, H.J. (2001). CarE - Ein Softwaretool zur Kognitiv-Ergonomischen Bewertung von Komponenten in Fahrzeugen. In D.P. Akademie (Ed.), *Psychologie am Puls der Zeit* (pp. 403 - 407). Bonn: Deutscher Psychologen Verlag.

Eilers, K., Nachreiner, F., & Hänecke, K. (1986). Entwicklung und Überprüfung einer Skala zur Erfassung subjektiv erlebter Anstrengung. *Zeitschrift für Arbeitswissenschaft, 40*, 215-224.

Green, P. (2000). *Potential expansion of the 15-second rule.* University of Michigan Transportion Research Institute, Human Factors Division.

Harbluk, J., Burns, P.C., Lochner, M., & Trobovich, P.L. (2007). *Using the lane-change test (LCT) to assess distraction: Tests of visual-manual and speech-based operation of navigation system interfaces.* Paper presented at the Fourth International Driving Symposium on Human Factors in Driver Assessment, Training and Vehicle Design. Retrieved from http://ppc.uiowa.edu/driving-assessment/2007/proceedings.htm

Hart, S.G., & Staveland, L.E. (1988). Development of a multi-dimensional workload rating scale: Results of empirical and theoretical research. In P.A. Hancock and N. Meshkati (Eds.), *Human mental workload* (pp. 139-183). Amsterdam: Elsevier.

Huemer, A., & Vollrath, M. (2008). Alkoholstandard für die Lane Change Task? - Wirkung von Alkohol auf die Leistung der Fahrer. In VDI (Ed.), *Integrierte Sicherheit und Fahrerassistenzsysteme* (pp. 253-261). Düsseldorf: VDI Verlag.

ISO/DIS 26022 (2007). *Road vehicles—Ergonomic aspects of transport information and control systems — Simulated lane change test to assess driver distraction.* Draft International Standard. Geneva: ISO.

Jahn, G., Rösler, D., Oehme, A., & Krems, J. (2004). Kompetenzerwerb im Umgang mit Fahrerinformationssystemen, *Bericht der Bundesanstalt für Straßenwesen* (Vol. F47). Bremerhafen: Wirtschaftsverlag NW.

Mattes, S. (2003). The Lane Change Test as a Tool for Driver Distraction Evaluation. In H.R.H.Strasser and H. Bubb (Eds.), *Quality of Work and Products in Enterprises of the Future* (pp. 57-60). Stuttgart: Ergonomia Verlag.

Reed-Jones, J., Trick, L.M., & Matthews, M. (2008). Testing assumptions implicit in the use of the 15-second rule as an early predictor of whether an in-vehicle device produces unacceptable levels of distraction. *Accident Analysis and Prevention, 40*, 628-634.

Salvucci, D.D. (2006). Modeling driver behavior in a cognitive architecture. *Human Factors, 48*, 362-380.

Salvucci, D.D. (in press). *Rapid prototyping and evaluation of in-vehicle interfaces.* To be published in: Transactions on Human-Computer Interaction. Retrieved 15.06.2009 from http://www.cs.drexel.edu/~salvucci/publications.html

Salvucci, D.D., Zuber, M., Beregovaia, E., & Markley, D. (2005). *Distract-R: rapid prototyping and evaluation of in-vehicle interfaces.* Paper presented at the Proceedings of the SIGCHI conference on Human factors in computing systems.

Sanders, A.F. (1983). Towards a model of stress and human performance. *Acta psychologica, 53*, 61-97.

Schmitz, M., Kebeck, G., & Parnow, A. (2003). *Verfahren zur Bewertung der kognitiven Ergonomie von Bedien- und Anzeigekonzepten im Fahrzeug.* Paper presented at the 45. Tagung experimentell arbeitender Psychologen (TEAP), Kiel.

Seifert, K. (2002). *Evaluation multimodaler Computer-Systeme in frühen Entwicklungsphasen. Ein empirischer Ansatz zur Ableitung von Gestaltungshinweisen für multimodale Computer-Systeme*. Dissertation: Technische Universität Berlin.

Stevens, A., Board, A., Allan, P., & Quimby, A. (1999). *A Safety Checklist for the Assessment of In-Vehicle Information Systems*. Crowthorne Berks, UK: TRL Library.

Timpe, K.P., Jürgensohn, T., & Kohlrep, H. (2002). *Mensch-Maschine-Systemtechnik. Konzepte, Modellierung, Gestaltung, Evaluation*. Düsseldorf: Symposion.

Totzke, I., Hofmann, M., & Krüger, H.-P. (2005). Alte Fahrer und Fahrerinformationssysteme: Ansätze zur Reduktion möglicher Alterseffekte. In VDI (Ed.), *Der Fahrer im 21. Jahrhundert* (pp. 129-150). Düsseldorf: VDI-Verlag.

Urbas, L., Heinath, M., & Leuchter, S. (2007). Bedienermodellgestützte Bewertung des Ablenkungspotenzials von Komfortsystemen im KFZ in frühen Phasen der Systementwicklung. *iCom - Zeitschrift für interaktive und kooperative Medien, 2/2007*, 21-29.

Wu, C., & Liu, Y. (2007). Queueing network modeling of driver workload and performance. *IEEE Transactions on Intelligent Transportation Systems, 8*, 528-537.

Evaluating the distractive power of secondary tasks while driving

Claire Petit[1], Antoine Clarion[2], Carolina Ramon[3], & Christian Collet[2]
[1] Renault Research Department, Guyancourt
[2] Claude Bernard University, Lyon
[3] INSA, Lyon
France

Abstract

Distraction is likely to increase the risk of crash during driving as it may elicit periods of divided attention from the main task, e.g. by glancing away from the road scene. Distraction may occur when a secondary task is performed concurrently with driving. However, as many skills are automated, the driver may keep the ability to perform secondary tasks. Besides, some of them are closely associated with driving actions (e.g. opening a window or switching the radio on). The aim of this study is to evaluate the distractive power of several day life tasks that can be performed while driving. The experiment took place on a private circuit on which 39 drivers were asked to drive as usual. At some predetermined times, additional tasks were requested by the experimenter. Drivers had to manage the dual-tasks in a counterbalanced order, considered the distractors: air-conditioning adjustment, mental arithmetic, money preparation for toll highway, CD-programming and response to questions. The extent to which each secondary task was distractive was assessed by recording electrodermal activity (EDA). EDA evolved as early as the dual-task demand increased. Preparation of money for toll highway had the most distractive power, whereas air-conditioning adjustment was the less distractive, the others eliciting intermediate distraction. Whereas managing two tasks simultaneously remains possible, the distractive power of a secondary task is dependent upon its degree of interference with driving. Tasks sharing cognitive and motor demand with driving were the most distractive.

Introduction

Contribution of driver inattention and distraction in crashes is known for many years. Treat (1980) identified inattention and internal distraction as major driver error causal factors. He defined inattention as "a non-compelled diversion of attention from the driving task", whereas an internal distraction was a "diversion of attention from the driving task that is compelled by an activity or event inside the vehicle". Inattention is more difficult to monitor, being a special state of the driver involved in his thoughts. Stutts et al (2001) listed several possible sources of driver distraction either due to outside or inside events, a contribution which was completed recently by Regan et al. (2009). Drivers are likely to be distracted with the increase of embedded systems in the car (radio, CD, audio and video tape, navigation system, cell phone …). Other sources of inside distraction can be the passengers, or driver's own activity e.g. drinking, eating, or moving objects in vehicle, causing interference with driving actions (Horberry et al., 2006).

As a whole, epidemiologic studies strongly suspected distraction as being responsible of about 23% of accidents or nearly accidents (Regan et al., 2009). In a report from Klauer et al. (2006) this proportion climbs up to 80%. Thus, there is an urgent need to develop standardized methods and metrics so that distraction can be objectively measured (Caird et al., 2005). There is no common basis for determining when an activity represents a distraction. Behavioural recordings using video may provide information about the degree to which the driver is distracted and many studies focused on driver actions and visual behaviour (Fairclough et al, 1991). For example, more erratic steering was interpreted as being caused by visual distractions during curve negotiation whereas auditory cues had a less distraction power in the same situation (Donmez et al., 2006). However, few studies were conducted with a physiological approach, i.e. using physiological variations as indices of distraction (Brookhuis et al., 1991). Variables from the autonomic nervous system (ANS) are good candidates since the orthosympathetic branch is very sensitive to external stimuli. Sokolov (1963) showed that stimulus novelty elicits high autonomic responses simultaneously with different bodily reactions, the most common being the orienting response. These are interpreted as focusing or diverting attention if a main task is already performed. The ANS is specialized in mobilizing energy resources in response to internal and external milieu demands (Wallin & Fagius, 1986). In turn, responding to stimuli makes subjects' arousal increasing. The ANS is known to give a close estimation of subjects' arousal and its variations especially through the orthosympathetic branch (Boucsein, 1992). The eccrine sweat glands system is innervated by sympathetic endings only and is more sensitive to psychologically significant stimuli than to thermo regulator stimuli, which results in changing the level of arousal. Consequently, changing the level of arousal results in changes of electrodermal activity (EDA), i.e. increase in skin conductance or decrease in skin resistance. Previous studies have shown that EDA varied as a function of the workload that the drivers encountered (Collet et al., 2003). More recently, Collet et al. (2009a), confirmed that EDA was a reliable index of the strain underwent by air-traffic controllers, the number of aircraft to be monitored being highly correlated with electrodermal changes. In a second study by Collet et al. (2009b) EDA was again taken as an index of arousal and was shown to separate unambiguously the periods of dual-task from those of single task performing. However, if increasing arousal can be an index of distraction, this must be associated with other indices from the electrodermal signal or from other sources, to provide a set of effective tools with high ability to detect periods of distraction. While driving, a secondary task has the potential to elicit distraction by increasing demand of resources. We expected that changes in the base line would attest that distraction need higher arousal to manage two tasks simultaneously. We supposed that performing the secondary task while driving should require the driver to process more information than during a normal driving period. Consequently, the amount of electrodermal responses should be higher during the dual-task period and should be closely related to the dispersion around the mean baseline value. Finally, we expected that the slope of the electrodermal curve could account for distraction, the steepness of the slope being expected to be related to distraction intensity. As redundant information is believed to increase reliability, the aim of the present study is to evaluate drivers' functional state changes that are thought to be associated with distraction by using several physiological indices from electrodermal activity recordings.

Material and methods

Participants

Thirty-nine male participants took part in this experiment. The mean age was 40.4 years old (SD=9.7, min: 23 and max: 59).

Experimental design

Participants were requested to drive three rounds on the private track, designed as a highway of about 5 km (15 minutes). During the first round, they only had to drive normally, with respect to speed limitations. During the next two rounds, they were involved in five distractive tasks while driving. The distractive tasks were controlled by an experimenter and performed at random, nevertheless, at particular places of the circuit, to guarantee that the driving conditions were comparable among dual-tasks. Data were processed as repeated measures as dual task sessions were compared to normal driving and as each participant was confronted to all dual-task conditions. Information related to dual-tasks was given by the experimenter from the back seat, and was separated from the next one by a period of normal driving allowing the driver to rest and the physiological variable to recover baseline.

Five additional tasks were selected for their distractive power and are described below.

Action on the ventilation system
The instruction was: "please select warm air intensity 2 directed both on your feet and on your face. Then orient manually the air on your face and activate the demisting". This task was requested on straight line only and lasted about 30 seconds.

Mental arithmetic
This was a counting back task in steps of 3 from 147. This task was purely cognitive but was nevertheless supposed to interfere with driving. Drivers were told that they could count at their own pace and that errors made were not taken into account.

Insertion of a CD
The instruction was to select the seventh track of the CD player while the radio was switched off and was pre selected on the MF as early as the driver switched it on.

Search for small change: money preparation for highway toll
The instruction was to gather € 2.50 among small change, placed in a space-pocket in front of the shift lever and then to give it to the co-pilot. If the task was performed in less than 40 seconds, the driver was asked to start again, up to the signal of end of task.

Questions
The experimenter asked questions to the subject and told him to answer in the most accurate way. The following list describes some of the questions asked to the participants, each sequence (question-response) lasting about 40 seconds:

1. What is the speed limit on highway when it is raining?
2. What is the speed limit in city?
3. Can you give me the name of a presenter of Television News?
4. How long you need to go to work in the morning?
5. What model of car do you drive mostly?
6. How many windows are there in your house?
7. At home, at what side of the front door is the handle?
8. Can you give me the name of a French river?
9. What is the colour of the vehicle you are driving right now?
10. What is the current date?

Physiological data recordings

Electrodermal activity was the dependent variable and was recorded continuously, as early as each participant started to drive. Skin resistance was measured with 5 µA DC and recorded using 50 mm^2 unpolarizable Ag/AgCl electrodes (E243, Clark Electromedical Instruments, Edenbridge, UK). The current density was thus 9.9 µA/cm². Sensors were placed with isotonic gel (Teca, ref. 822-201210 or GEL101, Biopac Systems, Inc.) and held by adhesive tape. Despite traditional recommendations (Fowles et al, 1981) sensors were placed on the distal phalanx of the index and third finger of the non dominant hand to prevent sensors contact on the driving wheel which would have denatured the signal and elicited artifacts.

As briefly described in the introductory section, several indicators were selected from the electrodermal signal and taken as indices of distraction: i) EDA was averaged during the time-interval of the secondary task and was the first dependent variable; ii) The distractive power of each secondary task was evaluated by its ability to generate supplementary information to be processed. This second index was evaluated by the dispersion around the mean electrodermal value. We hypothesized that a high dispersion was due to electrodermal responses elicited by the process of the secondary task that divided attention; iii) the mean slope of electrodermal activity during the same period should also represent the extent to which the driver is distracted by the secondary task. Finally, the distractive power of each secondary task was assessed using a set of 3 indices extracted from EDA. These are summarized in Figure 1.

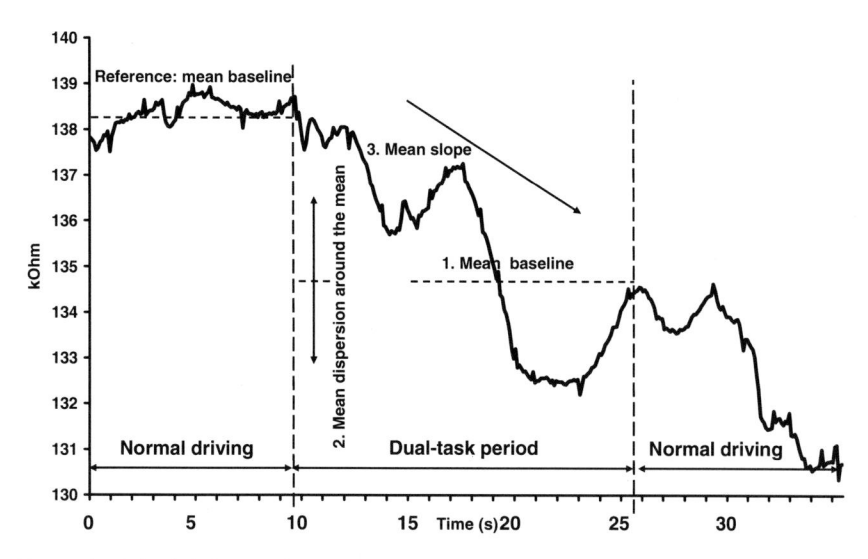

Figure 1. Method of data analysis: Example of Skin Resistance recording while performing a secondary task. The reference is the mean baseline just before the dual-task started. The duration of the reference period is the same to that of the dual-task. Thus, all indices computed during the dual-task period were compared to the same indices computed during the reference period. The mean baseline comparison gives evidence of changes in arousal. The mean dispersion around the mean is related to the difference in the amount of information that is processed during the two periods. The slope indicates the intensity of the distraction

Results

First, the time periods during which distraction was elicited were compared to those of normal driving. The three indicators extracted from the electrodermal signal gave evidence of significant differences between the distraction and the reference periods. The mean baseline value was lower during the dual-task period (66.19 kOhm) than during the reference period (68.74 kOhm), mean difference being 2.55, t=12.26, $p<.001$ (see Figure 2).

The mean dispersion around mean electrodermal baseline was higher during the distraction periods (1.89 kOhm) than during the reference periods (0.76 kOhm) as illustrated by Figure 3. The paired t-test was -9.66, $p<.001$, mean difference being 1.13.

Finally, the slope was negative during the distraction periods (-0.10 kOhm/s) whereas it remained almost constant during the reference periods (0.02 kOhm/s), as shown by Figure 4. The paired t-test was 2.01, $p<.05$, mean difference being 0.08.

To summarize, we defined a reference period for each distractive task to cancel the potential effect of time. Each task was different from the corresponding reference period (see Table 1). The present results showed that distraction was characterized by lower baseline values of EDA, higher dispersion around the mean and negative slope.

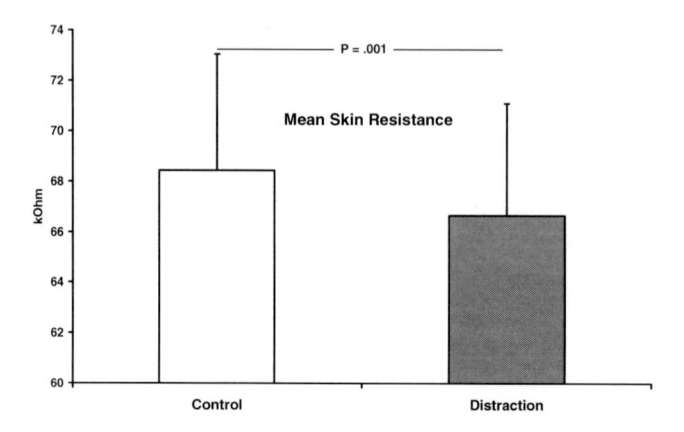

Figure 2. Mean Skin Resistance during the normal driving period (control) compared to the periods during which distraction was elicited by the secondary tasks. The electrodermal level is significantly lower under the distractive condition

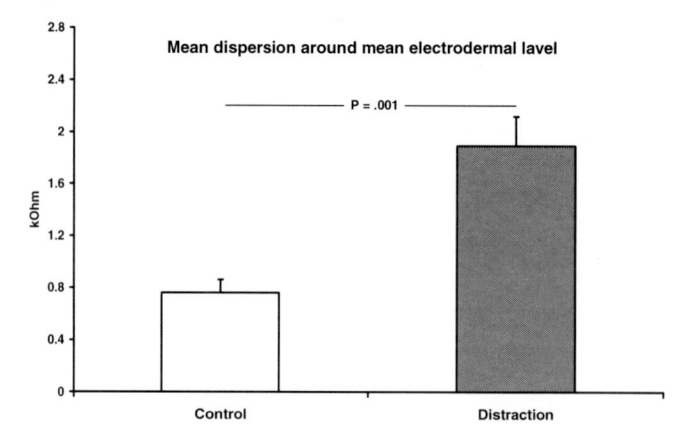

Figure 3. Mean dispersion around the Skin Resistance level during the normal driving period (reference) compared to the periods during which distraction was elicited by the secondary tasks. The dispersion is significantly higher under the distractive condition

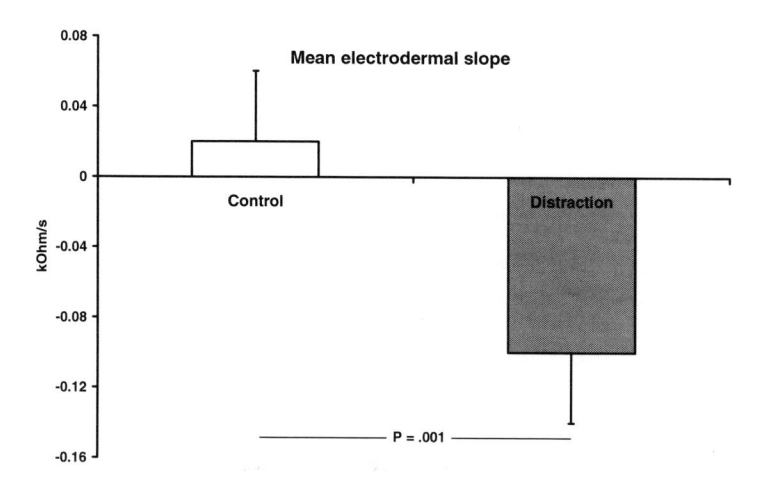

Figure 4. Mean Slope during the normal driving period (reference) compared to the periods during which distraction was elicited by the secondary tasks. The slope (kOhm/s) is close to zero during the reference period and negative during the distractive condition

Table 1. Summary of the distractive power of each secondary task. Each index from skin resistance recording (mean baseline, mean dispersion and slope) was compared with the same index computed during the reference period, just before the dual task started. At least two indices among three have the ability to differentiate each task from its own reference period. The slope is probably the less effective index whereas the two others clearly distinguished the distractive period from the reference

Electrodermal Variables	Mental Arithmetic / REF	CD / REF	Money / REF	Questions / REF	Ventilation / REF
Baseline (Mean level)	p < .001	p < .001	p = .005	p < .001	p < .001
Dispersion	p < .001	p < .001	p < .001	p < .001	p = .001
Slope	p = .03	NS	NS	p = .03	p = .001

We then used ANOVAs with repeated measures to compare the distractive power of all secondary tasks. This model showed a significant task effect in the three indices, if baseline values were compared ($F(4,190) = 12.1$, p<.001). The ANOVA for dispersion around the mean was $F(4,190) = 7.0$, p<.001 and $F(4,190) = 4.6$, p<.001, for the slope. The five distractive tasks (preparation of money, mental arithmetic, seeking a track on a CD, questionnaire and adjusting ventilation) were differentiated by at least one or several of the three dependent variables extracted from EDA (mean baseline, dispersion and slope) as summarized in Table 2.

Table 2. Differences among tasks obtained from mean level, mean dispersion and slope. At least one index is shown to differentiate each task from another, except when comparing mental arithmetic to CD manipulation. Comparisons were made using paired t-test, the p level being set at .05(/// NS).

	Mental arithmetic	CD	Money	Questions
CD	///	///	///	///
	///	///	///	///
	///	///	///	///
Money	///	///	///	///
	///	///	///	///
	Slope	Slope	///	///
Questions	Mean	///	///	///
	Dispersion	Dispersion	Dispersion	///
	Slope	Slope	///	///
Ventilation	Mean	Mean	///	///
	///	///	///	Dispersion
	///	///	Slope	Slope

Discussion

As expected, electrodermal activity was sensitive to distraction elicited by secondary tasks. The comparison of skin resistance, recorded when a secondary distractive task was performed, showed lower skin resistance values by comparison to that recorded during normal driving. Distractive stimulations elicited systematically a higher load and gave evidence of additional mobilization of mental resources as early as the secondary task started. The comparisons were made using a reference period taken just before the dual-task started thus guaranteeing that a potential effect of time was controlled. More detailed information was obtained as a higher dispersion around the mean baseline was also observed. Our hypothesis was that, during the dual-task, more information should be processed, thus eliciting more skin resistance responses by comparison with normal driving. The number of electrodermal responses could be considered an index of anxiety state (Haarmann et al., 2006). In the present study, the responses to the secondary task were due to external cues required to be processed simultaneously with those related to driving. Consequently, more dispersion around baseline was elicited that was considered as another index of distraction. This original index gave evidence of greater mental activity during dual-tasks management, which probably diverts attention from the road scene. To a lesser extent, the slope confirmed what the baseline change has previously evidenced. The reference period showed a slightly positive slope whereas the dual-task elicited a negative slope. Additional and redundant information is brought by this index which confirmed baseline changes.

The electrodermal activity thus provided accurate identification of periods during which drivers were distracted. We nevertheless acknowledge that the indices we computed are closely related to mental load and that their effectiveness also depends upon their ability to distinguish mental load from distraction. For this reason, the present analysis should probably be complemented by other methods, from driving performance measures, e.g. lane-keeping (Regan et al., 2009), and also by traditional questionnaires. The ability to process data both from the vehicle and from drivers' physiological activity in real time should contribute to build an integrated index of distraction. This requires obtaining specific behavioural and physiological patterns of activities correlated with the objective period of distraction state.

Classifying the dual-tasks as a function of their distractive power was made in this study, according to the potentiality of each secondary task to increase the mental load and, consequently, to divert the driver's attention from driving operations. Some of the tasks we tested in this experiment have a strong power of distraction. Among them, mental arithmetic showed high interference with the cognitive resources required to drive well. In the same way, preparing coins for toll-highway was highly distractive due to motor interference. This was probably the highest distractive task as this required to coordinate mental activity of arithmetic and motor activity of coins manipulation. Conversely, air-conditioning adjustment was more similar to a sensory-motor task implying a series of automated actions, closely linked with driving action and having little interference with it. Between these two categories, answering to questions elicited intermediate distraction, as well as choosing a track on a CD.

Despite the distractive power of each secondary task, the additional load remained relatively weak because of the highly automated action of driving on the one hand and because the drivers were not subjected to perform well, on the other. Several drivers might have reduced their speed, for example, to manage the two tasks under safety conditions. As further investigation, it would be interesting to evaluate to which extent the driving performance was influenced by additional task. The positive point is that this experiment was conducted under conditions close to the field. We should obtain a more powerful discrimination by associating indicators of electrodermal activity with other physiological variables, such as the analysis of cardiac and respiratory activities, as well as behavioural driving performance.

References

Boucsein, W. (1992). *Electrodermal activity*. Plenum Press, New York.
Brookhuis, K.A., de Vries, G., & De Waard, D. (1991). The effect of mobile telephoning on driving performance. *Accident Analysis and Prevention 23*, 309-316.
Caird, J.K., Lees, M., & Edwards, C., (2005). *The Naturalistic Driver Model: A Review of Distraction, Impairment and Emergency Factors*. UCB-ITS-PRR-2005-4 California PATH Research Report.
Collet, C., Petit, C., Priez, A., Brigout C., & Dittmar, A. (2003). Effects of a precision docking system on mental load in bus drivers during tight manoeuvre. Human Factors, 45, 539-548.
Collet, C., Averty, P., & Dittmar, A. (2009a). Autonomic nervous system and subjective ratings of strain in air-traffic control. *Applied Ergonomics, 40*, 23-32.

Collet, C., Clarion, A., Morel, M., Chapon, A., & Petit, C. (2009b, in press). Physiological and behavioural changes associated with the management of secondary tasks while driving. *Applied Ergonomics*.

Donmez, B., Boyle, L.N., & Lee, J.D. (2006). The impact of distraction mitigation strategies on driving performance. *Human Factors, 48*, 785-804.

Fairclough, S.H., Ashby, M.C., Ross, T. & Parkes, A. (1991). Effects of handsfree telephone use on driving behaviour. In *Proceedings of the 24th ISATA International symposium on automotive technology and automation* (pp. 403-409). Croydon, England: Automotive Automation Limited.

Fowles, D.C., Christie, M.J., & Edelberg, R. (1981). Publication recommendations for electrodermal measurements. *Psychophysiology, 18*, 232-239.

Haarmann A., Schaefer F., & Boucsein W. (2006). Adaptive automation using electrodermal activity during a simulated IFR flight mission. In S. Miyake and M. Trimmel (Eds.) *Proceedings of the 6th International Conference on PIE*. PIE-IEA Maastricht.

Horberry, T., Anderson, J., Regan, M.A., Triggs, T.J., & Brown J. (2006). Driver distraction: the effects of concurrent in-vehicle tasks, road environment complexity and age on driving performance. *Accident Analysis and Prevention, 38*, 185-191.

Klauer, S.G., Dingus, T.A., Neale, V.L., Sudweeks, J.D., Ramsey, D.J. (2006) *The impact of driver inattention on near-crash risk: an analysis using the 100-car naturalistic driving study data*. Report No DOT HS 810 594, National Highway Traffic Safety Administration, Washington DC.

Regan, M.A., Young, K.L., Lee, J.D., & Gordon, C.P. (2009). Source of driver distraction. In Regan, M.A., Young, K.L., Lee, J.D., Driver distraction; Theory, effects and mitigation (pp 249-279), CRC Press.

Sokolov, E.N. (1963). *Perception and the Conditioned Reflex*. London: Pergamon Press.

Stutts, J.C., Reinfurt, D.W., Staplin, L., & Rodgman, E.A. (2001). *The role of driver distraction in traffic crashes*. Washington, D.C.: AAA Foundation for Traffic Safety.

Treat, J.R. (1980). A study of the precrash factors involved in traffic accidents. *The HSRI Review, 10*, 1–35.

Wallin, B.G., Fagius, J. (1986). The sympathetic nervous system in man: aspects derived from microelectrode recordings. *Trends in Neurosciences, 9*, 63-67.

Menu interaction in Head-Up Displays

Natasa Milicic & Thomas Lindberg
BMW Group Research
Germany

Abstract

Introduced to the market several years ago, the Head-Up Display (HUD) has turned out to be one of the most innovative display technologies in the automotive industry. One of its most significant advantages is a minimized gauge reading time, compared to Head-Down Displays such as instrument clusters. However, certain cognitive effects are associated with the use of this display location. The most familiar effects are cognitive capture and perceptual tunnelling. These effects have to be considered carefully since advanced Human–Machine Interfaces could make a further step towards interaction concepts within the HUDs. Differences between Head-Up and Head-Down Displays, regarding interaction strategies and driver distraction are likely to be found. The question at hand is which types of interactions are suitable to be performed in a HUD. Therefore, a special driving simulator experiment was designed to examine these aspects. Within 120 min, 36 subjects had to perform several secondary tasks while driving in the BMW driving simulator. Each task had to be performed using the Head-Up Display as well as using the Central Information Display (CID) in the upper middle section of the dashboard. The HUD resulted in a more efficient operating time, better driving performance (lane keeping) and less cognitive and visual demand (measured by a Peripheral Detection Task). However, the interaction type and design remain crucial to glance time duration and to several driving performance parameters. Prolonged glance time has to be prevented by using an appropriate information scheme and by reducing interaction complexity to a minimum.

Introduction

Head-Up Displays (HUD) overlay the front scene with information. They were initially developed for use in military aircraft to enable pilots to see any relevant information whilst keeping their eyes on the flight scene. This technology has also been used in vehicles since the 1980s. In contrast to aircraft, where the information is shown in a combiner, HUDs in cars use the automobile windshield to reflect the virtual image in the driver's primary field of view. The construction is displayed in figure 1.

The most striking advantages of HUDs are a minimized gauge reading time (Gengenbach, 1997) and a significantly better recognition of sudden or unexpected events in the driving scene compared to Head-Down displays (HDD) like instrument clusters (Kiefer, 2000, Horrey, 2003).

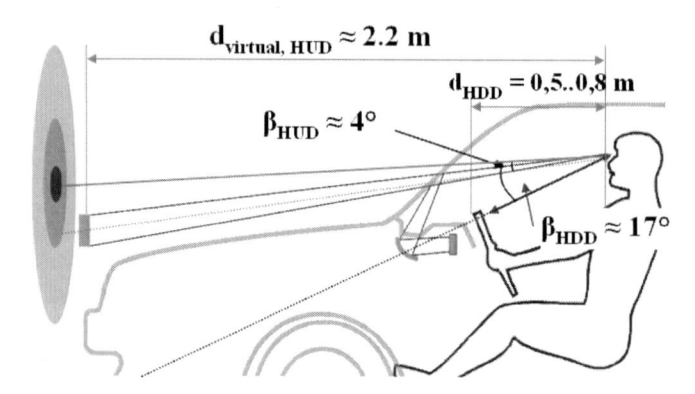

Figure 1. Virtual image of Head-Up Display

These effects are based on a smaller angle between the HUD and the street as well as the lower cost of optical adaption due to the same background brightness. A further reduction of accommodation effort comes with the virtual distance of the image in an HUD, which is about 2.20 m as shown in figure 1. The impact of all these aspects correlates with the driver's age (Kiefer, 2000; Adis, 1998). Older drivers especially appreciate enhanced comfort and safety by using HUDs.

On the other hand a virtual overlying image is eye-catching and might potentially distract the driver. Cognitive Capture is the most frequently mentioned disadvantage of HUDs (Gish & Staplin, 1995; Tufano, 1997; Weintraub & Ensing, 1992): the driver's attention lies on the HUD instead of lying on the driving scene. Thus when the driver switches from the HUD to the primary task, missed external targets or delayed responses can result. Hence former studies recommend the sole presentation of static information supporting the driver's primary driving task (Adis, 1998).

Finding the balance between the advantages of a HUD, i.e. safety and comfort, and the risk of cognitive capture is still in progress. The next step in the development of advanced Human-Machine-Interfaces on HUDs is the user's interaction with information. Here, driver distraction is a major topic of research and can be quantified by measuring attention demand, driving performance and visual behaviour. Thus, the study presented here investigates the effects of different types of interactions with a HUD compared to a common HDD – the Central Information Display (CID) at the top of the centre stack.

Questions

According to various standards, guidelines and directives, e.g., ESoP (2006), systems and performed secondary tasks in a car should not distract or visually entertain the driver. The aim is that interaction with system displays and controls remains compatible with the attention demands of the driving task. Cognitive and visual demands caused by an interaction, depend first of all on the type of interaction; such as scrolling down a list, setting up alphanumeric characters or adjusting analogue bars. Secondly, it depends on the construction of the menu (hierarchy orientated or object orientated structure), the number of functionalities as well as width and depth of the menu. Last but not least, design and amount of presented information are crucial. The structure of the menu is not supposed to be considered at this time.

An experiment was designed to measure the amount of caused distraction by different interaction types when using an HUD in contrast to a CID. For this reason, a wide spectrum of measures were considered:

1. The amount of visual and attention demand caused by different interaction types in the HUD is lower than in the CID, as the HUD is closer to the driver's line of sight and demands less accommodation time.
2. Depending on the type of interaction, the HUD causes different eye movements and glance behaviour. Drivers tend to look longer into the HUD than into the CID, as they feel safer using the HUD because the eyes are still on the road.
3. There is a difference in total task time and the number of interaction steps between the HUD and the CID depending on the interaction type.
4. Longitudinal and lateral driving performance is better under the HUD condition and depends furthermore on the interaction type.
5. In terms of acceptance, likeability etc, subjects prefer an interaction performed with the HUD.

Method

Evaluating secondary-task interactions can be done by using driving simulator scenarios like Lane Change Tasks (Mattes, 2003) or a Car Pursuing Tasks. For this study, a driving simulator course containing a simple pursuing task on a three-lane motorway was created. Keeping a constant distance of 50 meters, subjects had to follow a lead vehicle driving at a constant speed of 100 km/h in the middle lane while performing secondary tasks.

Apparatus and tasks

Secondary task
Subjects had to solve secondary tasks consisting of three basic principles of interaction. Each type of interaction was divided in two subtypes:

- *Scrolling through lists*: short (3-15 items) vs. long (15-30 items)
- *Adjusting*: analogue bars vs. digital values
- *Setting up alphanumeric characters*: numbers vs. letters by using a speller

To fulfil the task "scrolling through lists", participants had to choose the correct category (1 of 5) and scroll through the list to find the mentioned item. By pushing the control element, the secondary task was ended.

For the task "adjusting" the scale was presented by an analogue bar or just by digital values. Selecting one of 3 categories, e.g. climate, the user was asked to decrease the temperature to a certain scale. During "setting up numbers", participants were asked to make a call. Thus they dialed their own number on the interface via the control element. By turning the dial they could skip between the numbers and by pushing the button for selecting the number. Then finally the call was made by pushing to the right. To "set up letters" a speller was used. This means, choosing a letter the system creates a list with all matches. Consequently it is possible to choose the right item by scrolling through the list after only two or three letters.

The layouts of the scrolling and the adjusting tasks were particularly designed for the HUD. Interfaces were clearly arranged with a minimum presented information and details. Interface design for setting up alphanumeric characters was adopted from the iDrive menu of the current BMW 5. iDrive was introduced by BMW in 2001 and consists of a hierarchical menu with complex layouts and detailed information.

Display condition
The tasks were presented in both displays – Head-Up Display (HUD) and Central Information Display (CID). The CID was positioned in the upper middle section of the dashboard, whereas the HUD was shifted 12° to the right of the driver's field of vision and 4° below the horizon, as shown in Figure 2. The control element for interaction was a single turn-and-push dial in the centre console, which could be operated with one hand.

Experimental design and independent variables

The experiment was a 3x3x2 factor in-between design with the following conditions:

- First independent variable (*display condition*): HUD, CID and NO DISPLAY (baseline)
- Second independent variable (*type of interaction*): scrolling through lists, adjusting values, setting up alphanumerics
- Third independent variable (*subtype of interaction*): long vs. short lists, analogue vs. digital values, setting up alphanumerics with or without speller

Thus, it is possible to distinguish between the different types of interactions (second independent variable) on one hand and the effects caused by the specific design (interaction subtype: third independent variable) on the other hand. All tasks were repeated several times during one drive, each participant driving six experimental drives.

PDT interferes with normal eye behaviour. Thus, subjects had to solve each interaction task without an additional peripheral detection task (PDT) in two separated drives (one per display). In this condition the eye tracking system was used.

The conditions led to a design containing eight simulator drives. Two of them – one pure driving task without interaction and a second one with an additional PDT – were used as baselines. The remaining six drives where split into three drives for each type of display. These contained all tasks with all versions. Table 1 gives an overview of the experimental design.

Table 1. Experimental design

Number of drives	Interaction in HUD	Interaction in CID	No interaction = Baseline	Total
PDT	1 drive (3 types of interaction x 2 versions)	1 drive (3 types of interaction x 2 versions)	1	3
Without PDT	1 drive (3 types of interaction x 2 versions)	1 drive (3 types of interaction x 2 versions)	1	3
Total	2	2	2	6

Measures

In order to take a closer look at distractions caused by interactions with HUDs, several metrics were considered: longitudinal and vertical driving performance, time and number of steps required for the interaction, reaction time and number of missed PDT stimuli, number and mean duration of glances and eye movement behaviour. Subjective opinions on the workload, acceptance and likability were also added to the list of objective criteria.

Interaction activity
Among others, the quality of an interaction can be defined by the time and number of steps required for the interaction, as well as the number of operating errors.

Peripheral Detection Task
The PDT-method measures the driver's mental workload and visual distraction by other on-road vehicles. Drivers must respond to random optical cues presented in their field of view. The more demanding the task, more cues will be missed and PDT response times prolong. The results of Martens and Van Winsum (2000) suggest that PDT measures the variations in selective attention which increases with the driver's workload (cognitive tunnelling). In order to analyse differences in cognitive and visual workload between HUD and CID, PDT stimuli in form of red dots were presented randomly at seven different spots in the driving scene (exact positions are shown in figure 2). They appeared for 2 seconds with an inter-stimuli interval between three and five seconds. A button on the left side of the steering wheel had to be pressed every time the participant detected a PDT stimulus. The reaction time and the number of missed PDT stimuli are considered for analysis.

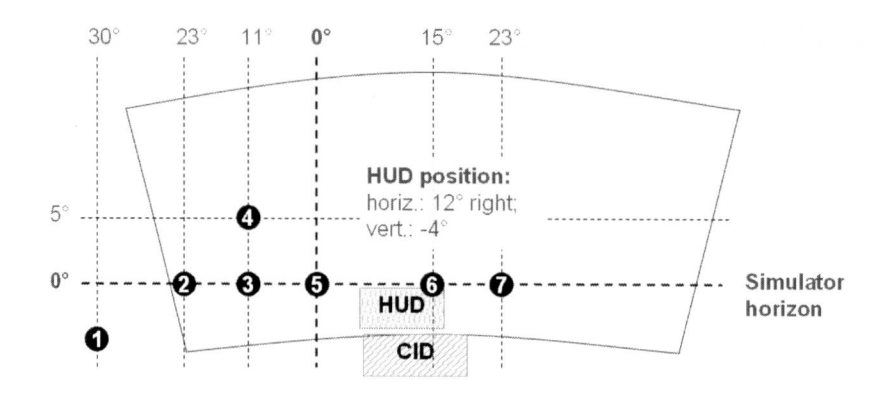

Figure 2. Positions of displays and PDT-Stimuli

Eye movements
In order to analyze different glance behaviour and duration of glances depending on display condition, a head-mounted eye-tracker with two cameras was used. One camera captured the images of the right eye; the other recorded the field of view. A gaze movie was subsequently created by the overlay of the simultaneously recorded eye video stream and the field-camera video stream.

PDT interferes with usual eye movement behaviour. For this reason, gaze movies of 21 participants were recorded without PDT (the remaining 15 movies could not be used due to technical difficulties). Afterwards, the duration of glances, mean number of glances, total glance times and glance behaviour were analyzed.

Table 2. Dependent variables

	Metric 1	**Metric 2**	**Metric 3**	**Metric 4**	**Metric 5**
Driving performance	SDLP	TTC	Keeping distance	Mean velocity	Variation in velocity
PDT	Reaction time	Hitrate			
Interaction activity	Total task time	Number of required steps	Number of operational errors		
Eye movements	Mean glance duration	Mean number of glances	Total glance duration	First look: mean gaze time	
Subjective questionnaires	DALI Workload	Acceptance	Likeability		

Subjective opinion
Objective measures of workload should always be verified by the subject's own opinion. The standardized questionnaire "Driving Activity Load Index" (DALI) is used to measure subjective workload in different categories like effort of attention,

visual demand or stress. It contains a six point rating scale for workload with a proven sensitivity (see Cherri et al., 2004, and Pauzié et al., 2007, for details).

Last but not least, acceptance and likeability are essential for automotive interfaces. Participants had the chance to give their opinion on whether the interaction in a HUD is reasonable or unsafe, which display they prefer and what type of interaction they would like to perform in a HUD. Furthermore, they had to estimate the distraction they experienced while interacting with both displays. They were also asked whether the HUD masked the driving scene in any way. Table 2 gives an overview of the applied methods of measurement.

Experimental procedure

Participants
36 participants (32 male, 4 female) took part in the experiment. The average age in the group was 38.8 years, with a standard deviation of 6.8. All subjects were licensed drivers and experienced users of HUDs and the BMW-iDrive. Thus, all subjects were used to perform several secondary tasks while driving.

Procedure
The study took place in July 2007 in the static driving simulator of BMW with an all-round field of view (figure 3). At the beginning of the experiment, the participants were to fill out a demographic questionnaire. Afterwards they were asked to follow the leading vehicle and keep a distance of 50 m. All secondary tasks were practiced while driving after a test run during which the participants familiarized themselves with the simulated testing environment and PDT.

Figure 3. BMW driving simulator

Subsequently, subjects continued with the experimental driving task (follow the leading vehicle at 100 km/h at 50 meters apart) and performed secondary tasks under all display conditions, with and without PDT. The type of display (HUD vs. CID) and type of secondary task were randomly permuted between the subjects. The first run was a baseline. After two drives with one display type, the subjects filled in the DALI and acceptance questionnaires. Before the next two drives using the other display, a second baseline was recorded. At the end of the experiment, DALI and questionnaires concerning acceptance and likeability were completed.

Results

Interaction activity: Total Task Time (TTT)

The Total Task Time is measured between the first step and the last step of the interaction. It contains all steps and errors made by the participants and indicates the demand caused by interaction.

Using the HUD, TTT is significantly shorter for scrolling down short lists (T(0.95, 28)= −2.96 p=.006), long lists (T(0.95, 28)= −2.37 p=.025) and for adjusting analogue bars (T(0.95, 27)= −2.35, p=.027). No difference was found between the two displays for the other tasks.

Table 3. Mean Total Task Time

Type	Scrolling down lists		Adjusting		Setting up alphanumerics	
Sub-type	short	long	analogue bars	digital values	numbers	letters
HUD	11.8 (sd.3.9)	18.8 (sd.4.6)	15.6 (sd.2.9)	17.2 (sd.3.4)	43.5 (sd.25.8)	30.7 (sd.10.5)
CID	12.4 (sd.3.6)	19.4 (sd.4.8)	17.8 (sd.6.1)	16.6 (sd.2.5)	41.8 (sd.12.5)	29.6 (sd.9.5)

Driving performance

Longitudinal metrics like Time to Collision (TTC), distance to the lead vehicle, mean velocity and variation in velocity did not differentiate between the two types of display. More demanding tasks like setting up numbers caused a higher mean variation in velocity for both displays.

A difference in driving performance was found in the Standard Deviation of Lateral Position (SDLP). SDLP is lower for scrolling down lists and adjusting under the HUD condition, but only shows significant differences for analogue adjustment (W(33)= −2.939, p=.003).

Table 4. SDLP – Standard Deviation of Lateral Position

Type	Scrolling down lists		Adjusting		Setting up alphanumerics	
Sub-type	short	long	analogue bars	digital values	numbers	letters
HUD	0.19 (sd.0.13)	0.2 (sd.0.21)	0.16 (sd.0.07)	0.25 (sd.0.22)	0.3 (sd.0.23)	0.34 (sd.0.42)
CID	0.34 (sd.0.23)	0.25 (sd.0.24)	0.3 (sd.0.22)	0.24 (sd.0.17)	0.26 (sd.0.13)	0.30 (sd.0.31)

PDT performance

The Peripheral Detection Task is a sensitive method for measuring workload and driver distraction (Martens & Van Winsum, 2000). Participants were to react on PDT-stimuli presented on seven different spots in the driving scene during 5 of 8 drives.

Reaction time

The mean reaction time referring to seven PDT stimuli positions while setting up numbers is presented in figure 4. Significant differences compared by a pair wise T-tests are marked by a star.

Significant differences are found for scrolling through lists, adjustment of values and setting up numbers. No differences were detected for setting up characters and adjusting analogue bars.

PDT reaction time for setting up numbers

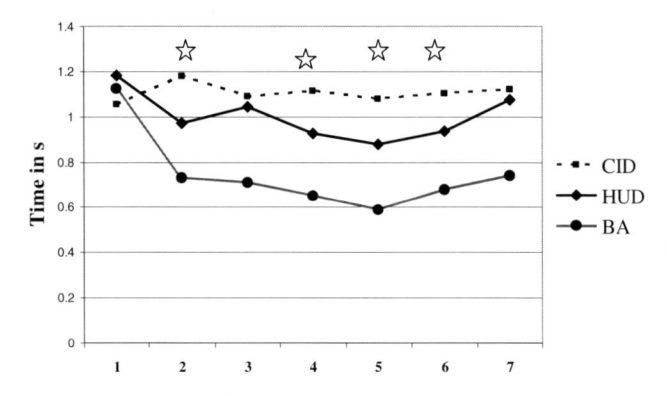

Figure 4. PDT reaction time. On the x-axis: PDT Stimuli positions

Compared to the baseline, the cognitive and visual demands caused by an interaction are considerable.

Misses

The missing rate is calculated by the number of missed targets divided by the total number of presented PDT-stimuli on a certain position during a task. The missing rate while adjusting digital values is shown in figure 5. Significant differences between HUD and CID are marked by a star.

The percentage of missed PDT-stimuli under a HUD condition while scrolling through lists and adjusting values is significantly lower compared to the usage of a CID. For setting up alphanumeric characters there is no difference between both display conditions.

PDT missing rate
while adjusting digital values

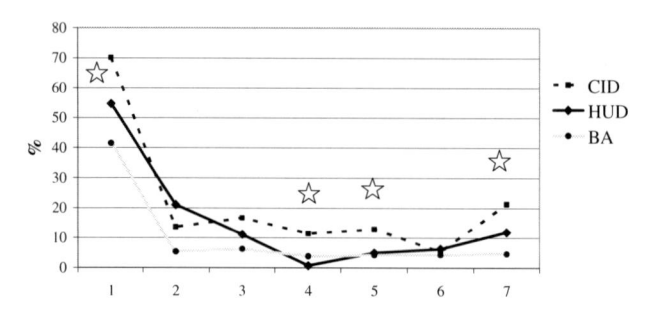

Figure 5. PDT Misses. On the x-axis: PDT Stimuli positions

Eye movements

Glance duration

The usual metrics of eye movement are mean glance time, mean number of glances and total glance time while performing secondary tasks. These metrics were considered and compared between the two displays for all interaction types. One criterion for interpreting eye movement as considered by AAM guidelines suggests an upper limit of 2 seconds for the 85th percentile of all glances caused by a secondary task in vehicles. Table 3 shows all mentioned glance metrics.

Table 5. Metrics of eye movement

Task	Display type	Mean glance time	Num. of glances	Median / Modus	85th percentile	Sig diff
Scrolling down lists	HUD	1.09(sd 0.31)	6.12	1.10 / 0.64	1.33	-
	CID	1.06(sd.0.3)	6.02	1.00 / 0.64	1.27	
Adjusting analogue bars	HUD	0.92(sd 0.34)	6.17	0.84 / 0.5	1.23	-
	CID	0.94(sd 0.36)	6.05	0.79/ /0.57	1.25	
Adjusting digital values	HUD	1.04(sd 0.32)	6.12	0.89 / 0.68	1.42	-
	CID	1.02(sd 0.47)	5.7	0.84 / 0.60	1.19	
Setting up characters with speller	HUD	1.85(sd 0.65)	13.36	1.85 / 0.83	**2.21**	X
	CID	1.3 (sd 0.43)	15	1.11 / 0.82	1.82	
Setting up numbers	HUD	1.62(sd 0.64)	14.1	1.43 / 0.8	**2.17**	X
	CID	1.22(sd 0.37)	16.1	1.16 / 0.82	1.51	

A significant difference between the two displays was found for setting up characters (T(0.95,14)=5.86, p<.001) and numbers (T(0.95,14)=4.60, p<.001). In addition to significantly longer mean durations of glance, these two tasks did not comply with the AAM 85th-percentile (marked in Table 3) criterion on the HUD condition. With CID, all mean glance durations are below the AAM Criterion.

Setting up characters was the most demanding secondary task and caused most of the measured distraction in all other metrics. Scrolling through lists and adjustment of values did not cause longer glance times at a HUD than at a CID.

The following two diagrams in figure 6 and figure 7 show histograms of all glance durations for adjusting analogue bars and setting up numbers. It is obvious that the type of interaction does not have any influence on eye movement behaviour while using a CID. Participants are aware of risks caused by eye-off-road time and they limit the interaction to a minimum. Using a HUD may make them believe that they can look longer at the display if the task is too demanding, given that they can continue to maintain an overview of the driving scene.

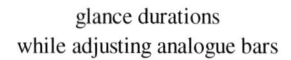

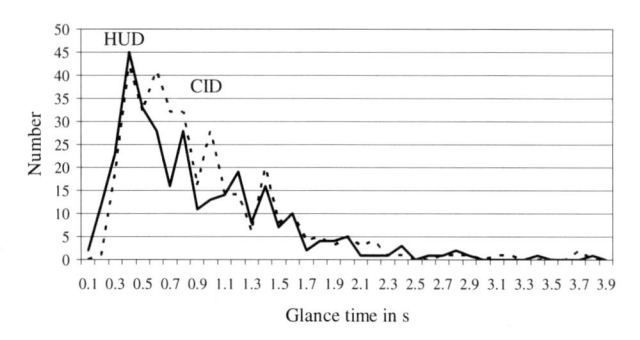

Figure 6. Histogram of glance durations while adjusting

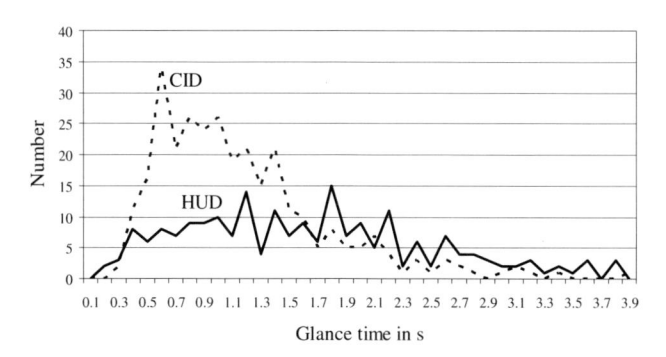

Figure 7. Histogram of glance durations while entering numbers

Eye movement strategies
Upon instruction for a secondary task, the glance at a display is always used to scan the display, realize the instruction and plan the next step. The subsequent glance

aims at the driving scene. During the next look at the display the interaction is started. Thus, the first glance at the display is used to recognize the presented information and the task. Its mean duration for all tasks was compared between the two displays and is shown in figure 8.

Mean duration of first look

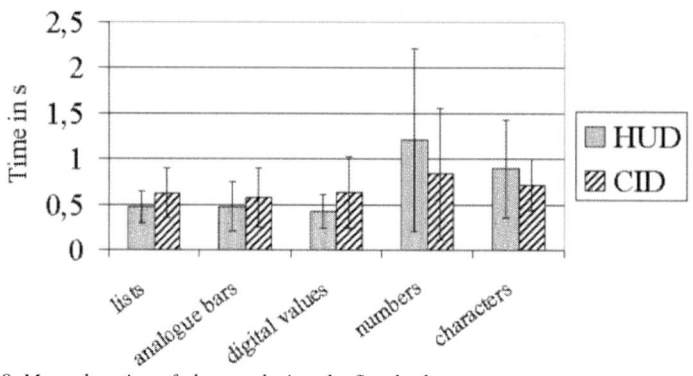

Figure 8. Mean duration of glances during the first look

Tasks were tested with a Wilcoxon test and revealed significant differences for setting up numbers (W(17) = –2.42, p= .016) and characters (W(19)= –1.95 p=.05). The first look at the HUD was longer than at the CID. No differences were found for all other types of interaction.

As a consequence, the design of the screen has a greater impact when presented in a HUD than when presented on a CID. Simple graphical layouts can be recognized faster in the HUD whereas complex layouts cause significantly longer glances.

Subjective opinion

Seventy-two percent of all participants preferred the HUD for all types of interaction. They also thought that the presentation of interaction was more reasonable in HUDs (T(0.95,33)= –4.881, p<.001) and less risky (T(0.95,33)= 4.7, p<.001) than in CIDs.

DALI identified a significantly lower workload for the HUD condition in the following categories: effort of attention (T(0,95,33)=4.31, p=.0011), visual demand (T(0,95,33)= 4.42, p<.001), stress (T(0,95,33)=4. p=.0012), temporal demand (T(0,95,33)=3.44, p<.001) and interference(T(0,95,33)=4.22, p<.001).

The subjects felt less distracted by the HUD and indicated that the driving task was less affected by interaction when using the HUD (T(0,95,33)=5.92, p=.00)).

Discussion

Interaction activity: Total Task Time is less for the HUD condition than for the CID condition for easy tasks. In particular, graphically simply designed user interfaces like analogue bars and lists are more suitable for being presented in HUDs than complex screen layouts.

Driving performance: Considering driving metrics, interactions with the HUD are not disadvantageous in comparison to CID. Especially for adjusting analogue bars, the HUD is to be preferred due to the significantly lower SDLP.

PDT: The advantages of HUDs mentioned in the introduction can be quantified by PDT measurement. Visual and cognitive demand caused by an interaction depends on the type of display and the layout of the screen. Scrolling through lists and making adjustments indicate lower cognitive and visual demand for interaction in the HUD compared to the CID. In contrast, setting up alphanumeric characters is the most demanding and distracting of all tasks, but does not result in a significantly lower miss rate for the HUD condition. All PDT results also confirm the thesis that scrolling through lists and making adjustments are types of interactions which are suitable for HUDs with quantified advantages.

Eye movements: Eye movement results indicate that the design and type of interaction have a greater impact on the driver's visual behaviour when using HUDs. As an attractive display, it can cause more distraction than a usual CID. Participants look longer at the HUD than at the CID when the task or the screen design is complex. However, with easy tasks and well structured information, it is possible to avoid long glances at a HUD and use all associated benefits.

Conclusion

The current study was intended to lay a foundation for future applications of the HUDs in vehicles. The effects caused by different types of interactions with a HUD were assessed by analysing a number of different metrics.

Concise and easy interactions like scrolling down lists and adjusting analogue bars are suitable to be performed in a HUD. Significant HUD advantages are a partly better driving performance (SDLP) and a more efficient interaction (TTT). Lower cognitive and visual demand measured by Peripheral Detection Task (PDT) and a lower subjective workload add to an overall positive result for the HUD. Using an appropriate design, the glance durations for the tasks of scrolling through lists and adjusting analogue bars can be reduced.

References

AAM - Statement of Principles, Criteria and Verification Procedures on Driver Interactions with Advanced In- Vehicle Information and Communication Systems Version 2.0 (2002). Driver Focus-Telematics Working Group

Adis - *Advanced Traveller Information Systems and Commercial Vehicle Operations Components of the Intelligent Transportation Systems* (1998). Head-Up Displays and Driver Attention for Navigation Information. Publication No. FHWA-RD-96-153. Georgetown, USA: U.S. Department of Transportation Federal Highway Administration Research Development..

Cherri C., Nodari E., & Toffetti A (2004). *Review of existing Tool and Methods*. Deliverable D2.1.1 of the EU-Project aide (adaptive integrated driver-vehicle interface). Orbassano, Italy: CRF

Pauzié A. & Manzano J. (2007). *Evaluation of driver mental workload gacing new in-vehicle information and communication technology*. Pater No. 07-0057, INRETS – National Research Institute on Transport and Safety, LESCOT – Laboratory Ergonomics & Cognitive Sciences in Transport, France.

ESoP - Commission of The European Communities (2006): *Commission Recommendation of 22 December 2006 on safe and efficient in-vehicle information and communication systems: Update of the European Statement of Principles on human machine interface*. Brussels.

Gengenbach, R.(1997). *Fahrerverhalten im Pkw mit Head-Up-Display: Gewöhnung und visuelle Aufmerksamkeit*. VDI Verlag, Düsseldorf.

Gish, K. & Staplin, L. (1995). *Human Factors Aspects of Using Head Up Displays in: A Review of the Literature*. Report No. DOT HS 808 320, Washington: U.S. Department of Transportation.

Horrey, W., & Alexander A.L., & Wickens C.D. (2003). *Does Workload Modulate the Effects of In-Vehicle Display Location on Concurrent Driving and Side Task Performance?* Michigan: DSC North America Proceedings.

Kiefer, R. (2000). Older Drivers' Pedestrian Detection Times Surrounding Head-Up Versus Head-Down Speedometer Glances. In A.G. Gale et al. (Eds.), *Vision in Vehicles -VII* (pp.111-118). Amsterdam: Elsevier.

Mattes, S. (2003). The Lane Change Task as a Tool for driver Distraction Evaluation. In H. Strasser, H. Rausch and H. Bubb (Eds.), *Quality of Work and Products in Enterprises of the Future*. Stuttgart, germany: Ergonomia Verlag.

Martens M. & Van Winsum W.(2000) *Measuring distraction: The Peripheral Detection Task*. Soesterberg, The Netherlands: TNO Human Factors

Tufano, D. (1997). Automotive HUDs: The overlooked safety issues. *Human Factors, 39,* 303-311.

Weintraub, D. & Ensing, M. (1992): *Human Factors Issues in Head-up Display Design: The Book of HUD*. Wright-Patterson AFB, OH: CSERIAC.

Representation of driver's mental workload in EEG data

Shengguang Lei, Sebastian Welke, & Matthias Roetting
Berlin Institute of Technology, Berlin
Germany

Abstract

Prior researches indicate that driver overload is one of the most important contributors to traffic accidents. The present study outlines the contribution of driving speed to driver's mental workload. A standardized Lane Change Task (LCT) was used as driving task. Participants were asked to perform the lane change task at different speed levels which induced different levels of task load.

Electroencephalography (EEG) technique was used to record the electrophysiological response invoked by LCT from 30 participants. The recorded data were divided into epochs related to the lane change commands and the analysis of Event Related Potentials (ERP) revealed that LCT evoked cognitive response with CNV, P2, N2, P3b, etc. Additionally, statistical analysis showed that the amplitude of P3b decreased with the task load. This finding indicates that the amplitude of ERP component could be used for representing driver's mental workload.

Introduction

In the last decades, a large number of researches have been conducted to investigate driver workload using different methods, such as subjective measurement (Pauzié & Pachiaudi 1997), performance measurement (De Waard, 1996), as well as physiological parameters, such as Electroencephalography (EEG) or Electrocardiography (ECG, Wilson et al., 1988; Piechulla et al., 2003; Chen et al., 2005). EEG as the measurement of brain electrical activity recorded from electrodes placed on the scalp provides a promising way for driver mental workload monitoring. Characteristic changes in the EEG and event related potentials (ERPs) that reflect levels of mental workload have been identified (Wilson et al., 1988; Gevins et al., 1998; Raabe et al., 2005). Raabe et al. (2005) revealed a decline of the amplitude of P300 evoked by a secondary oddball task when the primary task load or task difficulty increased. Other researchers calculated the power spectral density of EEG signals using fast Fourier transform (FFT) to examine the change of frequency characteristics. Such an approach allows understanding how the ratio of a specific frequency band (i.e. alpha band) changes when the mental work level changes (Gevins, et al., 1998).

In D. de Waard, J. Godthelp, F.L. Kooi, and K.A. Brookhuis (Eds.) (2009). *Human Factors, Security and Safety* (pp. 285 - 294). Maastricht, the Netherlands: Shaker Publishing.

Our group aims to develop new approaches of driver state monitoring based on brain signals. Assessing the driver mental workload is one of our objectives. This article presents our first step – a study of EEG measurement as an index of driver's mental workload in a stimulated driving task- Lane Change Task (LCT) (Mattes, 2003; Burns et al., 2005). The workload levels in LCT were manipulated by three operation speeds (low, moderate and high). Senders et al. (1967) and McDonald et al. (1975) both presented that the attention demand of the road increased with the operating speed. Consistent result was as well obtained in Cnossen's study that the mental workload increased according to the speed level (Cnossen et al., 2000). A pre-study verifying if the workload increases with speed in LCT was conducted using the subjective measurement, NASA-TLX (Hart & Staveland, 1988), and performance measurement.

In the formal EEG study, a four-block experiment was carried out. Workload levels were manipulated in two different ways – (1) variation of speed levels and (2) changes of difficulty of a secondary auditory task, the Paced Auditory Serial Addition Task (PASAT, Gronwall, 1977). However, this article focuses on the effect of operation speed on driver's mental workload.

Methods

Lane Change Task

The Lane Change Task (LCT) was initiated by the project ADAM (Advanced Driver Attention Metrics) as an easy-to implement, low-cost, and standardized methodology for the evaluation of the attention associated performing in vehicle tasks while driving (for details, see Mattes, 2003; Burns, et al., 2005).

Figure 1. Lane Change Task

Participants are required to repeatedly perform lane changes when prompted by road signs (Figure 1). The quality of these lane changes can be evaluated by the difference (mainly based on Mean Deviation) between a normative lane change path and the driver's actual lane change path, which is influenced by the driver's performance of detecting and responding to the road signs as well as their lateral control maintenance. Lane change performance with a secondary task (driving and using the telematics system of interest) is evaluated against a normative model of single task performance, which enables the possibility to evaluate the extent of distraction due to the secondary task.

Paced Auditory Serial Addition Task (PASAT)

The PASAT is a method for triggering cognitive function that assesses auditory information processing speed and flexibility, as well as calculation ability. It was developed by Gronwell in 1977. The PASAT is presented using audio cassette tape or compact disk to ensure standardization in the rate of stimulus presentation. Single digits are presented every 3 seconds (PASAT 3) and the participant must add each new digit to the one immediately prior to it. Shorter inter-stimulus intervals, e.g., 2 seconds or less have also been used with the PASAT but tend to increase the difficulty of the task. The digit is randomly arranged to minimize possible familiarity with the stimulus items when the PASAT is repeated over more than one occasion.

Pre-study: manipulating workload in the Lane Change Task

A pre-study, involving 18 participants ranged in age from 23 to 37 (M=30, S=12.3), was taken to examine the correlation between operation speed and workload. They were required to perform the LCT in three 20 minutes blocks. Each block involved three speed levels 60 km/h, 90 km/h and 120 km/h (Low, moderate and high) lane change trials in random order. After each trial, participants reported their subjective workload by answering the NASA-TLX questionnaires. Figure 2 shows the results of pre-study.

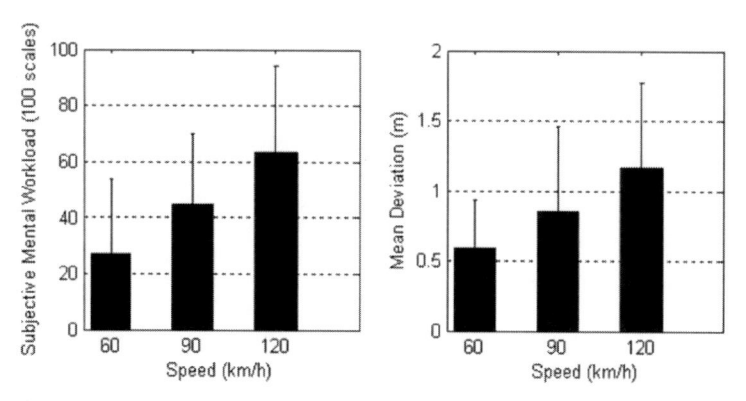

Figure 2 Correlation of subjective mental workload and performance to speed levels in LCT

Although the scores of workload and Mean Deviation differed significantly among individuals, there was an obvious speed-associated trend either for subjective task load or for the performance. The subjective mental workload steadily increased with the speed level (F (2, 51) = 27.35, p<0.001). The increase of mean deviation (F (2,

51) =84.96, p<0.001) indicated that there was a decline of performance with high speed. This finding demonstrated that the manipulation of task load by the speed level was successful.

Monitoring driver's mental workload based on EEG

Participants

Overall, 30 participants between the ages of 20-34 were assessed (M=26.1, SD=11.8). All individuals reported being free of neurological/psychiatric disorders and received a cash payment for their participation.

Experiment

The formal experiment involved four blocks. The first block was the primary driving task. Participants were requested to perform the LCT under three different speeds 60 km/h, 80 km/h, 100 km/h, which represented three task load levels respectively (low, moderate and high). (In the formal study, we transferred the experiment from the original LCT device employed in the pre-study to a driving simulator and the steering wheel was more sensitive to participants input. Thus, we slightly changed the speed settings.) The second block was PASAT under two paced conditions: slow and fast (the numbers were presented every 5 and 3 seconds). Participants were requested to calculate the numbers and report the results. The third block was the combination of the primary task and secondary task. Participants were requested to do the calculating while performing the LCT. However, they were instructed that the primary task was more important. The last block was another dual task. Participants were requested to press a button embedded in the steering wheel at the moment they started to take the action of changing.

While performing the LCT the brain activity was recorded with 32 Ag/AgCl impedance-optimized electrodes (ActiCap, Brain Products), referenced to the nasion, sampled at 1000 Hz and wide-band filtered. Electromyogram (EMG) was recorded from both forearms with two bipolar electrodes. The horizontal and vertical eye-movement was recorded using the Electrooculogram (EOG). For data recording the Brain Vision Recorder by Brain products was used. This article concentrates on the EEG data obtained in the first experiment block, the single task LCT with three different speeds.

Data analysis

EEG data analysis was performed using EEGLAB 6.03, a freely available open source toolbox running under Matlab 7.3.0. A detailed description about EEGLAB is provided by Delorme and Makeig (2004). For pre-processing, data was down-sampled to 500 Hz to save computation time, and was then digitally filtered using band pass filter (pass band 0.5 to 40 Hz) to minimize drifts and line noise. Then, data was average rereferenced to avoid the influence of an arbitrary chosen local reference (Nunez, 1981). Finally, data epochs were extracted from 2000 ms before stimulus, the command for lane change direction, until 2000 ms after stimulus and the average of time range [-2000ms, -1000] as the baseline was removed from every epoch.

EEG recordings involve plenty of artefacts, such as eye movements, muscle noises, cardiac signals and line noises, etc. In the present study, Independent Components Analysis (ICA) was used to improve the data quality and to identify the ERP components. ICA decomposes EEG data into temporally independent and spatially fixed components, which account for artefacts, stimulus and response locked events and spontaneous EEG activities. Recently, it has been considered as a powerful tool for EEG components identification and artefacts removal (Makeig, et al., 1996; Delorme & Makeig, 2004).

After the calculation of independent components (ICs), we followed the following procedures to remove artefacts. (1) The correlations coefficient of ICs and EOG, ICs and EMG were calculated and the highest correlated ICs were removed. (2) The correlations of the single-trial ERP and the mean ERP for each electrode site, condition and subject were calculated and the number of trials which high correlated to mean ERP was recorded (correlation coefficient was higher than the threshold 0.2). The ICs which had less than 20% trials with high correlation to the mean ERP were removed.

Analysis of Variance (ANOVA) was employed to verify the significance of difference in amplitude of the component of interest under three conditions and the paired conditions. The Post Hoc Test was used to offer a detailed view of the differences among the three conditions.

Results

Independent Components Analysis

ICA decomposes the EEG into temporally independent and spatially locked components. These components could either be the result of task events or stimulus, or account for the artefacts. Figure 3 shows some typical IC examples evoked by LCT.

As shown in Figure 3, some of these components showed a high coherence across over all trials and the peaks seemed to be time locked, e.g. (a) and (b), and they were likely evoked by the involved events. Empirically, we could conclude that Figure 3 (a) shows a P2 or a early P3a liked positive peak, since it happened around 250mm and was located in the frontal-central area, while Figure 3(b) demonstrates a P3b-liked positive peak located in the posterior area and with a latency of 500ms. Both of them showed a high coherence across all trials. However, some of them presented low coherence across over all trials and the peaks were randomly located at the time course, e.g. (c) and (d), and might account for artefacts. Figure 3 (c) shows the eye movement artefact and Figure 3 (d) shows eye blink artefact, which is characterized by a location of frontal area of the head and a strong power activity in the low frequency and inconsistent time course across trials.

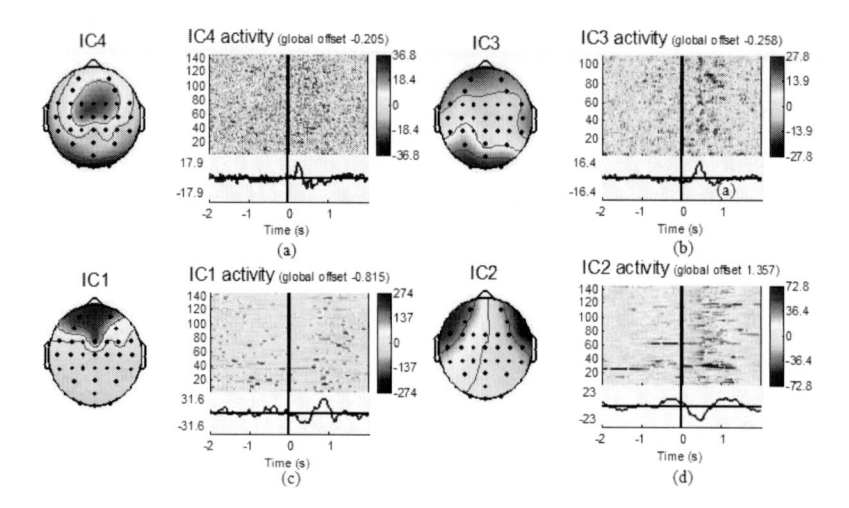

Figure 3. Independent components which demonstrate the typical components obtained in the present study (left: topography of the IC and right: the corresponding averaged signal). A colour image is available on http://extras.hfes-europe.org

ERP evoked by LCT

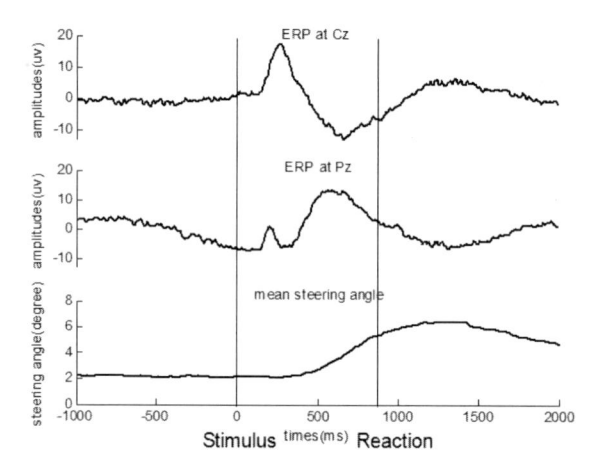

Figure 4. Artefacts-corrected ERP and Steering angle in LCT

Artefacts-corrected data showed a prominent positive peak around 250ms (P250) after the stimulus onset at the frontal-central area of the brain. This peak had its maximum (21.6 μv) at FCz and still existed at the parietal area but with smaller amplitude and earlier latency (200 ms, see Figure 4). Another prominent positive peak occurred at the latency of 550 ms (P550) at the parietal and occipital area (see Figure 4) and it had a maximum (20.2 μv) at POz. The steering angle recorded in LCT indicated that both of these two components happened before the participants took an action of lane change, which happened around 800ms after the stimulus onset. Additionally, there seemed still other components involved, such as the contingent negative variation (CNV), an increasing negative shift that is associated to an expected "go-signal", and N2, a negative maximum around 250 ms at the parietal and occipital area.

ERP and workload

Figure 4 shows the artefacts-corrected ERP under three speed conditions. It is well known that the amplitude and latency of some ERP components might be used for the index of workload levels. As shown in Figure 5, it seemed that there were no differences in the amplitude of frontal-central component P250 among these three task conditions, whereas there seemed a decrease in amplitude of the parietal-occipital component P550 as the driving speed increased.

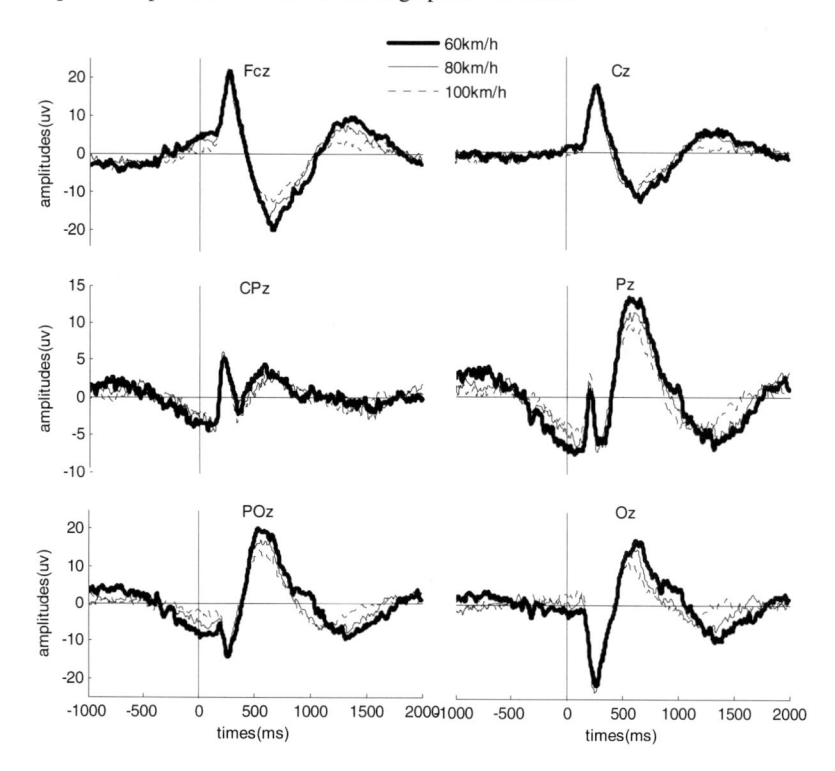

Figure 5. Artefacts corrected ERP. For the frontal-central P250, the amplitude was evaluated by the average of 200-300 ms and for the parietal-occipital P550, the amplitude was evaluated by the average of 450ms-650ms

Statistic analysis demonstrated a consistent finding. For P250, no differences were obtained in its amplitudes (F(2,23)<1, NS at Fcz; F(2,23)=1.36, NS at Cz). However, the differences in amplitudes of P550 ms under three speed conditions were significant at the parietal-occipital area (F(2,23)=12.83, p<0.0001 at P4;

$F(2,23)=6.66$, $p<0.01$ at POz; $F(2,23)=6.25$, $p<0.01$ at Oz). Post hoc test indicated there were significant differences between speed conditions 60 km/h and 80 km/h, 60 km/h and 100 km/h, but no difference between conditions 80 km/h and 100 km/h at the location of POz, Oz. But at location of P4, the differences between the three pairs of speed conditions were significant. Furthermore, statistic analysis showed no difference in the latency both for these two components. ($F(2,23)<1$, NS at Cz; $F(2,23)<1$, NS at Pz; $F(2,23)<1$, NS at Oz).

Discussion

What are these components? A Task Analysis

Lane Change Task is a typical two-stimulus task paradigm. In Lane Change Task, a white sign without any information is first presented to the participant and could be considered as a warning stimulus S1, which draws the participant's attention towards preparation for command perception. Later, the road sign is presented with some crosses and arrows, which indicate which lane the participant should change to, and this could be the second stimulus S2 allowing participant's perception of lane change direction and readiness for action taking. These two stimuli are both visually presented.

So far, a large variety of researches investigate the visual stimulus evoked EPRs. Generally, there is a late wave complex consisting of an enhanced P2 (200-250ms), a negative peak (designated in literature as N200 or N250, N2) and a large prominent positive peak (P300). (Halgren, et al. 1995; Giger-Mateeva, et al. 1999; Kubová, et al. 2002). P300 has been reported to consist of two sub-components-P3a and P3b. P3a exhibits a frontal/central maximum peak at the latency of 300 ms, whereas P3b is located in central-posterior area with the maximum around the latency of 500ms.

In the present study, both ICA and artefacts corrected ERP demonstrate the existence of a frontal/central positive maximum around 250ms and a positive central-posterior maximum around 550ms. The frontal-central P250 could be either a P2 or an early P3a (200-300ms). P3a seems involved in the non-target and involuntary attention, whereas P2 is usually observed after any visual stimulus and usually has its negative aspect at parietal and occipital area (N2). This highly consists with our finding in the present study. Thus, we rather conclude it as a P2 wave. The central-posterior peak around 550ms is a later P3b, which is reported to be involved in participant's voluntary attention. Additionally, CNV, which is a common cognitive response in the two-stimulus paradigm, is also obtained in the present study.

Effect of Task load on the amplitude of P300

Previous researches have reported that the amplitude of the P300 reflects the amount of attention resources allocated (Polich, 2004). A large number of Oddball paradigm researches indicate that the amplitude of P300 for the target is much larger than that of the non-target P300. That is there seems a positive ratio between the amplitude of P300 and the amount of attention allocated to stimuli. This theory also could be reasonable for the interpretation of results from some dual-task researches. Raabe et al. (2005) reported that the amplitude of P300 decreased with the task load in an oddball paradigm. The P300 was evoked the secondary auditory task including 96 frequent tones and 24 seldom tones. The primary task was set by two task load conditions, self-paced driving and car-following. Admittedly, the primary task in the higher workload level (car following) draws more attention of the subject than in the lower workload level (self-paced driving), which leads to the fact that less attention is allocated to the secondary task and therefore, the amplitude of P300 decreased. In the present study, the Lane Change Task could be also considered as a dual-task. The primary task is to keeping the vehicle in the lane and secondary task is to perceive the road sign for lane change. P3b is generated in response to the visual stimuli in the secondary task. As the speed increases, more attention is paid to the lane keeping and car controlling and less attention is allocated for the visual perception. Thus, the amplitude of P3b is reduced. Nevertheless, there seems no difference in P3b amplitude between conditions 80 km/h and 100 km/h. This might because that the difference of the mental workload under these two conditions is not distinguishable. Nevertheless, the amplitude of P3b, in general, decreases as the task load (speed level) increases in LCT and could be a good candidate for driver mental workload assessment.

References

Burns, P.C., Trbovich, P.L., McCurdie, T., & Harbluk, J.L. (2005). Measuring Distraction: Task Duration and the Lane-Change Test (LCT). *Proceedings of the 49th Annual Meeting of the Human Factors and Ergonomics Society* (pp. 1980-1983). Santa Monica, CA, USA: HFES.

Chen, W.H., Lin, C.Y., & Doong, J.L. (2005). Effects of interface workload of in-vehicle information systems on driving safety. *Transportation Research Record, 1937,* 73-78.

Cnossen, F., Rothengatter, T., & Meijman, T. (2000). Strategic Changes in Task Performance in Simulated Car Driving as and Adaptive Response to Task Demands. *Transportation Research Part F, 3,* 123-140.

Delorme, A. & Makeig, S. (2004). Eeglab: an open source toolbox for analysis of single-trial eeg dynamics. *Journal of Neuroscience Methods, 134,* 9-21.

De Waard, D.(1996). *The measurement of drivers' mental workload.* Ph.D Thesis. University of Groningen. Traffic Research Centre, Haren, The Netherlands.

Gevins, A., Smith, M.E., Leong, H., McEvoy, L., Whitfield, S., Du, R., & Rush, G. (1998). Monitoring working memory load during computer-based tasks with EEG pattern recognition methods. *Human Factors, 40,* 79–91.

Giger-Mateeva, V.I., Riemslag, F.C., Reits, D., Schellart, N.A., & Spekreijse, H. (1999). Isolation of late eventrelated components to checkerboard stimulation. *Electroencephalography and Clinical Neurophysiology Supplement, 50,* 133-149.

Gronwall, D. (1977). Paced Auditory Serial Addition Task: A measure of recovery from concussion. *Perceptual and Motor Skills, 44,* 367-373.

Halgren, E., Baudena, P., Clarke, J.M., Heit, G., Marinkovic, K., Devaux, B., Vignal, J.-P., & Biraben, A. (1995). Intracerebral potentials to rare target and distractor auditory and visual stimuli. II. Medial, lateral and posterior lobe. *Electroencephalography and Clinical Neurophysiology, 94,* 229-250.

Hart, S.G. & Staveland, L.E. (1988). Development of the NASA TLX: results of of empirical and theoretical research. In P.A. Hancock and N. Meshkati, *Human Mental Workload* (pp. 139-183). Amsterdam: North Holland.

Kubová, Z., Kremlácek, J., Szanyi, J., Chlubnová, J., & Kuba, M. (2002). Visual event-related potentials to moving stimuli: normative data. *Physiological Research*, 51, 199-204.

Makeig, S., Bell, A.J., Jung, T.-P., & Sejnowski, T.J. (1996). Independent component analysis of electroencephalographic data. *Advances in Neural Information Processing Systems*, 8, 145-151.

Mattes, S. (2003). The Lane Change Task as a Tool for driver Distraction Evaluation. *IHRA-ITS Workshop on Driving Simulator Scenarios*, Dearborn, Michigan.

McDonald, B., & Ellis, N.C. (1975). Driver workload for various turn radii and speed. *Transportation Research Record, 530*, 18-29.

Nunez, P. (1981). Electric Fields of the Brain: The Neurophysics of EEG. New York: Oxford University Press.

Pauzié, A. & Pachiaudi, G.. (1997). Subjective evaluation of the mental workload in the driving context. In T.Rothengatter & E. Carbonell Vaya (Eds.), *Traffic & Transport Psychology: Theory and Application* (pp. 173-182). Oxford: Pergamon.

Piechulla, W., Mayser, C., Gehrke, H., & Konig, W.(2003). Reducing drivers' mental workload by means of an adaptive man–machine interface. *Transportation Research, Part F, 6*, 233-248.

Polich, J. (2004). Clinical application of the P300 event-related brain potential. *Physical Medicine and Rehabilitation Clinics of North America, 15*, 133-161.

Raabe, M., Rutschmann, R.M., Schrauf, M., & Greenlee, M.W. (2005). Neural Correlates of Simulated Driving: Auditory Oddball Responses Dependent on Workload. In Schmorrow, D. D. (Eds.), *Foundations of Augmented Cognition* (pp. 1067-1076). Mahwah, NJ: Lawrence Erlbaum Associates.

Senders, J.W., Kristopherson, A.B., Levison, W.H., Dietrich, C.W., & Ward, J.L. (1967). The Attentional Demand of Automobile Driving. *Highway Research Record,195*, 15-32.

Wilson, G.F. & O'Donnell, R.D. (1988). Measurement of operator workload with the neuropsychological workload test battery. In P.A. Hancock and N. Meshkati (Eds.), *Human Mental Workload* (pp. 63-115.) Amsterdam: Elsevier Science Publishers.

Effects of a driving assistance system on the driver's functional near-infrared spectroscopy

Kazuki Yanagisawa[1], Hitoshi Tsunashima[1], Yoshitaka Marumo[1],
Makoto Itoh[2], & Toshiyuki Inagaki[2]
[1] Nihon University
[2] University of Tsukuba
Japan

Abstract

In this study driver's brain activity was measured by using functional near-infrared spectroscopy (fNIRS) when using driving assistance systems such as Adaptive Cruise Control (ACC) system and Lane Keeping Assistance (LKA) system. The fNIRS measurement system detects the radiated near-infrared rays, and measures relative variations of oxygenated hemoglobin (oxy-Hb) and deoxygenated hemoglobin (deoxy-Hb) based on those absorbencies.

Subjects followed a leading vehicle, which had a certain speed pattern including stop and go situations, with and without ACC system. The activity in the dorsolateral pre-frontal cortex (DLPFC) of the drivers without the ACC was modulated by accelerating and decelerating the car, whereas this was not the case for the drivers using the ACC. Also, the overall DLPFC as measured by fNIRS was lower when drivers were using the ACC compared to when they were not using ACC. These results may reflect a difference in mental workload between the two conditions.

Introduction

In recent years, various driving assistance systems have been developed to ensure safety by giving support and reducing driver workload. Examples include the Adaptive Cruise Control (ACC) system, which maintains a safe distance between the driver's vehicle and the vehicle ahead of it, and the lane-keeping assistance system, which keeps the car in a lane through steering support. However, it is also possible that, while driver workload is reduced, the driver's attention is also reduced, resulting in unexpected accidents. The undesirable behavioural adaptation was reported by Hoedemaeker and Brookhuis (1999) and Hjälmdahl and Várhelyi (2004). It is necessary to examine driver workload and attention from the viewpoints of cognitive engineering and human physiology. First, it is necessary to clarify the relationship between driver workload and brain activity, which includes recognition and judgment. It is then necessary to evaluate the driver's attention and to clarify the relationship between brain activity and driving performance.

Functional magnetic resonance imaging (fMRI), a non-invasive imaging diagnosis, has recently attracted attention. Using magnetism to detect local changes in the concentration of oxygenated hemoglobin (oxy-Hb) in the brain, fMRI has helped clarify higher brain functions from the viewpoints of linguistic and cognitive sciences. However, fMRI has many shortcomings in evaluating driving performance, because it requires the subject to lie in a narrow cylinder during evaluation and does not permit movement of the body, particularly the head.

Another diagnostic tool, functional Near-Infrared Spectroscopy (fNIRS), has gained attention in recent years. This non-invasive technique uses near-infrared light to evaluate an increase or decrease in oxy-Hb or deoxygenated hemoglobin (deoxy-Hb) in the tissue from the body surface. NIRS can detect the hemodynamic of the brain in real time while the subject is moving. Therefore, brain activity can be measured in various environments. Recent research has used functional Near-Infrared Light Spectroscopy (fNIRS) to measure brain activity in experiments focusing on mental calculation tasks (Kozima et al., 2007) or using train driving simulators (Tsunashima et al., 2004).

In this study the possibility of evaluating the reduction of driving workload with the above mentioned ACC from brain activity.

The principle of fNIRS

Using near-infrared rays, fNIRS non-invasively measures changes in cerebral blood flow. Its principle of measurement, which was developed by Jöbsis (1977), is based on measurement of oxygenation of hemoglobin in the cerebral blood flow.

In uniformly distributed tissue, incident light is attenuated by absorption and scattering. Therefore, the following expression, a modified Lambert-Beer law, was used (Tamura, 2002):

$$Abs = -\log(I_{out}/I_{in}) = \varepsilon dC + S$$

Here, I_{in} is the irradiated quantity of light; I_{out} is the detected quantity of light; ε is the absorption coefficient; C is the concentration; d is the averaged path length; and S is the scattering term.

If it is assumed that no scattering changes in brain tissue occur during activation of the brain, the change in absorption across the activation can be expressed by the following expression:

$$\Delta Abs = -\log(\Delta I_{out}/\Delta I_{in}) = \varepsilon d \Delta C(\Delta Hb_{oxy}, \Delta Hb_{deoxy})$$

Furthermore, if it is assumed that the change in concentration (ΔC) is proportional to the changes in oxy-Hb and deoxy-Hb, ΔHb_{oxy} and ΔHb_{deoxy}, the following relational expression can be obtained:

$$\Delta Abs(\lambda_i) = d[\varepsilon_{oxy}(\lambda_i)\Delta Hb_{oxy} + \varepsilon_{deoxy}(\lambda_i)\Delta Hb_{deoxy}]$$

The absorption coefficients of oxy-Hb and deoxy-Hb at each wavelength, $\varepsilon_{oxy}(\lambda_i)$ and $\varepsilon_{deoxy}(\lambda_i)$, are known; therefore, $d\Delta Hb_{oxy}$ and $d\Delta Hb_{deoxy}$ can be obtained by performing measurements with near-infrared rays of two different wavelengths and solving simultaneous equations. However, the physical quantity obtained here is the product of the change in concentration and the averaged path length, so care should be taken.

In general, the averaged path length d varies largely from one individual to another, and from one part to another. Therefore, caution must be exercised in evaluating the results.

The fNIRS signal and workload

To measure brain activity under workload using fNIRS, we used the workload of mental calculation. Mental calculation tasks were set to low, medium, and high levels as follows:

- Low-level task: Simple one-digit addition (e.g., 3 + 5)
- Medium-level task: One-digit addition of three numbers (e.g., 6 + 5 + 9)
- High-level task: Subtraction and division with decimal fraction (e.g., 234/(0.61 - 0.35))

The subject answered the questions displayed on the PC screen without speaking. fNIRS signals are illustrated in Figure 1. Results revealed that oxy-Hb increased and the brain was activated during mental calculation tasks. Furthermore, such changes became larger as the level of mental calculation task became higher. Figure 2 presents the WWL score determined by NASA-TLX. The workload became higher when the task level was higher. Comparison of brain activity detected by fNIRS (Figure 1) and the workload scores obtained from NASA-TLX (Figure 2) exhibited good correlation. This result confirmed the feasibility of evaluating workload using the signal of cerebral blood flow obtained from fNIRS.

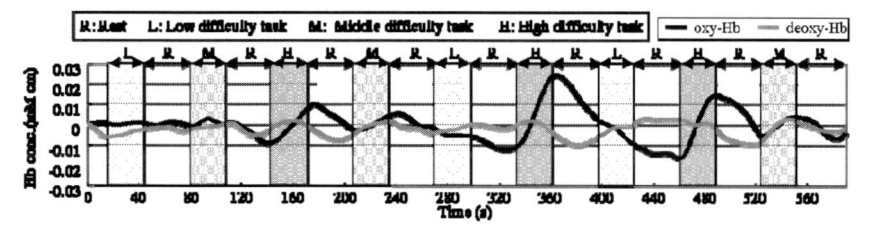

Figure 1. fNIRS signal in the mental calculation task

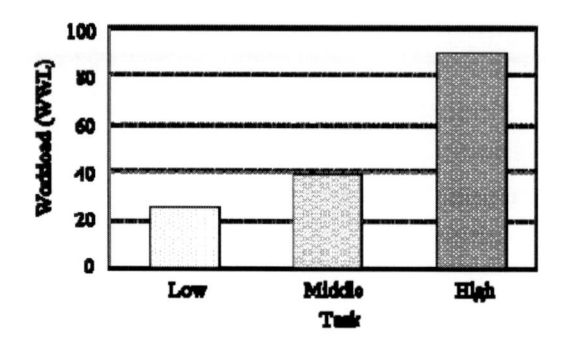

Figure 2. Workload evaluation by NASA-TLX

Method - experiment

Contents of the task

To verify that the driving workload reduction of ACC could be evaluated from brain activity, we conducted an experiment that involved the use of a driving simulator to follow a vehicle (Figure 3). Driving tests were conducted under two conditions: one involved following a vehicle by utilizing ACC, and the other involved following a vehicle while driving without ACC. Figure 4 presents the sequence of the experiment. The subject performed practice runs to become somewhat skilful in handling the driving simulator and then drove two times under each condition. Brain activity during one condition was compared with that during the other condition.

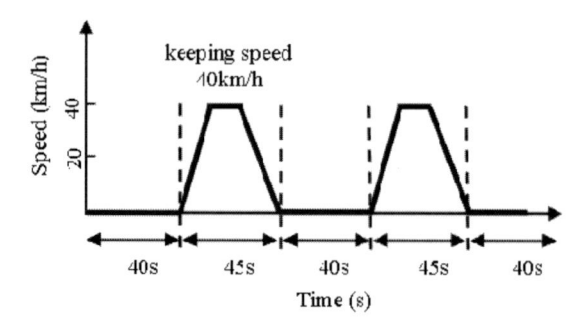

Figure 3. Speed pattern of leading car

Figure 4. The sequence of the experiment

Method-measurement

Brain activity in the frontal lobe was measured using fNIRS. The measuring instrument was a near-infrared imaging device, OMM-300, Shimadzu Corporation, Japan. Figure 5 illustrates the arrangement of optical-fiber units. The numbers between the light-emitting fiber unit and the light-receiving fiber unit denote the measurement channels; measurements were performed through a total 42 channels. Furthermore, driving performance was also recorded on the driving simulator while measuring brain activity. The four subjects were males who were in their 20s and who had a valid driving license.

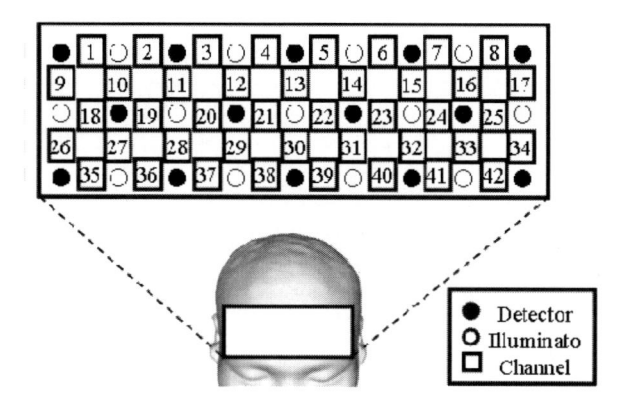

Figure 5. Position of optical fibers and channels

Analysis of fNIRS signals

Decomposition and reconstruction of fNIRS signals

The fNIRS signals include signals that are not related to brain activity (e.g., noise of the measurement instrument, influences of breathing, and changes in blood pressure). It was necessary to remove these unrelated signals to evaluate brain activity in detail. Therefore, the measured fNIRS signals were decomposed through MRA using discrete wavelet transform, and the components related to the driving task were reconstructed.

Figure 6 presents the MRA results for oxy-Hb in the 26 channels, where task-related changes were remarkable. The trend of the whole experiment was extracted on the approximated component (a_{10}). Here, d_1 and d_2 had a relatively large amplitude. It is possible that these were changes in blood flow, based on heart rate, and measurement noises. Because the interval of repetition of tasks and rests was 85sec, the d_8 component was the central component of task-related changes. Therefore, signals were reconstructed by adding the d_7, d_8, and d_9 components.

Statistical processing after conversion into z-scores

The fNIRS signal expresses the quantity of relative changes using the start point as the reference; however, comparisons of measurements between subjects or statistical processing of measurements of all subjects cannot be implemented using this signal as it is.

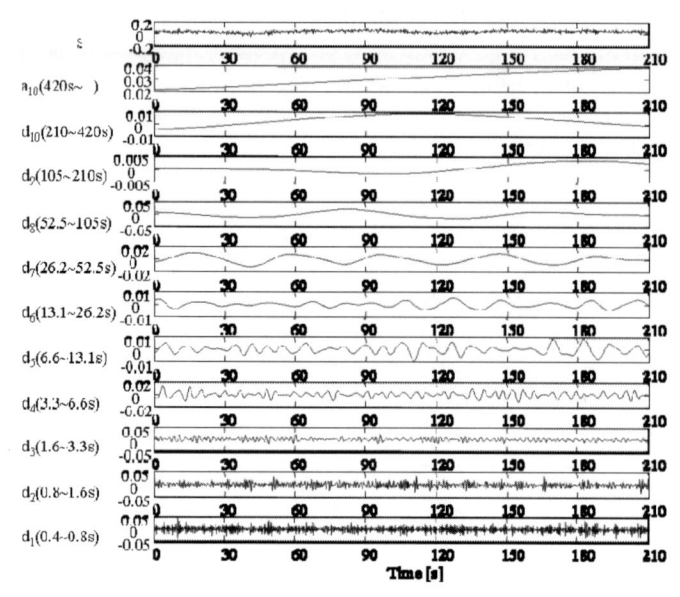

Figure 6. Decomposition of fNIRS signal by multi-resolution analysis (channel 26)

Therefore, we propose a method for converting data of oxy-Hb and deoxy-Hb reconstructed by MRA into z-scores, so that the mean value is 0 and standard deviation is 1.

Results

Figures 7 and 8 depict the relationships between brain activity when the subject followed the lead vehicle manually without using ACC and that when the subject used ACC (26 channels at the outer right portion of the frontal lobe) and vehicle speed.

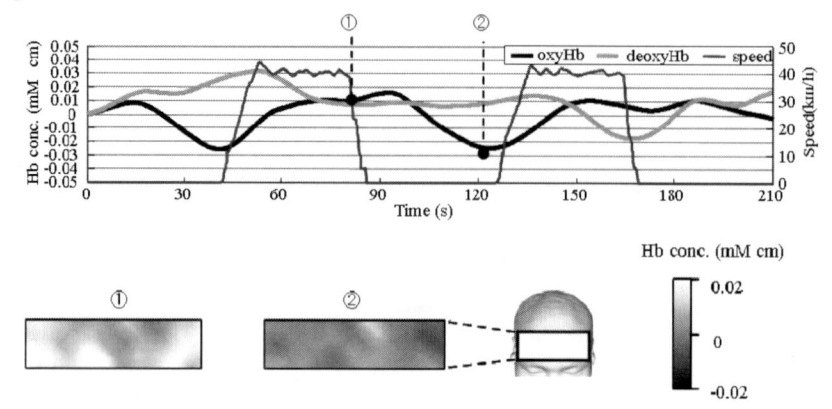

Figure 7. Result of following a leading vehicle without ACC system. Top (a): Task-Related fNIRS signal and vehicle speed (channel 26), Bottom (b): Functional brain imaging

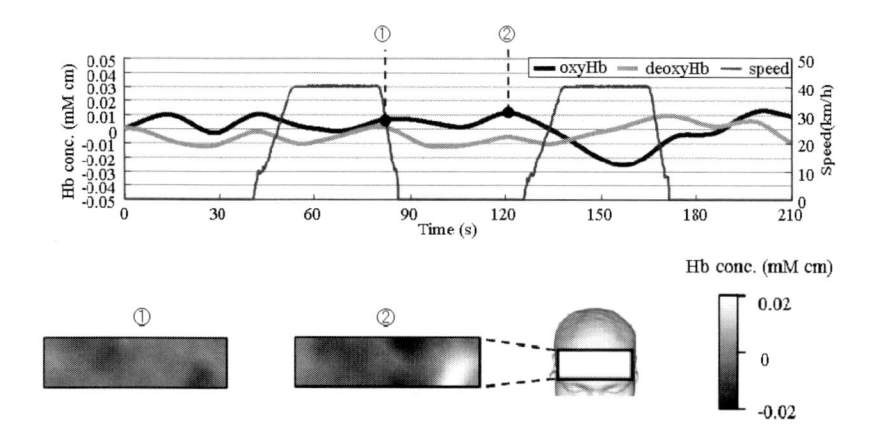

Figure 8. Result of following a leading vehicle with ACC system. Top (a): Task-Related fNIRS signal and vehicle speed (channel 26), Bottom (b): Functional brain imaging

Activity related to driving performance could be confirmed in both outer portions of the frontal lobe. Figure 7(a) reveals that oxy-Hb and deoxy-Hb changed in connection with vehicle speed when ACC was not used. After the start of acceleration, oxy-Hb started to increase and exhibited a high value during deceleration. Because changes in oxy-Hb were highly correlated with changes in regional cerebral blood flow (rCBF) (Hoshi et al., 2001) and an increase in rCBF reflected an increase in neural activity (Jueptner & Willer, 1995), we focused on oxy-Hb and created brain-activity images from the reconstructed oxy-Hb signals (Figure 7(b)). Figure 7(b) reveals that oxy-Hb increased in both outer portions of the frontal lobe during deceleration. This result indicated that both outer portions of the frontal lobe were active during manual driving without using ACC.

When ACC was used, however, oxy-Hb and deoxy-Hb in Figure 8(a) exhibited changes not related to vehicle speed. Furthermore, the brain activity images in Figure 8(b) indicate that oxy-Hb did not increase during deceleration.

The fNIRS signals after MRA were converted into z-scores, and the z-scores of the four subjects were added and averaged under two conditions: only driving and driving with the use of ACC. Although some problems were involved (e.g., the measurement points differed slightly from one subject to another, and driving performance differed among participants), addition and averaging was a very effective method for evaluating the reduction of driving workload by the use of ACC, based on brain activity.

Figure 9 presents the result of addition and averaging with the subject driving without ACC, and Figure 10 presents the result with ACC. Figure 9(a) confirms that oxy-Hb increased when the subject drove without ACC and exhibited a high value in the latter half of the task. The brain function imaged in Figure 9(b) confirms that, as common brain activity, both outer portions of the frontal lobe became active during deceleration.

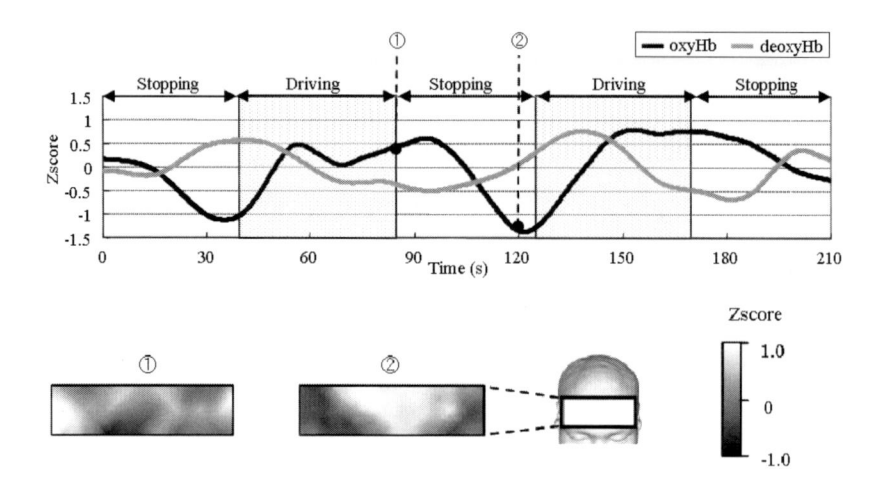

Figure 9. Averaged result of following a leading vehicle without ACC system. Top (a): Averaged fNIRS signal (channel 26), Bottom (b): Functional brain imaging

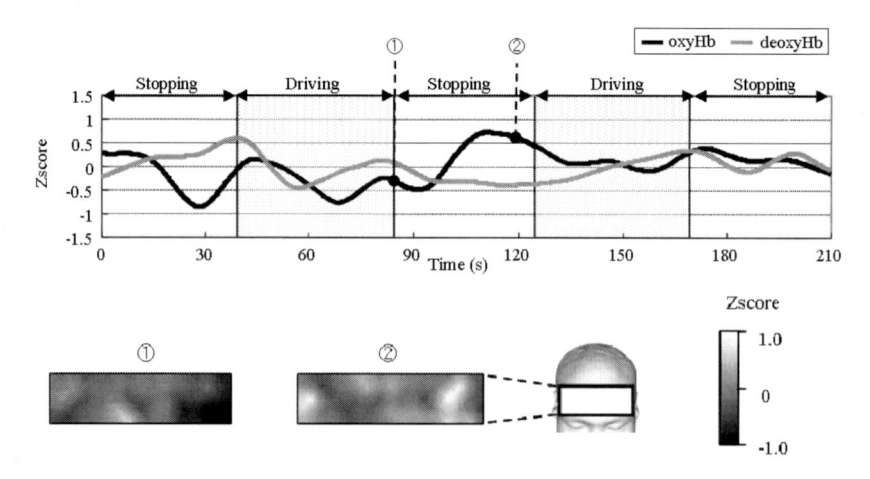

Figure 10. Averaged result of following a leading vehicle with ACC system. Top (a): Averaged fNIRS signal (channel 26), Bottom (b): Functional brain imaging

Figure 10(a) indicates that oxy-Hb did not increase while driving with the use of ACC. Also, the brain function image in Figure 10(b) reveals that the frontal lobe was less active than when the subject drove without ACC. This result may reflect the reduction of driving workload by ACC.

Conclusions

Whether the reduction of driving workload by ACC can be evaluated from brain activity was evaluated through pilot experiments using a driving simulator. MRA revealed that while the outer portions of the frontal lobe were active in connection with driving performance when the subject drove without ACC, it indicated no activity related to driving performance with the use of ACC. Furthermore, adding and averaging the z-scores of all subjects confirmed a common trend that the frontal lobe was not as active while using ACC as it was when the subject drove without ACC and therefore may reflect reduction of driving workload by ACC. These results suggest the possibility of evaluating driving assistance systems through evaluation of the driving workload from measurement of brain activity using fNIRS.

In the future, we will increase subjects and handle statistical analyses. Furthermore, the result considers whether it differs significantly with and without ACC system.

The goal of our research is to evaluate the driver's attention and intention from brain activity. Though this study was only the evaluation of the driving workload, we are planning to evaluate driver's attention and intention from activated region.

Acknowledgement

This work was supported by Grant-in-Aid for Scientific Research (A) 18201031 of the Japanese Ministry of Education, Culture, Sports, Science and Technology (MEXT).

References

Hoshi, Y., Kobayashi, N., & Tamura, M. (2001). Interpretation of nearinfrared spectroscopy signals, A study with a newly developed perfused rat brain model. *Journal of Applied Physiology, 90*, 1657-1662

Hjälmdahl, M., & Várhelyi, A. (2004). Speed regulation by in-car active accelerator pedal Effects on driver behaviour. *Transportation Research Part F, 7*, 77-94

Hoedemaeker, M., & Brookhuis, K.A. (1998). Behavioural adaptation to driving with an adaptive cruise control (ACC). *Transportation Research Part F, 1*, 95-106

Jöbsis, F.F. (1977). Non-invasive infrared monitoring of cerebral and myocardial oxygen sufficiency and circulatory parameters. *Science, 198*, 1264-1267

Jueptner, M., & Willer, C. (1995). Dose measurement of regional cerebral blood flow reflects synaptic activity?- implications for PET and fMRI. *Neuroimage, 2*, 148-156

Kojima, T., Yanaginuma, T., Tsunashima, T., Hirose, S., Shimizu, T., Shiozawa, T., Taira, M., & Haji, T. (2007). Measurement of Higher Brain Function with Workload by Using Functional Near-Infrared Spectroscopy (fNIRS) (in Japanese with Engish Summary) (pp.19-22). *JSAE annual congress*, No.60-07.

Tamura, M. (2002). Functional near-infrared spectoroscopy. *Advances in Neurological Science, Series C, 47*, 891-901

Tsunashima, H., Kojima, T., Shiozawa, T., & Takada, H. (2004). Development of Train Simulator for Evaluation of Human Factors and Application for Measurement of Brain Activity. *Journal of Reliability Engineering Association of Japan, 26*, 617-626

Go Ahead – I will follow you!
Social pull-factors in driving manoeuvres

Jessica Seidenstücker, Ernst Roidl, & Rainer Höger
Leuphana University of Lüneburg
Germany

Abstract

In Germany more than 26% of all road accidents outside towns involving personal injury occurred in curves and thereby 1,322 people died. This study concentrated on a person's steering abilities and velocity in curves with regard to social risk factors precipitated by social pressure from other road users. The assumption is that drivers orient themselves toward the driving characteristics of the preceding car (pull-hypothesis). Thirty-nine participants (divided in groups of low/highly experienced drivers) participated in a simulator-based study using a rural road scenario. Two conditions were simulated: a sharp bend with (1) and without (2) a fast vehicle in front of the participant.

The results showed that the speed of the preceding vehicle is principally contributory to the total number of accidents (e.g. car going off the road). The comparison between the two conditions indicated that accidents occur significantly more often in the first condition. Furthermore the results indicate that less experienced drivers produce more accidents than experienced drivers. This result led to the conclusion that the ability to steer in curves may underlie a social pull-effect. This social pull-effect is caused by their adjusting to the inappropriate speed of the preceding car, regardless of the possible dangers within the road scenario.

Introduction

Accident studies suggest that misperceiving and therefore misestimating of traffic hazards lead to critical moments. Especially young and inexperienced drivers tend to the unrealistic self-assessment in safe-driving by believing themselves to be more capable and skilful than the average driver (Gregersen, 1996). Schönbach (1996) investigated social interaction in foggy traffic conditions. Driving in fog involves a lot of elements of uncertainty and that is why a typical phenomenon in these traffic circumstances is that drivers tend to follow a car without maintaining an appropriate distance to the preceding car (Schönbach, 1996). The absence of a safety clearance can cause serious accidents or even multiple pileups. Schönbach explains this with the overestimating of the own capabilities paired with a high degree of subjective uncertainness. Drivers seam to *glue* to the backlight of the car in front of them, which is interpreted with a pull-hypothesis. One of the reasons for this pull-effect can be described with the social comparison theory. Human behaviour depends on social comparison processes which mean that people evaluate their abilities, skills and options. If objective, non-social aspects are missing or not available at the critical moment, individuals evaluate their abilities and options by comparing themselves with others and based on this evaluation they make specific decisions (Festinger, 1954). The present paper tries to carry this theory over into road traffic and accident analysis by investigating a special form of social comparison processes with regard to a driver's steering abilities and speeding behaviour in curves.

In order to obtain a high external validity the survey used as adequate stimuli a traffic scene which represents an accident hotspot on German roads: curves. Because of the special roadway arrangement, curves implicate (similar to fog) a high degree of complexity paired with a limited visibility of the following course of the track. In 2007 68% of all accidents happened on urban and 32 % on extra-urban roads (6% on motorways). The typical road design for rural roads is poorer than motorways and hence drivers feel often too comfortable, underestimate the risk to be in an accident and speed up too much. With regard to the accident statistics rural roads implicate a high perilous potential, because the speed on those streets is always higher and therefore the accidents are often more serious. In the last year 4,949 persons died on German roads and most of them (61%) died on streets out of town (Destatis, 2008; De Waard et al., 1995). From these statistics it follows that the risk to die on extra-urban roads is three times higher than on urban sites. With a total number of 25,000 crashes and a body count of 1,322 (36% of all traffic deaths) curves display one main place of accidents outside cities (Vorndran, 2006).

Based on the findings of Schönbach and on Festingers social comparison theory the survey follows the aim to investigate if those processes also can be found in cornering tasks on rural roads. The assumption is that drivers speeding up too much, because they orient themselves toward the negative driving characteristics of a preceding, excess speeding car and overlook their own abilities and ignore the appropriate velocity. Therefore two variants of a specific extra-urban scene (with and without a social factor) were tested against each other.

Method

Participants

Thirty-nine drivers took part in this study, of which twenty-one were male and eighteen were female. The participants were aged between seventeen and sixty-five (mean = 29.5), with an average of 11.9 years driving experience. Based on a special index the participants were divided into experienced and inexperienced drivers. To generate this two-level variable for *experience* all participants were assigned to a certain experience group. The assignment was performed using an index calculated as follows: for each participant the amount of kilometres covered in the last 12 month was multiplied by a frequency code. The frequency code is a value for the utilisation of the car per week and month respectively. The resulting index can be understood as a measure of the current or present experience in car driving. Subsequently a dichotomy for the participants was performed to obtain two contrast groups. The following table shows the descriptive statistics for both groups to see how "separated" these groups are.

Table 1. Descriptive statistics for the participants (separated according to the experience groups)

	sex		age	kilometers/year	mileage in total
experienced drivers	m:12	f: 5	40.2	65,588.24	817,647.06
inexperienced drivers	m: 9	f:13	21.3	1,052.27	19,931.59

All participants were paid for their participation and informed that they could stop participation at any time.

Stimuli

The simulator scenery consisted of a 3700 meters winding rural route including signs, trees and oncoming traffic. After 2382 meters covered distance an extreme left turn had to be passed at an appropriate velocity, otherwise the possibility to produce a potential danger or even an accident arised. The scene existed in two different versions: a) just the critical sharp bend without any external influence and b) with another car overtaking the participants' car just before the curve. In the following the condition with the overtaking car is called *negative prime*, because a negative influence of this speeding car on the participants driving behaviour is assumed. In this *negative primed version* the sports car passes after a covered distance of 1800 meters with a velocity of 10 metres per second faster than the participants' car. Thirty metres in front of the curve the car finished the overtaking process. In the following the preceding car drove through the left turn with a speed of 25 meters per second faster than the participant. These velocity differences between both vehicles were necessary to ensure that the triggered car is able to complete the passing manoeuvre.

Figure 1. Left: car overtakes the participant, right: lead car during the curve run

The exact position of the turn was between 2382 and 2516 meters distance down the road. The curvature had a radius of 66 meters with an overall distance of 100.4 meters and the width of the street is 4 metres each lane.

Furthermore a special evaluation of the conception and design of the simulator roads where performed by traffic experts of the German automobile club – ADAC to achieve a maximal convergence between the simulator surrounding and the reality on German roads.

Apparatus

The study was conducted with the driving simulator StiSim W100 by System Technology Inc. The simulator includes a racing force feedback wheel, accelerator pedal and brake pedal (all by Logitech) mounted in a wooden vehicle cabin and plugged to a Dell desktop computer. During the simulation drive the StiSim W100 registered all driving activities like accelerating, braking or steering behaviour at a sample rate of 10 Hz. The presentation of the scene was realized by an Optoma EP 725 projector on a white painted area of 300 cm x 225 cm. To obtain an appropriate sharpness and picture rate the resolution was set up at 1024x768 to provide a constant frame rate of 60 frames per second.

Design

The experimental designs consisted of two independent factors and both were varied in two steps. In this 2x2 between subjects factorial design the first factor *prime* is manipulated by the two different traffic scenes and the second factor *experience* is manipulated by the factor experienced and inexperienced drivers.

Procedure

All participants were tested individually. Firstly, they had to complete a questionnaire concerning their demographic and driving history. As an introduction to the graphical presentation of the virtual StiSim environment and to get used to the handling of the steering and speed controls a training scene was prefixed to the actual driving task. Within this 6 minutes familiarization run the participants had to drive on a huge parking site, as well as a rural highway and an urban street. During this preparation drive, the investigator gave the verbal introduction for the following experiment. The participants briefing was to move straight on the defined route until other advises follow. They were instructed to act like an ordinary car driver by obeying the general German Highway Code. After the training the participants were randomly assigned to one of the two experimental conditions.

Results

Data generation of the steering-performance and the crash variable

The steering ability is a good predictor for driving errors and accidents (Hatakka, Keskinen, & Hernetkoski, 2003; Lin, Popov, & McWilliam, 2003). Therefore the steering or lane-keeping ability need to be analysed to distinguish between good and bad performance. To operationalise this lane-keeping ability the exact position on the track while driving through the curve was detected. Initial points for this were the individual distances to the centre line and the exact position down the road independently of the driven speed at these measurement spots. Adapted from these StiSim simulator generated data a graphical lane plot for each participants was created via the TecPlot 360 2008 software. The software produce two different layers within a specific coordinate system (layer 1: course of the track; layer 2: participants distance to the centre line). The Figure 2 shows the prototypical shape of different lane-keeping occurrences from the ideal racing line to skidding off the road.

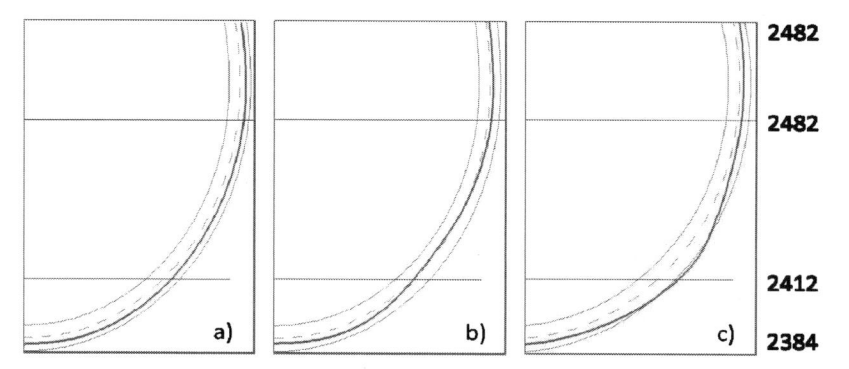

Figure 2. Geometric plot to display types of lane-keeping: a) racing line; b) cut a corner and c) skid off the road (including the position marker in meters down the road on the right side).

Because of the used time- (and not distance-) triggered measurement the number of position values is not constant between the participants. Hence it was not possible to create a mathematical derived value for the driving performance. That is why a behavioural monitoring was performed by two autonomous and trained observers. As shown in figure 2 the curve was separated in three different sectors. The observers had to evaluate each sector of the plots according to accurately pre-defined criteria, displayed in Figure 2. Afterwards the level of inter-rater agreement was calculated via Cohen's Kappa. The Kappa-values of the sectors are all highly significant and range between .77 and .89. Subsequent to the expert evaluation two different variables for the driving performance were created: 1) an *accident* variable (crash vs. no crash) and 2) a variable for the *quality of steering ability*. Thereby an accident is any erroneous behaviour that could cause an accident by leaving one's own lane. All participants' curvatures were classified to one of those two parameter values. The variable for *quality of steering ability* exists in the two variants a) good performance and b) bad performance. Other than the mentioned *crash* variable *bad performance* is additionally characterised by near accidents (i.e. erroneous behaviour without the proper lane line crossing).

The influence of the prime version on the velocity and the driving performance in a curve

Before the effect of the driven speed on the *quality of steering ability* and the *accident rate* is analysed, the general velocity within the curvature should be contemplated. It is assumed that the particular velocity has the main impact on good and bad performance when passing the curve. In order to obtain a first overview on the driven speed a velocity plot for the two priming versions differentiated for the crash occurrence should be designed.

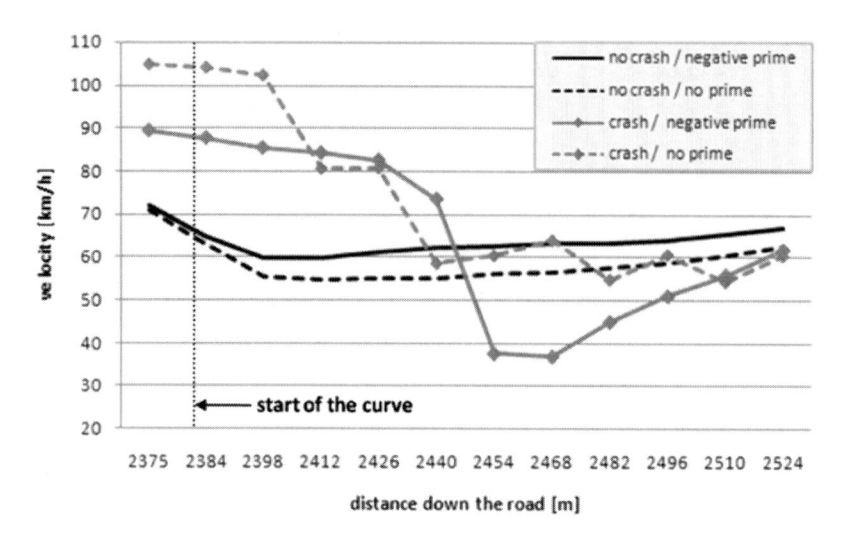

Figure 3. Velocity plot for the two priming versions differentiated for the crash occurrence

Within the course of the bend there are different measurement spots to record the current driving speed via the StiSim simulator. Based on these data the graph (Figure 1) was calculated. The pictured outcome shows the mean velocity plot of all participants within the bend including the marker for the start of the curve at a distance of 2482 meters down the road represented by the dashed vertical line.

To investigate a possible relation between the driving speed and the steering performance an analysis of correlation was performed. The analysed measuring points are between 2375 (seven meters in front of the curve) and 2454 meters down the road. Independently of the prime condition the correlation analysis indicates for the curve-sector between 2412 and 2440 meters down the road a strong influence of the driven *speed* on the *quality of the steering ability* (t-value=3.57; r=.50; p< .01) and the *crash-rate* (t-value= 3.90; r=.53; < .01). The faster the participants drove the worse the quality of driving performance was (i.e. bad lane-keeping performance; more (near) accidents). Apparently within the continuative course of the curve the next measure point (at a distance of 2454 meters down the road) constitutes a breach. The analysis shows a negative correlation with the velocity in this spot and the *accident rate* at this spot (r=-.31; p < .05). The reason is that the participants drive abruptly slower between 2440 and 2454 meters. The change of velocity is an indicator for the location where the most crashes occurred and hence in the presented experiment this point is the accident hotspot. These results show that the speed in the sector between a covered distance of 2412 and 2440 meters seems to be the most relevant to distinguish between good and bad performance and also crashes. Therefore a general speed value for this *critical sector* was developed. A correlation analysis with *speed in the critical sector* and the prime results that participants drove significant faster in the negative prime version (r=.42; p<.05). In average the participants drove 12.80 km/h faster in the negative prime condition than in the condition without any prime.

A correlation analysis revealed a significant influence of the prime on the crash occurrence: in the negative primed traffic scene the participants produce more accidents than in the scene without any external prime (r=.32; p<.05). Figure 4 points out the percental relation between the crash rate and the experimental prime condition.

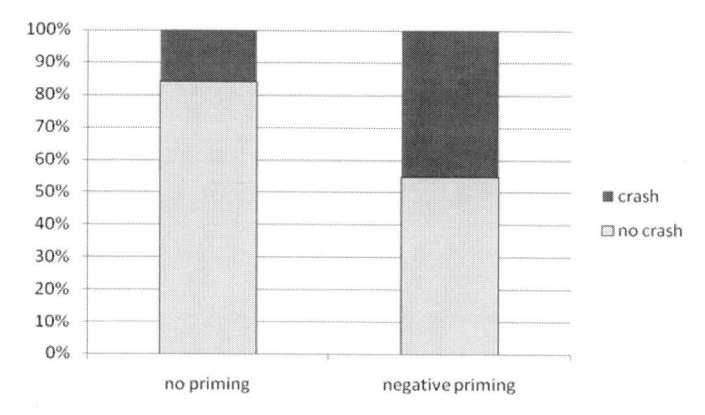

Figure 4. Percental disposition of the crash occurrence for each prime version

The influence of the prime version and the driving experience on the velocity and the driving performance in a curve

One of the main objectives of the survey was to find out if the driving experience has an influence on the steering abilities in curves. The frequency distribution leads to the assumption that inexperienced drivers produce show a worse steering performance then experienced drivers. Figure 5 illustrates the (percentage) distribution for good and bad performance for the two experimental groups.

To determine if there is a significant difference between the expected frequencies and the observed frequencies of good or bad steering performance for the dichotomized driving experienced a chi-square test was performed. The results indicate that there is a significance difference between the two groups of experienced drivers regarding their steering abilities within the curve-task (χ^2= 4,5; p<.05).

In the next step a two-way repeated-measurement analysis of variance (ANOVA) with the factors *experience* and *prime* for the speed in the *critical sector* (shortly before the accident hotspot between 2412 and 2440 meters down the road) was conducted. The analysis revealed significant effects for the velocity in the critical sector on the independent variable *prime* ($F(1,36) = 5.4$; p<.05) and furthermore an interaction of the factors *prime* and *experience* ($F(1,36)$ 4.4; p<.05, Figure 6).

	good performance	bad performance
experienced drivers	12	5
inexperienced drivers	8	14

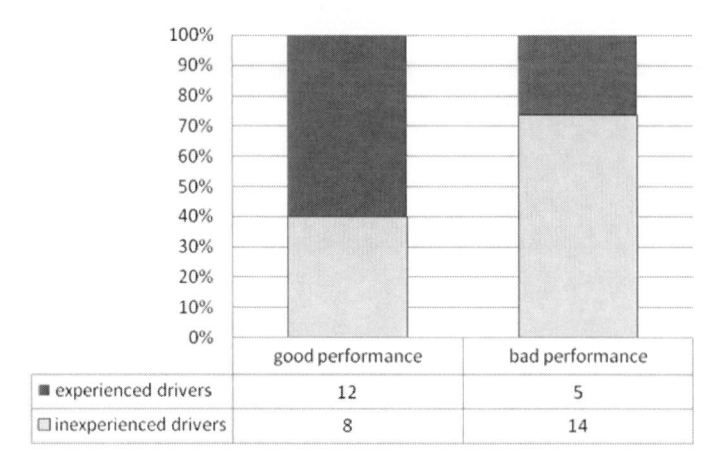

Figure 5. Frequency and percental distribution of the steering performance for inexperienced and experienced participants

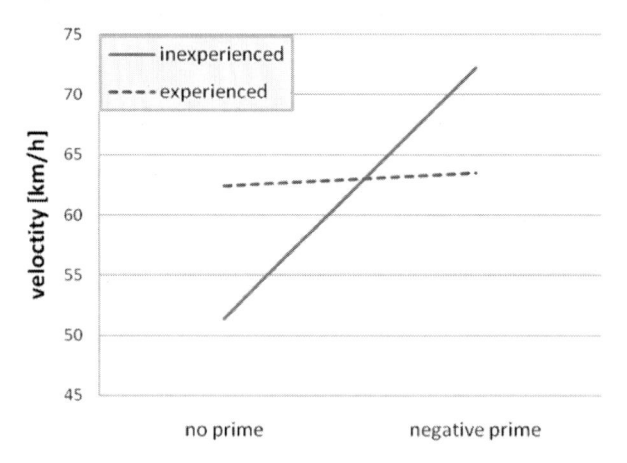

Figure 6. Mean values for the velocity in each prime version separated by the driving experience

The results depicted in figure 6 show that inexperienced participants drive faster in the experimental condition with the negative prime stimuli than experienced ones. Inexperienced drivers are more capable of being influenced by a negative social factor in this turn-task.

Discussion and conclusion

The present study was designed to observe the effect of social pull-factors in driving manoeuvres. Therefore two version of an extra-urban bend scene were tested against each other concerning to the steering abilities and the velocity of the car drivers. The analysis of the velocity revealed that the sector shortly before most accidents occurred (at a distance of 2454 meters down the road) is the profoundly critical for a safe drive. If the participants drive too fast in this critical sector, the possibility to produce a crash is much higher than at other parts within the bend. Furthermore the experiment shows the following, recapitulatory results:

1. The velocity in the critical sector of the bend has a highly significant influence on the quality of the steering abilities.
2. The velocity in the critical sector of the bend has a highly significant effect on the accident rate.
3. The presence of a social pull-factor has a strong negative effect on the steering ability.
4. Inexperienced drivers are more manipulable by the pull-effect of the preceding car. They show higher velocities within the critical sector of the bend in the social prime version than experienced drivers.
5. Independently of the prime condition the driving experience has an effect on the crash occurrence. The less the driving experience the faster the participants go and the more accidents are produced.

Previous studies confirmed the existence of a positive effect on the driving performance of specific external traffic objects. There it was demonstrated that participants responded in a better way to emerging events if corresponding concepts were activated previously. This was interpreted as a positive prime-stimuli (Seidenstücker & Höger, 2008; Höger & Seidenstücker, 2007). In the present paper not the positive, but the negative influence on a traffic-relevant pre-activation was detected by identifying a social pull-factor in a sharp bend scenery.

The premise for a safe drive is the correct estimation of the traffic circumstances including the course of the track and imported external factor like the other road users. Furthermore the accurate estimation of the own opinion and ability is of paramount importance. The higher the level of subjective and perceptional uncertainties the more humans are motivated to compare themselves with others (Festinger, 1954). In a situation like in the presented curve-task the preceding car drives too fast and this leads to a form of uncertainness towards the own speed selection. The results show that in the condition with a possible social pull- or comparison-factor the participants drive significant faster and produce more accidents than in the control version. Possible reasons for this behaviour could be seen in the following aspects:

- Participants try to compensate their own misestimation of the possible danger of the curve by orientating themselves toward the inappropriate behaviour of the preceding car.
- When a motive to self-improve the own behaviour is activated, individuals tend to compare themselves with others who perform better. In the present cornering-task the preceding car seems to pass the bend better, faster or more effective and therefore the overtaking manoeuvre was viewed as good driving. The participant tries to follow this (negative) example by speeding up too much.
- Participants overestimate their own steering abilities by putting their skills on a level with the other driver.

The lack of objective, non-social aspects in the critical moment bring the participants toward a comparison with others and this could lead to a fatal decision. This fatal decision is ignoring the speedometer (as an objective reference to the physical world) and just look about the behaviour of another member of the social traffic group. Road traffic can be understood as a special form of social interaction and therefore drivers perceive other drivers as a member of a specific reference group (Hyman, 1942) and they probably use this group as a benchmark for their self-evaluation in critical situations.

Acknowledgement

This research project was supported by a grant (F.A.-Nr. 2006.743) from the 'Arbeitsgemeinschaft Innovative Projekte' (AGIP) of the Ministry of Science and Culture, Lower Saxony, Germany. The realisation would not have been possible without the support and assistance provided by colleagues (especially Cornelia Hack and Swantje Robelski) and the participation of the ADAC Nord in Embsen/Germany.

References

Destatis (2008). *Unfallgeschehen im Straßenverkehr 2007* [Road traffic accidents 2007]. Wiesbaden, Germany: Statistisches Bundesamt.

De Waard, D., Jessurun, M., Steyvers, F.J.J.M., Raggatt, P.T.F., & Brookhuis, K.A. (1995). Effect of road layout and road environment on driving performance, drivers' physiology and road appreciation. *Ergonomics, 38*, 1395-1407.

Festinger, L. (1954). A Theory of Social Comparison Processes. *Human Relations, 7*, 117-140.

Gregersen, N.P. (1996). Young drivers' overestimation of their own skill – an experiment on the relation between training strategy and skill. *Accident and Prevention, 28*, 243-250.

Hatakka, M., Keskinen, E. & Hernetkoski, K. (2003). Goals and contents of driver education. In L. Dorn (Ed.), *Driver behaviour and training Vol. I* (pp.309-316). Aldershot (UK): Ashgate.

Höger, R. & Seidenstücker, J. (2007). Driving and activation of mental concepts. *Advances in transportation Studies an international Journal, 2007 Special Issue*, 91-96.

Hyman, H.H. (1942). The psychology of status. *Archives of Psychology, 269*, 5-91. Reprint in H. Hyman & E.Singer (1968, Eds.), *Readings in reference group theory and research* (pp. 147-165). New York: Free Press.

Lin, M., Popov, A.A., & McWilliam, S. (2003). Sensitivity analysis of driver characteristics in driver-vehicle handling studies. In L. Dorn (Ed.), *Driver behaviour and training, Vol. I* (pp.263-276). Aldershot (UK): Ashgate.

Miller, R.L. (1977). Preferences for social vs non-social comparison as a means of self-evaluation. *Journal of Personality, 45*, 343-355.

Schönbach, P. (1996). Massenunfälle bei Nebel [Group accidents during fog]. *Zeitschrift für Sozialpsychologie, 27*, 109- 125

Seidenstücker, J. & Höger, R. (2008). Effects of preactivated mental representations on driving performance. In D. de Waard, F.O. Flemisch, B. Lorenz, H. Oberheid, and K.A. Brookhuis (Eds.) *Human Factors for Assistance and Automation* (pp. 129-139). Maastricht, The Netherlands: Shaker Publishing.

Vorndran, I. (2006). *Unfallgeschehen im Straßenverkehr 2006 - Weniger Unfälle auf deutschen Straßen* [Road accidents in 2006 –Less accidents on German roads]. Wiesbaden, Germany: Statistisches Bundesamt.

Attitudes of public bus drivers towards technology integration in the dashboard design

Emine Nazli Özer, Gulsen Töre, & Çigdem Erbuğ
Middle East Technical University
Ankara, Turkey

Abstract

Integration of technological features, such as navigation and information displays, automation of manual controls and digitalization of displays, into the public bus can be promising in terms of providing solutions for the critical contextual problems of bus drivers, such as undesired interactions with passengers and increased cognitive workload caused by the necessity of checking time frequently for complying with the schedule. However, if these systems are not accepted by bus drivers, the solutions they provide become useless and they may even constitute critical threats for safety. Therefore, in this paper the aim is to investigate bus drivers' views towards technology integration in the bus, to provide technological functions, which correspond better to their needs. To achieve this aim, (1) drivers' attitudes towards general technology, (2) their attitudes towards current technology integration in the bus dashboard, and (3) degree of involvement with the current technological products were questioned with a structured questionnaire that was delivered to 30 respondents in interview format. The paper outlines the findings of the study by describing bus drivers' attitudes towards technology, identifying design features that lead to perception of technology in the dashboard, and making associations between the drivers' characteristics and their attitudes towards technology integration.

Introduction

Due to stressors which exist in the work environment of the bus driver, such as heavy traffic conditions, inflexible time schedules, interaction with passengers and crowdedness, etc., both cognitive and physical workload dramatically increase. In order to reduce the impact of these stressors, apart from the organizational improvements, workspace of the bus driver should be properly designed (Grosbrink & Mahr, 1998, Tse et al., 2006). The problem is attempted to solve by integrating technological equipments in the cabin design, e.g. navigation and information systems, which are designed to reduce the workload of the driver by providing information about conditions of the bus, the route, passenger load, etc. However integration of technological systems may ease the driver's tasks, if it can be successfully used by the driver. He should accept these systems, in order to utilize them and engage with them, while carrying out his tasks.

In D. de Waard, J. Godthelp, F.L. Kooi, and K.A. Brookhuis (Eds.) (2009). *Human Factors, Security and Safety* (pp. 317 - 329). Maastricht, the Netherlands: Shaker Publishing.

In the project related to which the study at issue is conducted, it is aimed to design a dashboard and workspace in a bus, which corresponds better to the needs and preferences of users. At the previous stages of the project, exploratory research has been conducted in order to (1) understand actual working conditions, (2) figure out the physical interactions with the workspace, and (3) find out general expectations and preferences of drivers with respect to an ideal bus. The results were mainly about ergonomic aspects and expectations of drivers regarding the bus interior and technical properties. Moreover, a list of criteria for evaluating a bus was formulated based on the opinions of the drivers, which were stated during the semi-structured interviews. The results of the interviews also indicate a variance in perception of technology among drivers, although no direct question about technology was asked. While evaluating a bus, some of them considered electronic controls and displays as a positive factor, whereas some others approached to the electronic system in a negative manner because these systems do not allow the driver to detect and fix malfunctions by himself. Since attitudes towards technology will be a crucial factor in the final outcome of the project, investigating this topic has great importance for the usability of the product.

Theoretical constructs of technology acceptance

Performance gains are often obstructed by users' (un)willingness to accept and use technological systems (Bowen, 1986). Following the rapid growth of investment in information technology by organizations, user acceptance of information technologies has received much attention. There are a vast number of approaches in information technology acceptance (e.g. Davis, 1989; Moore & Benbasat, 1991; Taylor & Todd, 1995; Venkatesh, 2000; Venkatesh & Davis, 1996, 2000; Venkatesh et al., 2003). The technology Acceptance Model (TAM) of Davis (1989) has emerged as a predominant model and has been used widely as a theoretical basis for many studies on technology acceptance. This model focuses on two theoretical constructs as fundamental determinants of system use, perceived usefulness –the extent the users believe that the new system will help them perform their job better– and perceived ease of use –the evaluation of the effort of using the application.

Venkatesh and Davis (1996) examined the determinants of perceived ease of use. The results show that computer self efficacy has a constant effect on perceived ease of use, while objective usability effects the perceived ease of use only after direct experience. Venkatesh (2000) proposed a framework which integrated constructs that are the determinants of perceived ease of use, from other user acceptance models into TAM. The results indicate that internal control (conceptualized as computer self-efficacy), external control (conceptualized as facilitating conditions), intrinsic motivation (computer playfulness) and emotion (computer anxiety) serve as anchors that determine the initial perceptions of ease of use. Meanwhile, objective usability, perceptions of external control and perceived enjoyment from system use become effective with increasing experience with the system. Besides the studies of Venkatesh and Davis (1996), Karahanna and Straub (1999) examined the effect of perceptions of social presence, social influence, accessibility and training on perceived usefulness and perceived ease of use.

A number of studies have examined the validity and applicability of TAM to different contexts; e.g. study of Hu et al. (2005) and Hu et al. (1999). TAM has been proven to be a powerful and robust model in understanding and explaining IT usage (Legris et al., 2003; King & He, 2006). Nevertheless the model has several shortages highlighted by the studies that analyze and extend TAM by introducing new factors other than perceived usefulness and perceived ease of use. These factors are subjective norm (Hu et al., 2005; King & He, 2006), organizational and social factors (Legris et al., 2003), social pressure (Kwon & Chidambarm, 2000), task-technology fit (Dishaw & Strong, 1999; King & He, 2006), enjoyment/fun (Kwon & Chidambarm, 2000), individual factors such as prior experience (King & He, 2006; Agarwal & Prasad, 1999; Taylor & Todd, 1995), computer self efficacy (King & He, 2006), tool experience (Dishaw & Strong, 1999), gender (King & He, 2006; Kwon & Chidambarm, 2000), age (Kwon & Chidambarm, 2000; Burton-Jones & Hubona, 2005), education level (Agarwal & Prasad, 1999; Kwon & Chidambarm, 2000; Burton-Jones & Hubona, 2005), professional orientation (Kwon & Chidambarm) contextual factors such as task characteristics (Dishaw & Strong, 1999) and culture (King & He, 2006).

Table 1. Summary of the determinants of technology acceptance

Technological factors	Contextual factors
Perceived usefulness	Facilitating conditions
Task-technology fit	Training
Job relevance	Task characteristics
Relative advantage	Voluntriess of use
Performance expectancy	
Perceived ease of use	
Effort expectancy	
Technological complexity	
Objective usability	
Output quality	
Result demonstrability	
Affective quality	
Compatibility	

Individual factors	Social factors
Gender	Subjective norm
Age	Social influence
Education level	Social pressure
Professional orientation	Image
Intellectual capacity	Visibility
Computer self-efficacy	
Personal innovativeness	
Perceived enjoyment/fun	
Cognitive absorption	
Computer playfulness	
Computer anxiety	
Tool experience	
Prior experience	
Cultural background	

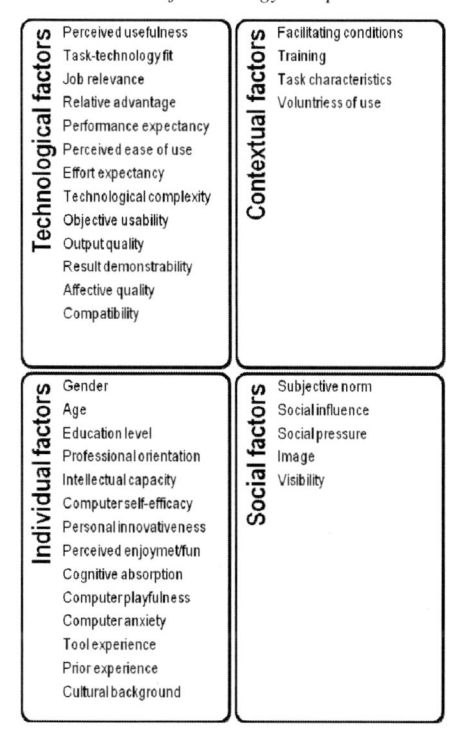

Besides TAM, other approaches from Moore and Benbasat (1991), Compeau and Higgins (1995), Goodhue and Thompson (1995), Venkatesh and Davis (2000),

Agarwal and Karahanna (2000), Venkatesh et al. (2003), Sun and Zhang (2004b), Han et al. (2006), Baron et al. (2006), and Zhang et al. (2005, 2006) are considered. Subsequently, all the determinants of usage proposed by the mentioned researchers are clustered and categorized under four main groups:

- Technological factors
- Contextual factors
- Individual factors
- Social factors

Table 1 provides a summary of all the determinants.

Methodology

Definition of variables

The research is aiming at (1) describing bus driver's attitudes towards technology, (2) determining the perceptions of technological appearance in a dashboard design, and (3) making associations between the driver's characteristics and the research variables. Constructs that are derived from literature review and relevant information from the previous exploratory research has been taken into consideration and variables are determined as shown in Table 2.

Table 2. Research variables

Technological factors	
	performance of the vehicle
	driving safety
	passenger safety
	driving comfort
	durability and endurance
perceived usefulness	cost of maintenance and maintainability
perceived ease of use	anticipated ease of use
	perceived stylishness
affective quality	perceived comfort
	pleasure of driving

Individual factors	
	age
demographic characteristics	education level
	attitude towards general technology
personal traits	involvement with current technological products

Since there is no concrete technological product to investigate and the aim is to search for the design elements that are relevant for acceptance and usage behavior, contextual and social factors of acceptance are beyond the scope of this study.

Technological factors from the literature are condensed under three headings; perceived usefulness, perceived ease of use, affective quality. *Perceived usefulness* is the variable that spans perceived usefulness, task technology fit, job relevance, and relative advantages, which are shown in Table-1. Perceived usefulness is the extent to which a driver thinks an application helps him perform his job better. In order for the driver to find a system useful, the benefits should be in accordance with the concerns of the driver. From the findings of the previous research, the main concerns of drivers are found out to be "performance of the vehicle, driving safety, passenger safety, driving comfort, durability and endurance, cost of maintenance and maintainability". Perceived ease of use is the variable that spans performance expectancy, perceived ease of use, effort expectancy, technological complexity, objective usability shown in Table 1. Perceived ease of use is the driver's perception that using the system will be effort-free. Electronic functioning of the system provides several facilities like automated features and ease of diagnostics due to the detailed vehicle monitoring. Nevertheless, the interfaces of these systems may introduce cognitive difficulties with the interaction. Affective quality is an individual's perception that an object has the ability to change his/her affective state. The driver may find driving a bus with technological features very privileged, if he thinks these feature adds prestige to the vehicle. This kind of positive attitude will contribute to technological acceptance. Not only technological functions, but the 'technological look' in the vehicle is important in this respect. Therefore it is necessary to question aesthetic preferences of drivers about technological appearance of the dashboard. Perceived enjoyment, perceived stylishness and perceived comfort* are the criteria for evaluating this determinant.

Individual factors are considered in two headings; demographic characteristics and personal traits. "Attitude toward general technology" refers to drivers' affinity for technology, that is, how much closer the driver feels himself to general technology. In this statement, general technology means technological products or services, which the driver can identify as technological. Obtaining information about drivers approach towards general technology is important, since attitude towards general technology is the key determinant for adoption of technology integrated products, rather than demographic attributes of the consumer (Modalh, 1999). In order to measure this variable, affinity for technology scale, which is proposed by Edison and Geissler (2003) is translated to Turkish and employed in this study. This variable is supposed to span the determinants such as computer self-efficacy, personal innovativeness, perceived fun, computer anxiety. *Involvement with current technological products* refers to the degree of the driver's familiarity with products that have recent technology and it is the extent to which he is knowledgeable about the usage of these products. The products which are considered in this study consist of cellular phone, computer, MP3 player and digital camera, since they have digital interfaces that propose similar interaction with the systems that can be integrated into the dashboard.

---

*The meaning of comfort in 'driving comfort' is as Dumur et al. (2004) defines: the pleasant and satisfying feeling of being physically or mentally free from pain and suffering, or something that provides this feeling. On the other hand, the term comfort used in 'perceived comfort' means material well-being, conveniences that make life easier and more pleasant. This term is related to added values regarding the concepts like luxury, pleasure, expansiveness, etc.

The sample

In the study, 30 bus drivers in Ankara were interviewed, 12 of which were public bus drivers, 11 were private public bus drivers and 7 were service bus drivers. The age of the respondents varied from 23 to 61 (mean 39.9); 8 of the 30 drivers were primary school graduates, 14 of them were secondary school graduates and 8 of them were high school graduates.

Data collection

The research is conducted in a questionnaire format and it is delivered to the respondents on a face to face basis, since it was observed in the previous studies that these respondents had difficulty in reading and comprehending written statements.

In the questionnaire, after age and education level has been asked, a technology affinity scale was applied. Drivers are asked to evaluate the items on a 5 point likert scale, 3 being the neutral. Then they are asked about their involvement with their current technological products and their features accordingly. As the next step, drivers were asked to define the features of a 'technological' dashboard in a public bus. This study was conducted before a finalized design and actually it is supposed to constitute a basis for decision of integration and design of any particular feature. Therefore no concrete feature was questioned. Instead the aim was to get an idea of what the drivers have in mind regarding this issue.

Following this step, they evaluated the features they defined according to the variables that are derived from the literature (perceived usefulness, perceived ease of use, affective quality). This evaluation is supposed to reflect the positive or negative attitudes of drivers toward the features.

In the last step, photographs of seven different dashboards were presented and the drivers were required to rank order them considering the technological appearance. The features and design components that were important for the drivers while determining technological appearance were questioned in detail. The reason of presenting visual stimuli is to help drivers to focus deeper on the situation and express their ideas more freely.

Analysis and results

The results of the 'technology affinity scale' (Figure 1) reveal that attitude toward technology is rather positive (mean score 3.81, with S.D. 0.58), Only one driver has scored below 3 (in a 5 point likert scale) and 13 drivers are found to have very positive attitudes towards technology, having scored at least 4 in the scale. No relationship is apparent between the attitude towards technology and demographic characteristics.

As seen in Figure 2, 3 and 4; 14 out of 30 drivers do not use a computer, one of the drivers does not even use a cellular phone and 7 of the others use it only for calling, they do not engage in other features, not even SMS. Although the drivers do not attitudes towards dashboard design 323

seem to be practically close to technology, their attitude is not negative considering the scores in 'technology affinity scale'.

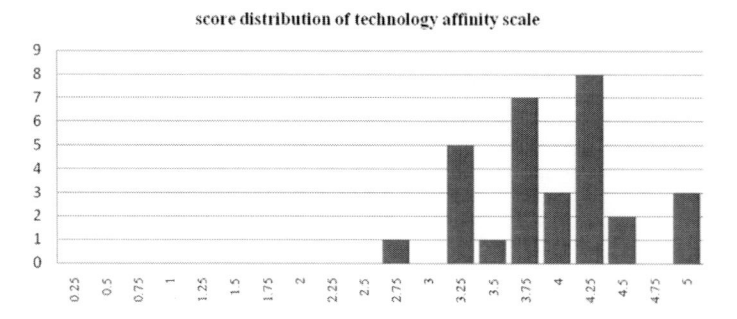

Figure 1. Results of the technology affinity scale

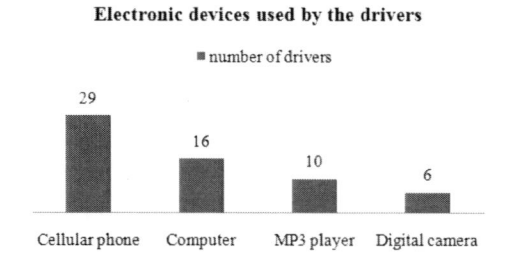

Figure 2. Electronic devices used by the drivers

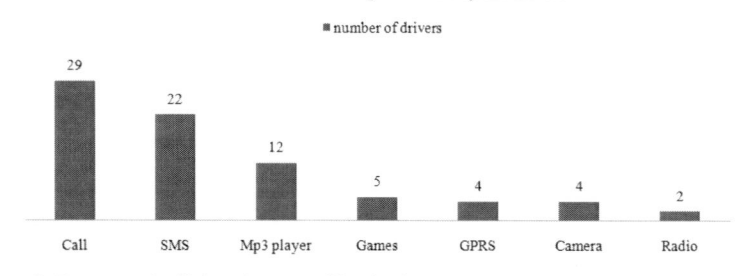

Figure 3. Features of cellular phone used by the drivers

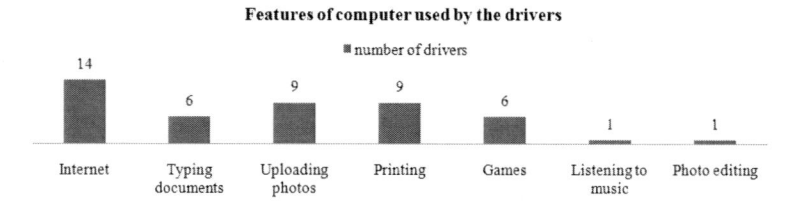

Figure 4. Features of computer used by the drivers

The aspects of technology perception regarding the dashboard are drawn out and certain categories of technology perception are defined by clustering the features of 'technological' dashboard that are stated by the drivers. These features also include the design components, which were defined by the driver during the rank ordering stage. Accordingly, driver's perception of technology in the bus can be categorized as follows:

Aspect 1: Defining technology with *features improving mechanical functioning* of the vehicle (Electronic processing of the system, brake systems such as ABS, ASR, retarder,...)

Aspect 2: Defining technology with *job assistive systems* that support the tasks of a public bus driver – regarding both the driving activity and interaction with the passengers (Information systems that warn the driver about the defects and malfunctions or about the route and the bus stops to stop by, video camera that allow the driver to control the passengers effectively, audible warning signals,...)

Aspect 3: Defining technology with *features that provide physical comfort of use* (Automatic transmission or gearshift that is positioned close to the steering wheel, automatic controls, voice controls, adjustable steering wheel,...)

Aspect 4: Defining technology with *features that* are not directly related to drivers' task, but *provide comfort* (meaning material well-being) (Radio, stereo, air conditioner,...)

Aspect 5: Defining technology with *'digital' features that have a 'technological look'* (Digital displays, digital screens,...)

The number of drivers that have mentioned about each aspect is seen in Figure 5. Accordingly, perception of technology regarding the dashboard is dominated mainly with job assistive systems (aspect 2) and digital features (aspect 5).

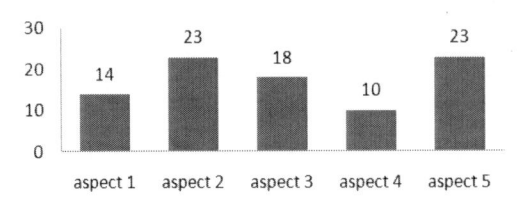

Figure 5. Drivers' frequency of mentioning each aspect

The relationships between technology perception and age, education level, technological affinity are checked. Education level and having scored higher or lower on technology affinity scale have no affect on perception. However, the results show some differences with age. As given in Figure 6, the drivers aged between 23 and 41 have mentioned more about *features that provide physical comfort of use, features that provide comfort* and *'digital' features that have a 'technological look'* more than the drivers who are aged between 42 and 61; whereas the latter group has referred the *features improving mechanical functioning* more than the former group.

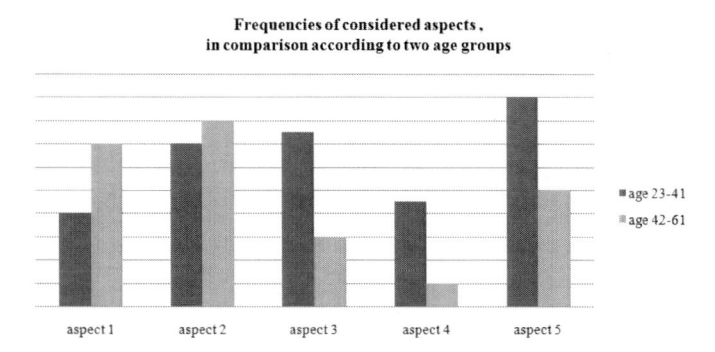

**Frequencies of considered aspects,
in comparison according to two age groups**

Figure 6. Comparison of technology perception according to two age groups

The features defined by the drivers are grouped according to these five categories of perception. Initially, the variables which are seen related to the defined features are checked out, since the scale had 'this item is unrelated to the feature' option. This phase gave a basic idea about the perception of the driver. Figure 7 shows categories regarding the perception of technology in the bus and the variables that they are mainly found related.

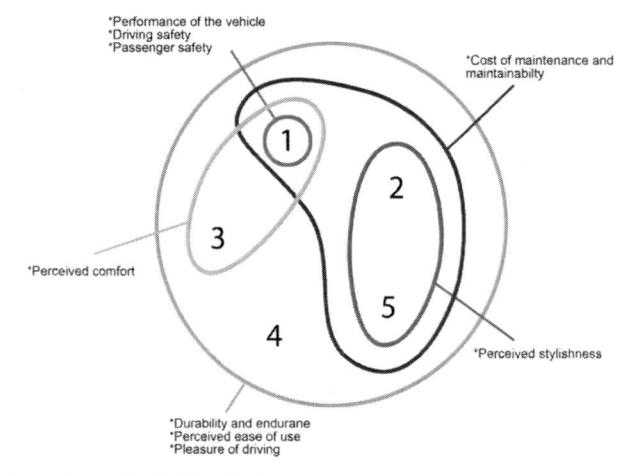

1- features improving mechanical functioning
2- job assistive systems
3- features that provide physical comfort of use
4-features that are not directly related to drivers' task, but provide comfort
5-'digital' features that have a 'technological look'

Figure 7. Related aspects of technology perception and acceptance variables

As the next step, the scores of evaluation are analyzed, with the aim of understanding how the defined categories are conceived according to the variables of technology acceptance. As seen in Table 3, technological features received the highest evaluation in regard to affective quality, whereas they are negatively evaluated with respect to two variables, durability and endurance and cost of maintenance and maintainability.

Table 3 Evaluation of five aspects of perception according to acceptance variables

		1- features improving mechanical functioning	2- job assistive systems	3- features that provide physical comfort of use	4- features that are not directly related to drivers' task, but provide comfort	5- 'digital' features that have a 'technological look'
perceived usefulness	performance of the vehicle	++	o	o	o	o
	driving safety	+	+	++	o	o
	passenger safety	+	++	++	o	o
	driving comfort	+	+	++	+	+
	durability and endurance	-	--	-	-	--
	cost of maintenance and maintainability	--	--	-	-	-
perceived ease of use	anticipated ease of use	+	++	++	+	+
affective quality	perceived stylishness	++	++	++	++	++
	perceived comfort	+	+	+	++	+
	pleasure of driving	++	++	++	++	++

- ++   higher score in evaluation
- +   high score in evaluation
- o   not related enough
- -   low score in evaluation
- --   lower score in evaluation

The drivers' attitude towards the issue of 'perceived ease of use' was quite positive; 8 drivers out of 30 evaluated the technological systems negatively regarding perceived ease of use. While responding to this question, quite a number of the drivers stated that, the use of a technological function or a system is easy:

- 'after that he has learned',
- 'to a person who already knows how to use',
- 'after he gets used to it',
- 'if he receives a training before',
- 'because this issue is related to being accustomed', etc.

These statements show that this issue has two dimensions; perceived ease of use at the very beginning and perceived ease of "learnability". The drivers seem to think that, technological features, such as navigational screens, may be difficult to use at the beginning, but they suppose they will learn it and get used to it easily, and evaluate it as easy to use accordingly.

Conclusion

Drivers having positive attitudes towards technology in general and towards the technology integrated into the dashboard implies that no serious resistance is expected against a design including technological features. However, data regarding the low involvement of drivers with current technological products such as cellular phone and computer imply that there are considerable numbers of drivers, who are not familiar with menu-structure, which is commonly used in vehicle information systems. In previous studies the drivers have also stated that they received no training on specific features of the dashboard before they begin driving a bus. Therefore, the design of the interface of these systems requires elaborate consideration.

Based on the result that technological features are evaluated negatively only with respect to durability and cost of maintenance, it can be concluded that these aspects should also be in the scope of main consideration. The technological features to be integrated should give a durable feel, so that the attitude of drivers is more positive. Comprehending the perceived durability necessitates further study.

As a last point, physical appeal has been found to have great importance in the formation of positive attitudes. Similarly, in the previous research there were participants, for whom aesthetic appeal of displays, controls and the dashboard in general was one of a main concern, in determining the general criteria for the design. Some of them even consider about self confidence that the driver gains in front of the passengers, when driving a bus with an aesthetic, technological, and appealing control and display panel. As long as new technological features to be integrated have the preferred physical appeal, they are supposed to be accepted easier. These features are to be considered in detail with further studies.

Acknowledgements

The project, which is the knowledge source of this paper, is funded by TUBITAK - The Scientific and Technological Research Council of Turkey- and TEMSA, which is a Turkey based automotive manufacturer company.

References

Agarwal, R., & Karahanna, E. (2000). Time flies when you're having fun: Cognitive absorption and beliefs about information technology usage. *MIS Quarterly, 24*, 665-694.
Agarwal, R., & Prasad, J. (1999). Are individual differences germane to the acceptance of new information technologies? *Decision Sciences, 30*, 361-391.

Baron, S., Patterson, A., & Harris, K. (2006). Beyond technology acceptance: Understanding consumer practice. *International Journal of Service Industry Management, 17*, 111-135.

Bowen, W. (1986). The Puny Payoff from Office Computers. *Fortune,* May 26, 20-24.

Burton-Jones, A., & Hubona, G.S. (2005). Individual differences and usage behavior: Revisiting a technology acceptance model assumption. *The DATA BASE for Advances in Information Systems, 36*, 58-77.

Compeau D.R., & Higgins, C.A. (1995). Computer self-efficacy: Development of a measure and initial test. *MIS Quarterly, 19*: 189-211.

Davis, F.D. (1989) Perceived usefulness, perceived ease of use and user acceptance of information technology. *MIS Quarterly, 13,* 319-340.

Dishaw, M.T., & Strong, D.M. (1999). Extending the Technology Acceptance Model with Task-Technology Fit Constructs. *Information and Management, 36*, 9-21.

Dumur, E., Barnard, Y., & Boy, G. (2004). Designing for comfort. In D.de Waard, K.A. Brookhuis, and C.M. Weikert (Eds.). *Human Factors in Design.* (pp.111-127). Maastricht: Shaker Publishing.

Edison, S.W. & Geissler, G.L. (2003) Measuring attitudes towards general technology: Antecedents, hypotheses and scale development. *Journal of Targeting, Measurement and Analysis for Marketing, 12*, 137-156.

Goodhue, D.L., Thompson, R.L. (1995). Task-technology fit and individual performance. *MIS Quarterly, 19*, 213-236.

Grosbrink, A. & Mahr, A.C. (1998). 102 Transport Industry and Warehousing.Ergonomics of Bus Driving. In J.M. Stellman (Ed.) *Encyclopaedia of Occupational Health and Safety - 4th ed.* Vol.3. Geneva: International Labour Office.

Han, S., Mustonen, P., Seppanen, M. & Kallio, M. (2006). Physicians' acceptance of mobile communication technology: an exploratory study. *International Journal of Mobile Communications, 4*, 210-230.

Hu, P.J., Chau, P.Y.K., Sheng, O.R.L., & Tam, K.Y. (1999). Examining the technology acceptance model using physician acceptance of telemedicine technology. *Journal of Management Information Systems, 16*, 91–112.

Hu, PJ, Lin, C., & Chen, H. (2005). User acceptance of intelligence and security informatics technology: A study of COPLINK. *Journal of the American Society for Information Science and Technology, 56*, 235–244.

Karahanna, E, & Straub, D.W. (1999). The psychological origins of perceived usefulness and ease-of-use. *Information & Management, 35*, 237-250.

King, W.R., & He, J. (2006). A meta-analysis of the technology acceptance model. *Information & Management, 43*, 740–755.

Kwon, H.S., & Chidambaram, L. (2000). A test of the technology acceptance model. The case of cellular telephone adoption. In *37th Hawaii International Conference on System Sciences Proceedings.*

Legris, P., Ingham, J., & Collerette, P. (2003). Why do people use information technology? A critical review of the technology acceptance model. *Information & Management, 40*, 191–204.

Modalh, M. (1999). *Now or Never: How companies must change today to win the battle for internet consumers.* New York: Harper Business.

Moore, G.C., & Benbasat, I. (1991). Development of an instrument to measure the perceptions of adopting and information technology innovation. *Information Systems Research, 2*, 192-222.

Sun, H., & Zhang, P. (2004a). A Methodological Analysis of User Technology Acceptance. In *37th Hawaii International Conference on System Sciences Proceedings.*

Sun, H., & Zhang, P. (2004b). The role of moderating factors in user technology acceptance. *International Journal of Human-Computer Studies, 64*, 53-78.

Szajna, B. (1996). Empirical evaluation of the revised technology acceptance model. *Management Science, 42*, 85-92.

Taylor, S., Todd, P. (1995). Assessing IT usage: The role of prior experience. *MIS Quarterly,* December 1995, 561-570.

Tse, J.L.M., Flin, R., & Mearns, K. (2006). Bus driver well-being review: 50 years of research. *Transportation Research Part F, 9*, 89–114

Venkatesh, V. (2000). Determinants of perceived ease of use: Integrating control, intrinsic motivation, and emotion into the technology acceptance model. *Information Systems Research, 11*, 342-361.

Venkatesh, V., & Davis, F.D. (1996). A model of antecedents of perceived ease of use: Development and test. *Decision Sciences, 27*, 451-481.

Venkatesh, V., & Davis, F.D. (2000). A theoretical extension of the technology acceptance model: Four longitudinal field studies. *Management Science, 46*, 186-204.

Venkatesh, V., Morris, M.G., Davis, G.B., & Davis, F.D. (2003). User acceptance of information technology: Toward a unified view. *MIS Quarterly, 27*, 425-478.

Zhang, P., Li, N. (2005). The importance of affective quality. *Communications of the ACM, 48*, 105-108.

Zhang, P., Li, N. & Sun, H. (2006). Affective quality and cognitive absorption. Extending technology acceptance research. In *39th Hawaii International Conference on System Sciences Proceedings.*

Public Surveillance

EYE MOVEMENTS DURING CCTV MONITORING

In D. de Waard, J. Godthelp, F.L. Kooi, and K.A. Brookhuis (Eds.) (2009). *Human Factors, Security and Safety* (p. 331). Maastricht, the Netherlands: Shaker Publishing.

Searching for threat: factors determining performance during CCTV monitoring

Christina J. Howard[1], Tomasz Troscianko[1], Iain D. Gilchrist[1], Ardhendu Behera[2], & David C. Hogg[2]
[1]*University of Bristol*
[2]*University of Leeds*
UK

Abstract

Monitoring closed-circuit television (CCTV) for security purposes is a task requiring sustained attention and the processing of many complex, constantly changing visual elements. Studies of performance in such tasks reveal a high level of workload and rapid loss of performance as workload is increased. Similarly, laboratory based experimental paradigms suggest that performance in CCTV monitoring is extremely dependent on the complexity and number of video screens monitored. We suggest that measuring eye movements during CCTV monitoring might provide a novel and rich source of data to illuminate the question of how CCTV monitoring is performed. In the psychological literature, two influences on attention are traditionally considered: the ability of events in the world to capture our attention regardless of our current goals (stimulus-dependent salience) and the ability to direct our attention towards stimuli relevant to the task we are trying to perform (goal-based relevance). Stimulus-dependent salience and goal-based relevance together determine the human fixation priority assigned to scene locations (Fecteau and Munoz, 2006, TICS 10, 382-390). Tests of the stimulus-dependent salience component of this process tend to look for regions in the image that are consistently fixated and link this to the underlying image properties. However, when the task is common across observers, consistent fixation location can also indicate that that region has high goal-based relevance. By examining the eye movements of multiple expert observers, we may start to characterise features of the moving video stimulus that are predictive of events likely to be judged as suspicious.

Eye movements in the real world

Studies of naturalistic task performance have used eye movements as a measure of how attention moves around the visual field, or 'attentional deployment' (e.g. Hayhoe & Ballard, 2005; Land, 1999; Underwood, Chapman, Brocklehurst, Underwood, & Crundall, 2003; Findlay & Gilchrist, 2003). The way people use eye movements to do various everyday tasks has been investigated; including steering a racing car (Land & Tatler, 2001), hitting a cricket ball (Land & MacLeod, 2000), making tea (Land, Mennie, & Rusted, 1999) and sandwiches (Hayhoe, 2000). The pattern of eye movements recorded suggest that the eyes do not perform a pure 'forward-planning' role for actions, nor do they wait for actions to occur before fixating relevant parts of a scene. Instead they provide visual information 'just-in-time' for action and lead actions by between around half a second to a second (Land & Hayhoe, 2001). In addition, where observers look in a scene is not determined solely by saliency derived from low level features of the kind described by Itti and Koch in their model (Itti & Koch, 2000). Instead, fixation 'priority' is instead determined by the dual influences of stimulus-dependent low-level salience and goal-based relevance i.e. whether the stimulus relates directly to the task at hand (Fecteau & Munoz, 2006). In the current programme of work we measure eye movements to investigate such attentional deployment for closed-circuit television (CCTV) monitoring. We hypothesise that eye movements are likely to lead responses by a buffer of the approximate magnitude suggested by Land and his colleagues (e.g. Land & Hayhoe, 2001). It remains an open question, however, whether the level of cognition required for a task such as CCTV might extend the length of this temporal buffer while cognitive processing of fixated events takes place.

Closed-Circuit Television (CCTV)

In CCTV control rooms, the principal duty of operators is to monitor a bank of screens through which about 50-60 images selected from up to 600 cameras are streamed, with the task of anticipating and detecting critical events. Each operator also has a desktop screen (spot monitor) through which, when something from the wall-mounted screens attracts attention, they can select chosen video images for display on their spot monitor and over-ride the pre-set programme. Despite the complexity of this task, the prediction of incidents by the operators appears to be possible (Troscianko et al. 2004). The main difficulty in this task is the degree of cognitive and visual overload both in terms of monitoring what may be complex and constantly changing scenes, and in terms of allocating attention effectively to many screens at once. Despite the familiarity of operators with the scenes and likely occurrences presented to them, as well as their ability to group images in terms of their geographical locations, this is still clearly a very demanding task. The paucity of research, and therefore, guidance on this topic has long been a subject of concern to CCTV operators and managers, and calls for an investigation into this issue are still being made by the CCTV surveillance community (e.g. Donald, 2005). Tickner and Poulton (1973) found that typical levels of perceptual load are greater than the average individual can process effectively. They had observers monitor 4, 9 or 16 screens, and recorded detection rates of 83%, 84% and 64%, respectively. More recently in a survey, 82% of operators reported feeling able to monitor a maximum of 16 cameras or fewer, with more than half of operators reporting a maximum of as little as one to four cameras (Wallace, Diffley & Aldridge, 1997). Clearly the task is a demanding one even for trained operators, and a deeper understanding of the way in which CCTV monitoring is performed may help us to design and implement systems that maximise human monitoring performance, bearing in mind that the number and type of camera images will depend on the type of surveillance task.

Figure 1. Scenes similar to those typically captured by CCTV. Even in this relatively sparse display of only four static scenes, cognitive and perceptual load is clearly high, and may result in missed incidents or late detection. Colour images can be viewed at http://extras.hfes-europe.org

Laboratory tasks relevant to CCTV monitoring

The literature on theoretically-driven tasks in the laboratory predicts that CCTV will place a heavy load on the perceptual and cognitive systems, with performance decreasing rapidly as the complexity and number of video streams is increased. This is certainly in accordance with the reports of CCTV operators and goes some way to explaining the nature of the task faced during CCTV monitoring.

From the laboratory, there is a wealth of evidence pointing towards the detrimental effects of load of performance. In visual search tasks, the speed with which search targets are detected is typically dependent on the number of objects in the search array. Although search for some features appears to exhibit 'pop-out', or to display relatively little effect of set size on reaction time (such as a target defined by its unique colour in the display), search times for targets defined by conjunctions of features, or search for an object defined by the presence of both of two or more features (e.g. a red, rightward-tilted object), exhibits large decrements with set size (Treisman & Gelade, 1980; Treisman, Sykes & Gelade, 1977). This is of particular relevance to CCTV monitoring tasks since even in a single CCTV monitor, there are many objects, people and environmental features all competing for attention. The target, i.e. potentially suspicious behaviour, appears likely to be defined by a conjunction of features such as suspicious patterns of movement, location in the scene and physical appearance. Furthermore, search for disjunctions of features (e.g. a red object or a rightward-tilted object) has been shown to cause a performance cost compared to search for a single feature in applied contexts. Manneer, Barrett, Phillips, Donnelly and Cave (2007) showed that for airport security screening, searching for two potential target types frequently caused a loss in accuracy, and in some cases an increase in search time, compared to searching for a single object type. Clearly, CCTV monitoring is likely to involve some similarity both to conjunction and disjunction searches, since there can be no single prototypical search target. Rather, operators must search simultaneously for many potentially criminal events, as well as accidents or other incidents requiring intervention.

Another field of laboratory work is particularly relevant to the task of CCTV monitoring, namely the multiple object tracking (MOT) task and its variants. The MOT task is reminiscent of the traditional fairground 'shells' game, whereby a shell is hidden underneath one of a number of visually identical cups. The cups are then shuffled in a manner requiring close and sustained attention. The observer must keep track of the cup that contains the shell to win a prize. In a typical MOT task, a number of identical moving objects are presented on a screen, some of which are designated as targets for tracking. Observers must attempt to keep track of the target objects and to continue to distinguish them from non-targets, so that at the end of the trial, when queried, they may say whether a probed object was a target or not. Typically in these tasks, a capacity limit of around four objects is reported (e.g. Pylyshyn & Storm, 1988) although more recently a flexible limit has been proposed that depends on objects' speed (Alvarez & Franconeri, 2007) and the degree of precision required (Franconeri, Alvarez & Enns, 2007; Howard & Holcombe, 2008). These results are likely to be applicable to the applied task of CCTV monitoring since the motion of many individuals around one or many screens is likely to be a powerful cue to their potential level of perceived suspiciousness. Many factors which are extremely prevalent in CCTV images have been shown to increase the difficulty of MOT tasks. For instance, performance decreases if objects change their shapes in a non-rigid fashion (vanMarle & Scholl, 2003) as real people would do as their form appears on the screen, or if they pass behind occluders (Scholl & Pylyshyn, 1999) such as would be expected from people moving behind each other, behind trees or buildings, for instance.

A third field directly pertinent to CCTV monitoring is that of divided attention or 'dual' tasks. In such experiments, observers are required to perform two tasks at once, and this has been repeatedly shown to cause a marked decrement in performance in both tasks (e.g. Pashler, 1995; Schneider & Shiffrin, 1977). It seems likely that these results apply at least on occasions to CCTV monitoring, since operators may need to make a cognitive or perceptual judgement about activity in two or more areas of screens simultaneously. Indeed they may also need to continue to monitor many events whilst making decisions about the level of appropriate action required.

Future directions

We believe that measuring eye movements will provide a new window into our understanding of how CCTV monitoring is performed. This is likely to have implications for the ways in which CCTV monitoring is managed and carried out, as well as for control room systems design and for training programmes. The method can be used to study systematically how the number, arrangement and content of screens affect performance. We hope that this method will also create a new synergy between laboratory-based models and their applications.

The results will help to test and constrain laboratory-based models of human performance in CCTV monitoring and in similar tasks. Most laboratory studies use simplified and carefully controlled tasks and stimuli in order to make claims about the underlying perceptual processes involved. The advantage of these methods is of course the ease with which conclusions can be drawn about perceptual processes. Testing in applied scenarios, however, is likely to help generalise these results to tasks that are more relevant to the real world with its inherent complexity. In particular, the visual complexity of CCTV images, and their dynamic nature are likely to influence performance in a manner that is very different from the static, simple stimuli most frequently used in the laboratory.

In addition, we hope that measuring eye movements during monitoring will provide an invaluable new method for studying responses to video stimuli drawn from real-world scenarios outside of CCTV monitoring. Eye movement data will provide a novel and rich source of data for a variety of applied tasks. There are many cases in which a moving image must be consistently attended, and this method could be applied to any scenario, perhaps relating to other fields of safety and security such as air traffic control, or relating to navigation in the real world, to sports behaviour or to social perception.

Figure 2. Sample fixation data shows the position of the eyes for eleven observers at a single point in time. The left-hand panel shows a time at which the between-observer consistency is low. The right-hand panel shows a time at which between-observer fixations are highly consistent, indicating an event of significant interest for observers. Colour images can be viewed at http://extras.hfes-europe.org

We plan to use this human eye movement data in future to inform and constrain a computer leaning algorithm, and to work towards automatic detection of elements of CCTV footage that are likely to be judged as potentially suspicious by human observers. We will study two distinct forms of the CCTV viewing situation. Firstly, to establish a 'ground truth', we will study the decisions of expert observers when looking at a single screen in a rested state, with no competing visual distracters. Secondly, we will study a realistic control-room situation in which there is a bank of multiple screens and a high perceptual and cognitive load. In this multiple screen scenario, we will be able to study the interactive effects of multiple screens on eye movements and attention, as well as strategies used by operators to spread their attention over multiple screens. Eye movement data from the single screen scenario will be used to extract features from the moving video image which are likely to be judged as suspicious. Eye movements are likely to precede behavioural responses and are therefore well placed to best inform artificial detection processes. In addition, when eye movement data is combined across different observers (see Figure 2) and with behavioural responses, it provides a measure of the direction of goal-directed attention.

Acknowledgements

This work is supported by an EPSRC Cognitive Systems Foresight grant. Many thanks to Kate Rennicks, David Walsh and Ian Townley at Manchester City Council CCTV Control Room for their invaluable contributions to this project.

References

Alvarez, G.A. & Franconeri, S.L. (2007). How many objects can you track? Evidence for a resource-limited attentive tracking mechanism. *Journal of Vision, 7*(13), 1-10.

Donald, C. (2005) How many monitors should a CCTV operator view? In *CCTV Image*. STL Publishing, London, England, pp. 355.

Fecteau, J.H. & Munoz, D.P. (2006) Salience, relevance, and firing: a priority map for target selection. *TRENDS in Cognitive Sciences, 10*, 382-390.

Franconeri, S.L., Alvarez, G.A., & Enns, J.T. (2007). How many locations can be selected at once? *Journal of Experimental Psychology: Human Perception and Performance, 33*, 1003-1012.

Findlay, J.M. & Gilchrist, I.D. (2003) *Active vision: The psychology of looking and seeing.* Oxford: Oxford University Press.

Hayhoe, M. (2000) Vision using routines: A functional account of vision. *Visual Cognition, 7*, 43-64.

Hayhoe, M. & Ballard, D. (2005) Eye movements in natural behavior, *TRENDS in Cognitive Science, 9*, 188-194.

Howard, C.J. & Holcombe, A.O. (2008) Tracking the changing features of multiple objects: Progressively poorer perceptual precision and progressively greater perceptual lag. *Vision Research, 48,* 1164-1180.

Itti, L. & Koch, C. (2000) A saliency-based search mechanism for overt and covert shifts of attention. *Vision Research, 40*, 1489-1506.

Land, M. (1999) Motion and vision: why animals move their eyes. *Journal of Comparative Physiology, 185*, 341-352.

Land, M.F. & Tatler, B.W. (2001) Steering with the head: The visual strategy of a racing car driver. *Current Biology, 11*, 1215-1220.

Land, M. & Furneaux, S. (1997) The knowledge base of the oculomotor system. *Philosophical Transactions of the Royal Society: B. Biological Sciences, 352,* 1231 – 1239.

Land, M.F. & Hayhoe, M. (2001) In what ways do eye movements contribute to everyday activities? *Vision Research, 41*, 3559-3565.

Land, M.F. & MacLeod, P. (2000) From eye movements to actions: how batsmen hit the ball. *Nature Neuroscience, 3*, 1340-1345.

Land, M., Mennie, N. & Rusted, J. (1999) The roles of vision and eye movements in the control of activities of daily living. *Perception, 28*, 1311-1328.

Menneer, T., Barrett, D.J.K., Phillips, L., Donnelly N. & Cave, K.R. (2007) Costs in searching for two targets: dividing search across target types could improve airport security screening. *Applied Cognitive Psychology, 21*, 915-932.

Pashler, H. (1995). Divided visual attention. In S. Kosslyn (Ed.) *Visual Cognition: Invitation to Cognitive Science* (pp. 71-100). Cambridge, MA, USA: MIT Press.

Pylyshyn, Z.W. & Storm, R.W. (1988). Tracking multiple independent targets: evidence for a parallel tracking mechanism. *Spatial Vision, 3*(3), 1-19.

Schneider, W. & Shiffrin, R.M. (1977). Controlled and automatic human information processing: 1. Detection, search, and attention. *Psychological Review, 84*, 1-66.

Scholl, B.J. & Pylyshyn, Z.W. (1999). Tracking multiple items through occlusion: clues to visual objecthood. *Cognitive Psychology, 38*, 259-290.

Tickner, A.H. & Poulton, E.C. (1973) Monitoring up to 16 synthetic television pictures showing a great deal of movement. *Ergonomics, 16*, 381-401.

Treisman, A.M. & Gelade, G. (1980). A feature integration theory of attention. *Cognitive Psychology, 12*, 97-136.

Treisman, A., Sykes, M. & Gelade, G. 1977. Selective attention and stimulus integration. In *Attention and performance VI*, S. Dornic (Ed.), Hillsdale, NJ, USA: Lawrence Erlbaum, pp. 333-361.

Troscianko, T., Holmes, A., Stillman, J., Mirmehdi, M., Wright, D., & Wilson, A. (2004). What happens next? The predictability of natural behaviour viewed through CCTV cameras. *Perception, 33*, 87-101.

Underwood, G., Chapman, P., Brocklehurst, N., Underwood J., & Crundall, D. (2003) Visual attention while driving: sequence of eye fixations made by experienced and novice drivers. *Ergonomics, 46,* 629-646.

vanMarle, K. & Scholl, B. J. (2003). Attentive tracking of objects versus substances. *Psychological Science, 14*, 498-504.

Wallace, E., Diffley, C., & Aldridge, J. (1997). Ergonomic Design Considerations for Public Area CCTV Safety and Security Applications, *International Ergonomics Association Congress*, July 1997.

Expertise and strategies in the detection of firearms via CCTV

Iain Darker, Alastair Gale, Anastasssia Blechko, & Marc Whittle
Applied Vision Research Centre
Loughborough University, UK

Abstract

Illicit firearms are frequently carried on the person, on the streets of the UK, in both concealed and unconcealed guises. The UK's considerable CCTV infrastructure offers a suitably distributed network of sensors for the detection of these firearms. However, the capacity of CCTV operators for surveillance is limited. The present research aimed to make explicit the expertise of CCTV operators in the detection of firearms. The overall intention is to inform the design of image processing algorithms which would automate the monitoring of CCTV cameras for firearms. The abilities of CCTV operators (n = 16) and lay people (n = 16) to detect concealed and unconcealed firearms in staged CCTV footage were quantified within a signal detection framework. The visual search strategies employed by each participant were elicited by self-report and indexed against efficacy. Search strategies were also verified empirically in a separate eye-tracking study (n = 24). CCTV operators had greater sensitivity to concealed firearms than lay people, dependent upon type of firearm and camera angle. The results of the study are discussed in terms of effective monitoring strategies, applicability to real-life CCTV monitoring, automatic firearm detection algorithms, and future research with real-life CCTV footage of gun crime.

Introduction

There is evidence that firearms are frequently carried in public places, in both concealed and unconcealed guises; the UK's considerable CCTV infrastructure offers a means by which such firearms can be detected (Darker, Gale, Ward, & Blechko, 2007; Darker, Gale, Ward, Blechko, & Purdy, 2007). CCTV operators spot a majority of incidents directly, through proactive surveillance (Gill et al., 2005). Thus, a key element in the efficacy of CCTV as a crime reduction tool is the skill of the CCTV operator. Therefore, it is desirable to make explicit the nature of expertise in the detection of mal-intent via CCTV. This information might then be used in the design of automatic image processing algorithms for the detection of firearms via CCTV.

Prior research has concentrated on the detection of the precursors of overtly violent behaviour, via CCTV (Troscianko, Holmes, & Stillman, 2004). However, the carrying of firearms need not be accompanied by overtly violent behaviour. There are indications that firearms are frequently carried on the street, in a casual fashion (Hales, Lewis, & Silverstone, 2006). Consequently, there is a need to address potentially subtle visual cues which might indicate that a CCTV surveillance target is carrying a firearm in a casual manner, either concealed on their person or held in plain view in an inconspicuous fashion; both modes of carrying a firearm have been observed in CCTV footage recorded at UK CCTV control rooms (Darker, Gale, Ward, & Blechko, 2007).

If the detection of firearms via CCTV involves learnt skill or expertise then CCTV operators should perform better at this task than lay people. Thus, a signal detection experiment was conducted to provide a robust quantification of the sensitivities of CCTV operators and lay persons in the detection of firearms. Additionally, it was posited that any influence of expertise might vary according to a number of parameters: concealment of the firearm, size of the firearm, and camera view. The experimental stimuli addressed the scenario in which a firearm is carried in a casual fashion. Surveillance targets were viewed in staged CCTV footage, whilst walking towards or away from the camera and carrying objects of various sizes, either concealed on their person or held in plain view in an inconspicuous manner. In an individual CCTV clip an object could be innocuous or it could be a firearm. CCTV operators and lay people were required to make a judgement after each clip regarding whether or not a firearm was present. Further, if expertise were found to confer an advantage in the detection of firearms via CCTV then the inspection strategies employed by CCTV operators might facilitate the reliable detection of firearms and they might inform the design of image processing algorithms to automate the detection of firearms via CCTV. Thus, CCTV operators and lay people were also asked to rate the frequency with which they used a series of strategies in the detection of firearms. However, it is noted that this self-report method for eliciting inspection strategy has limitations. Consequently, a follow-up experiment assessed inspection strategy in an empirical fashion using eye-tracking technology.

Experiment 1

Method

Participants
Thirty eight participants were recruited to the study. Data from six participants were discarded as, for each of these participants, the detection model did not account for all the variance in the data; the recommendation is to reject such data (Harvey Jr, 2007). The data of the remaining 32 participants were retained (28 men, age range 18 - 63 years, $M = 42$, $SD = 12$). Half of these participants were CCTV operators (all were engaged in public space surveillance; years of experience: 0.33 - 19, $M = 5$, $SD = 5$; hours per shift spent monitoring cameras: 1 - 12, $M = 8$, $SD = 4$), the other half were lay people with no security industry training or experience.

Stimuli
The stimuli were 2-second video clips of individual actors (hereafter referred to as surveillance targets) walking against a neutral backdrop. In a number of video clips

the surveillance target was required to carry an object: either a firearm or an innocuous object matched to the firearm for approximate size, weight, and colour (a plastic drinks bottle filled with a dark-coloured liquid). The firearm or bottle was either held in plain view in the hand, or it was concealed on their person, in their clothing. In either case the object was positioned between the surveillance target's body and the camera used to record the footage. The camera was positioned so to produce a view that might be obtained from a CCTV camera: an elevated, three-quarter angle view of the walkway at a height of 2.8 metres, approximately 9.8 metres from the centre of the walkway along an axis subtended at an angle of 50° to the backdrop. The camera view was panned out to show 13.5 metres of walkway, although during the 2-second clip surveillance targets appeared only within the central third of the walkway. To capture the surveillance targets in full stride the clip featured 1 second of walking either side of the centre point of the walkway. Seven hundred and ninety two video clips were used in the experiment: 72 clips of people carrying unconcealed firearms; 72 clips of people carrying concealed firearms; 216 clips of people carrying unconcealed bottles; 216 clips of people carrying concealed bottles; 216 clips of people carrying neither a firearm nor a bottle. Twelve different surveillance targets were filmed to produce these video clips and the numbers of each type of video clip were distributed equally across them. The clips involved three sizes of firearm: a .32 calibre revolver (small), an automatic pistol (medium), and a sawn-off shotgun (large). Each firearm was matched to one of three sizes and weights of bottle allowing gun detection performance to be assessed with respect to object size (small, medium, and large). Surveillance targets were filmed walking towards and away from the camera to produce ventral and dorsal views of the surveillance target, respectively. These variations were distributed equally across surveillance targets, concealment conditions (concealed, unconcealed), and views of the surveillance target (ventral, dorsal), and in the appropriate proportions across object types (ratio: 1 firearm: 3 bottle: 3 no additional object) at each object size (small, medium, large). At a viewing distance of 450mm, each surveillance target subtended approximately 4° visual angle and the firearms and bottles subtended between 0.1° and 0.5° visual angle, depending on type.

Firearm detection performance
Firearm detection performance was analysed within a signal detection framework (Harvey Jr, 2003) and indexed using ROC parameters. The signal detection analysis measures the ability of a sensor to detect a signal against a background of noise. In this instance the sensor is the human visual system and the signal is a firearm (and the firearm's influence upon the person carrying it). The noise condition tested here involved an innocuous object (and the innocuous object's influence upon the person carrying it). The innocuous object was a plastic drinks bottle which was matched to the firearm for approximate size, weight, and colour; this was in order that the experiment should test the detection of a firearm, specifically, rather than simply the detection of any object. The measure of detection sensitivity adopted for the present analysis was d_a which was obtained by means of a maximum likelihood algorithm (RSCORE 5.5.7) based on a Gaussian probability distribution (Harvey Jr, 2007).

Design
Participants were pseudo-randomly assigned to view either concealed or unconcealed objects within a stipulated design: 8 CCTV operators and 8 lay people viewed the concealed condition; 8 CCTV operators and 8 lay people viewed the unconcealed condition. Each participant saw 504 experimental video clip stimuli which were determined by the concealment condition to which they had been assigned: Concealed: 72 clips of people carrying concealed firearms, 216 clips of people carrying concealed bottles, 216 clips of people carrying no additional object (neither a firearm nor a bottle); Unconcealed: 72 clips of people carrying unconcealed firearms, 216 clips of people carrying unconcealed bottles, 216 clips of people carrying no additional object (neither a firearm nor a bottle). Influences on firearm detection performance were measured within a full-factorial, mixed design: expertise (CCTV operator, lay person) and concealment (concealed, unconcealed), between-participants; object size (small, medium, large) and camera view (ventral, dorsal), within-participants. Influences on firearm detection strategy were measured in a full-factorial, between-participants design: expertise (CCTV operator, lay person) and concealment (concealed, unconcealed).

Procedure
Stimulus presentation and data collection were automated by computer. Each participant was exposed to 8 practice video clips, followed by 504 experimental video clips. Following the presentation of each video clip, participants were prompted for a response: Was a firearm present in the video clip that you have just viewed? Responses were made on a six point scale of confidence: Definitely no; Probably no; At a guess, no; At a guess, yes; Probably yes; Definitely yes. The response was recorded automatically and the next video clip was initiated when the response had been submitted. The video clips were presented in a pseudo-randomised order which intermingled and balanced the presentation of the various experimental conditions evenly across the duration of the experiment. After the experiment, participants completed a questionnaire regarding the strategies that they had used to decide whether or not a firearm had been present in the video clips. CCTV operators then completed a questionnaire on CCTV camera handling and the use of CCTV to detect crime.

Firearm detection strategy
Detection strategy was assessed after participants had viewed all 504 experimental video clips in the concealment condition to which they had been assigned. Participants were asked: In the video clips that you have just seen, how did you decide whether or not a person was carrying a gun? In response, participants rated a series of 17 potential task strategies on a 5-point scale which reflected the frequency with which they had employed that strategy: 1) not at all; 2) very little; 3) sometimes; 4) a lot; 5) all the time. The strategy statements were derived from feedback to a pilot study (n = 8) which employed the same staged CCTV clips, as well as on the basis of literature reviews regarding the use of CCTV, surveillance and inspection strategies, and the influences of firearms on behaviour (Darker, Gale, Ward, & Blechko, 2007; Darker, Gale, Ward, Blechko, et al., 2007). Strategy statements were tailored to the concealment condition to which that participant had been assigned.

Both concealed and unconcealed conditions: I guessed; I looked at how they walked; I looked at how fast they were walking; I looked at the length of their stride; I looked at how they swung their arms; I looked at the expression on their face; I looked at their posture; I looked at their demeanour or overall behaviour; I looked at the shadow on the wall; I looked at how their clothing hung; I looked at how their clothing moved. Concealed only: I looked for a bulge in their clothing; I looked for a gun-sized bulge in their clothing; I looked for a gun-shaped bulge in their clothing; I looked at where they put their hands; I looked at whether their hands were hidden or not; I looked for possible hiding places in their clothing. Unconcealed only: I looked for a gun-sized object in the hand; I looked for a gun-shaped object in the hand; I looked for a gun-coloured object in the hand; I looked at how they held the object in their hand; I looked at how the object moved in their hand; I looked at how light reflected off of the object in their hand.

Results

The data violated assumptions underlying a parametric analysis and were subjected to non-parametric analyses.

Between-participants groups
Four participant groups were created through the combination of concealment condition (concealed, unconcealed) and expertise (CCTV operator, lay person). The four groups did not differ significantly in terms of age. The CCTV operator groups did not differ significantly in terms of years of experience as a CCTV operator or hours per shift spent proactively monitoring cameras.

Sensitivity to the firearm
Do CCTV operators and lay people differ in their abilities to detect firearms, and if so, under what circumstances? The influences of concealment and expertise on overall sensitivity to the firearm were assessed. Then the sensitivities of CCTV operators and lay people were compared for each unique combination of concealment condition, camera view, and object size.

Mann-Whitney U-tests were conducted between sensitivities in the concealed and unconcealed groups within each expertise condition, and also between the sensitivities of CCTV operator and lay person groups within each concealment condition (significant at $\alpha = .05$). The influence of concealment on sensitivity to the firearm was, perhaps, unsurprising. For CCTV operators, sensitivity to the firearm was significantly lower in the concealed condition (Mdn = -0.31) than in the unconcealed condition (Mdn = 2.59; $U = 0.00$, $n = 16$, $p < .001$, $r = -0.84$). Similarly, for lay people, sensitivity to the firearm was significantly lower in the concealed condition (Mdn = -0.47) than in the unconcealed condition (Mdn = 2.50; $U = 0.00$, $n = 16$, $p < .001$, $r = -0.84$). With respect to the influence of expertise, the sensitivities of CCTV operators and lay people did not differ significantly in either the concealed ($U = 16.00$, $n = 16$, $p = .105$, $r = -0.42$) or unconcealed ($U = 30.00$, $p = .878$, $n = 8$, $r = -0.05$) conditions, although in the concealed condition an advantage for CCTV operators tended towards significance.

Next, a series of Mann-Whitney U-tests were conducted to compare the sensitivities of CCTV operators and lay people for each unique combination of object size and camera view, separately for unconcealed and concealed firearms. For each of the two batches (concealed, unconcealed) of six comparisons the alpha-level was adjusted to avoid an inflation in the probability of a type II error (significant at α = .009). Only one of these comparisons revealed a significant difference: CCTV operators were significantly more sensitive than lay people to medium-sized concealed firearms, when given a ventral camera view (U = 7.00, n = 16, p = .007, r = -0.66; all other ps > .328). These results are summarised in Figure 1.

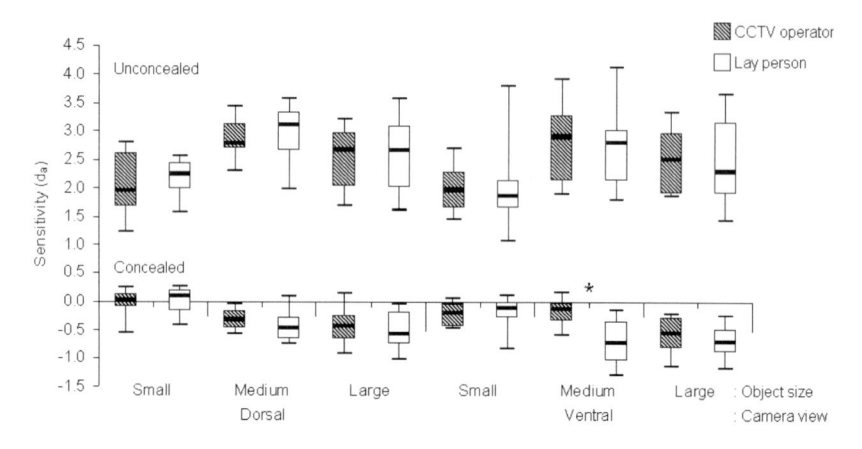

Figure 1. Box plots: The sensitivities of CCTV operators and lay people to concealed and unconcealed firearms of small, medium, and large sizes, viewed from the ventral and dorsal aspects.

Self-reported strategies in firearm detection
Do CCTV operators and lay people differ in their use of firearm detection strategies? Mann-Whitney U-tests were conducted to compare between CCTV operators and lay people in terms of the frequency of use reported for each strategy, separately for the concealed and unconcealed conditions. For each of the two batches (concealed, unconcealed) of seventeen comparisons (one for each strategy) the alpha-level was adjusted to avoid an inflation in the probability of a type II error (significant at α = .004). It was found that CCTV operators and lay people did not differ significantly in the frequencies of use reported for each strategy in either the concealed or concealed conditions.

The relative frequencies of use for each strategy were assessed within each of the CCTV operator and lay person groups, separately for each concealment condition (significant at α = .05). Friedman's ANOVAs revealed that the frequency of use ratings reported for each strategy varied within each of the four participant groups: CCTV operator, concealed ($\chi^2(16) = 49.45$, n = 8, p < .001); lay person, concealed ($\chi^2(16) = 64.84$, n = 8, p < .001); CCTV operator, unconcealed ($\chi^2(16) = 60.02$, n = 8, p < .001); lay person, unconcealed ($\chi^2(16) = 79.98$, n = 8, p < .001). Within each of the four groups, the strategies receiving the highest frequency of use ratings were defined on the basis of the critical differences (Siegel & Castellen, 1988) between the ranks of the strategies within each Friedman's ANOVA as those strategies whose frequency of use differed significantly from the lowest ranked strategy. CCTV operators had a smaller group of most frequently used strategies than lay people in both the concealed and unconcealed conditions (see Figure 2).

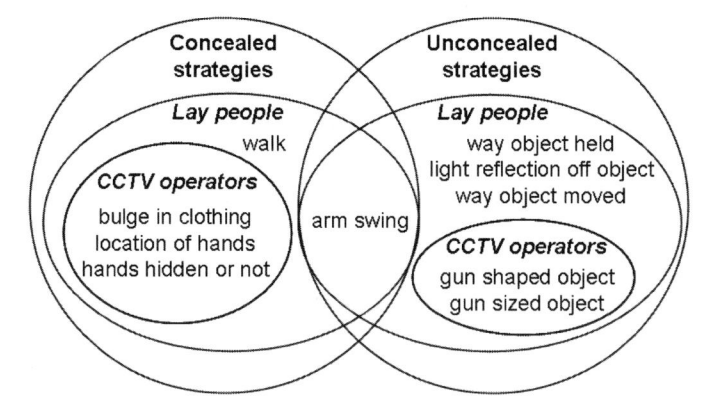

Figure 2. Euler diagram: Strategies falling in the high frequency of use category for CCTV operators and lay people in the concealed and unconcealed conditions.

Finally, the frequency of use for each strategy was correlated with sensitivity to the firearm across participants, within each of the four participant groups. This involved seventeen Spearman's correlations per group (one for each strategy) so the alpha-level was adjusted to avoid an inflation in the probability of a type II error (significant at α = .004). Only one correlation achieved significance: for CCTV operators in the unconcealed condition, a higher frequency of looking at demeanour or overall behaviour correlated with higher sensitivity to the firearm (r_s = .91, n = 8, p = .001; d_a range: 1.92 – 3.02; frequency of use rating range: 2 - 5; all other ps > .037). Other correlations may have had a restricted opportunity to arise on account of a limited range of values in the sensitivity or frequency of use variables. However, it is noted that a correlation involving demeanour or overall was not significant for lay people in the unconcealed condition over similar ranges of frequency of use and sensitivity (r_s = .28, n = 8, p = .506; d_a range: 1.82 – 3.73; frequency of use rating range: 2 - 4).

Discussion

CCTV operators had a greater sensitivity to firearms than lay people, but only for surveillance targets viewed from the ventral aspect and carrying a medium-sized firearm concealed on their person. With regard to the detection of unconcealed firearms, sensitivity was universally high despite the relatively small size of the objects in the footage and despite suboptimal visual conditions. Under these less testing conditions expertise was not observed to influence performance. Thus, expertise can increase reliability in the detection of firearms, but only under certain circumstances when the firearm is concealed. However, in the concealed condition sensitivity was principally below zero and, therefore, accuracy was mainly below chance; even for CCTV operators.

CCTV operators and lay people did not differ directly in their use of strategy. However, CCTV operators appeared to concentrate on a narrower range of strategies for detecting firearms in both the concealed and unconcealed conditions. This might reflect a common factor in the content of the training or experience of CCTV operators. The narrower focus of the CCTV operator group appeared to pay dividends, but only in the concealed condition. CCTV operators concentrated on looking for a bulge in the clothing or looking at the surveillance targets' hands, whilst lay people also diverted their attention to how the surveillance targets walked and to how their arms swung. In the unconcealed condition a narrower focus did not afford an advantage, perhaps reflecting that the task was not sufficiently challenging. CCTV operators tended to concentrate on the size and shape of the object in the hand, whilst lay people also diverted their attention to they way the object was held, how it moved and, again, to how the surveillance targets swung their arms.

For CCTV operators only, and only in the unconcealed condition, reporting a higher frequency of looking at demeanour or overall behaviour correlated with higher sensitivity to the firearm, whilst the same relationship did not approach significance for lay people. Although the correlation is merely an association, it might be inferred that, as a result of training or experience, CCTV operators can use the demeanour or overall behaviour of a surveillance target to improve their reliability in firearm detection. It is further noted that the self-report procedure did not reveal more explicit information about the nature of any visual cues that might subserve the use of demeanour or overall behaviour. Thus, if present, these cues could be quite subtle (as they relate to the firearm when carried in a casual fashion) and they might operate at a subconscious level. Such cues might relate to the *weapons effect*: firearms are thought to prime automatically, aggressive behaviour (Berkowitz & Lepage, 1967; Klinesmith, Kasser, & McAndrew, 2006), possibly via semantic association and the activation of aggression-related behavioural scripts (Anderson, Benjamin, & Bartholow, 1998). Thus, even those carrying a firearm in a casual fashion might reveal that fact through aggression in their body language. Indeed, the moods of those who participated in the filming of the staged CCTV footage were measured empirically, in a counterbalanced design, whilst they carried firearms and bottles: it was found that they experienced higher levels of aggression-related affect when carrying a firearm than when carrying a bottle. It is noted that the use of demeanour or overall behaviour did not correlate with sensitivity to the firearm in the concealed condition. It is not certain why such cues were not indicated in the detection of concealed firearms. Potentially, CCTV operators may have been unwilling to place much emphasis on what might amount to a "gut feeling" without the presence of stronger evidence in the stimulus. If any cues relating to demeanour or overall behaviour can be made explicit then they might also facilitate greater reliability in the detection of concealed firearms.

The self-report procedure for strategy has a number of drawbacks. For instance, participants may not fully appreciate the strategies that they used, or their reports may not reflect faithfully their behaviour. Thus, it is desirable to obtain an empirical assessment of firearm detection strategy. In a second experiment, eye-tracking was used to determine inspection strategies and loci of attention in a firearm search task. Lay people viewed still frames that were captured from the staged CCTV footage used in experiment 1, whilst their eye gaze was recorded. The stimuli consisted of two frames presented side-by-side, simultaneously. One frame featured a firearm, the other a bottle. The participant was required to indicate which frame contained the firearm in a two-alternatives forced-choice (2-AFC) task. Again, concealed and unconcealed conditions were tested. Ideally this task would have employed moving footage and would have tested CCTV operators alongside lay people. Even so, experiment 2 offers a means of verifying a portion of the self-report strategy data obtained in experiment 1.

Experiment 2

Method

Participants
Twenty four participants were recruited to, and took part in, the study (14 men, age range 19 – 44 years, $M = 28$, $SD = 8$). All participants were staff or students from Loughborough University.

Stimuli
Stimuli were constructed from still frames that were selected from the staged CCTV footage used in Experiment 1. Each stimulus comprised two images of the same person, presented side-by-side: one featured that person carrying a firearm, the other featured that person carrying an innocuous object. As in experiment 1, the innocuous object was a bottle matched approximately to the firearm for size, weight, and colour. Depending on the experimental condition, the firearm and innocuous object were either both concealed or both unconcealed. In total, 12 stimuli were constructed with concealed firearms and bottles and 12 stimuli were constructed with unconcealed firearms and bottles. Stimuli were viewed at a distance of 600mm. Each stimulus subtended $42.8°$ x $29.0°$ in visual angle. Each of the two instances of a person featured within a stimulus subtended approximately $5.7°$ x $11.4°$ in visual angle. (The experiment also featured stimuli comprised of still frames from real-life CCTV footage, but results pertaining to these stimuli are not reported here).

Firearm detection performance
Firearm detection performance was defined as accuracy (proportion correct) in the two-alternatives forced choice task.

Eye-tracking and inspection strategy
A Tobii x50 eye-tracker (Tobii Technology, Stockholm) was used to track eye-movements. A conservative estimate of this system's accuracy is $1°$ of visual angle. A single fixation was defined as eye-gaze falling exclusively within a region of radius $2°$ visual angle for at least 200 ms. Areas of interest (AOI) were defined prospectively for each stimulus: head, body, arms, legs, hands. Inspection strategy

was quantified in terms of the number of fixations on each AOI, averaged across stimuli.

Design
Influences on firearm detection performance were assessed in a between-participants design: concealment (concealed, unconcealed). Influences on inspection strategy were assessed in a full-factorial, mixed design: concealment (concealed, unconcealed), between-participants; AOI (head, body, arms, legs, hands.), within-participants.

Procedure
Stimulus presentation and data collection were automated by computer. Participants were pseudo-randomly assigned to a concealment condition with the stipulation that half should participate in each concealment condition. Participants were positioned in front of the screen with the eye-tracker sited directly underneath the screen at a distance of 600mm and the eye-tracker was then calibrated for that participant. They received stimuli appropriate to the concealment condition to which they had been assigned. Stimuli were presented in a randomised order. The side on which the firearm was present (left or right) was determined pseudo-randomly such that if one participant saw the firearm on the right for a given stimulus, another paired participant from the same concealment condition saw the firearm on the left for that stimulus. Whilst viewing each stimulus, participants indicated which image they thought featured the firearm, left or right, by pressing the left or right arrow key, respectively. Participants were asked to guess if unsure. The eye-tracker recorded eye-movements against each stimulus.

Results

The data violated assumptions underlying a parametric analysis and were subjected to non-parametric analyses.

Accuracy
A Mann-Whitney U-test confirmed that accuracy was significantly lower when the firearm was concealed (Mdn = 41.67%) than when it was unconcealed (Mdn = 83.33%; $U = 11.00$, $n = 24$, $p < .001$, $r = -0.73$).

Inspection strategy
Friedman's ANOVAs revealed that the number of fixations on each AOI varied significantly in both the concealed ($\chi^2(4) = 19.81$, $n = 12$, $p < .001$) and unconcealed conditions ($\chi^2(4) = 18.16$, $n = 12$, $p < .001$). Within each of the concealment conditions the AOIs receiving the highest numbers of fixations were defined on the basis of critical differences (Siegel & Castellen, 1988) between the ranks of the AOIs within each Friedman's ANOVA as those whose frequency of fixation differed significantly from the lowest rank. In the concealed condition the body AOI was fixated most frequently, whilst in the unconcealed condition the hand AOI was fixated most frequently. These results are summarised in figure 3. This was also true when the analysis was based on the duration of fixation data.

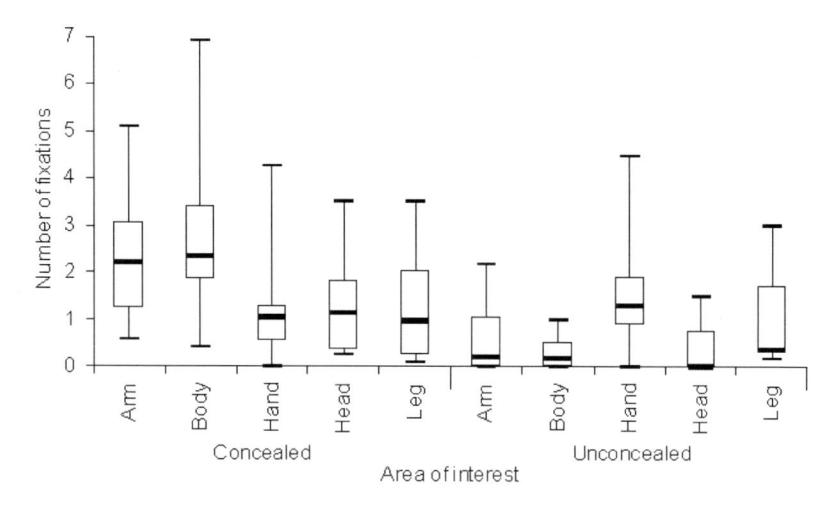

Figure 3. Box plots: The frequency of fixation on AOIs in the detection of concealed and unconcealed firearms.

Discussion

The influence of concealment on performance in the 2-AFC firearm detection task with still-frames (Experiment 2) was similar to that observed in the firearm detection task that involved moving footage (Experiment 1). With regard to firearm detection strategies, in the concealed condition the most frequently fixated AOI was the body, which is consistent with looking for a bulge in the clothing as in the self-report data of Experiment 1. However, the eye-tracking data did not reveal the special attention to the hands that was reported in Experiment 1. Additionally, strategies involving the walk and arm swing of the surveillance target could not be assessed in the still frames of Experiment 2. In the unconcealed condition the most frequently fixated AOI was the hand, which included the object held. This is in accordance with a focus on the physical properties of the object in the hand that was also apparent in the self-report strategies of Experiment 1. Strategies relating to the use demeanour or overall behaviour could not be assessed in the still frames of Experiment 2. Overall, the empirically determined inspection strategies of Experiment 2 are in broad accordance with the self-report strategies of Experiment 1. However, the use of still frames in Experiment 2 prevented the assessment of key strategies involving the surveillance target's movement, such as the use of demeanour or overall behaviour.

General discussion

When a firearm was carried in plain view, in an inconspicuous fashion, it was possible to spot that firearm, via CCTV, with a high degree of reliability. Detection did not benefit from expertise. The strategies used to detect unconcealed firearms revolved principally around the visual properties of the object in the surveillance target's hand. Conversely, when a firearm was concealed upon a surveillance target, reliable detection was not possible. Expertise augmented reliability in the detection of concealed firearms under certain circumstances and expert strategies in the detection of concealed firearms related to looking for a bulge in clothing. However, even amongst experts, the level of performance in the detection of concealed firearms was still near chance.

A further strategy that was linked to expertise in the present study was the use of the demeanour or overall behaviour of a surveillance target in the detection of unconcealed firearms. However, this inference relies upon an association that is based on self-report; it was not possible to further test this inference through eye-tracking in the still-frames of Experiment 2. Also, as the footage used in both Experiments 1 and 2 was staged it is likely to include only impoverished cues relating to demeanour or overall behaviour (even if those filmed did feel more aggressive whilst carrying a firearm). Cues relating to real intent, as well as situational cues surrounding the carrying of a firearm, are probably lacking. At present, the collection of a body of real-life footage involving firearms, from CCTV control rooms in the UK, is underway. Real-life footage might include cues relating to demeanour or overall behaviour that are strong enough to facilitate the detection of concealed firearms.

The ultimate goal of this body of work is to inform the design of image processing algorithms to automate the detection of firearms, via CCTV, and then alert CCTV operators when necessary. This is desirable as, typically, CCTV control rooms are served by more cameras than CCTV operators are able to monitor simultaneously and constantly (Gill et al., 2005). With respect to the detection of concealed firearms, the findings suggest that human strategies are unlikely to offer inspirations that will prove effective. However, potentially effective strategies can be derived for the detection of unconcealed firearms: an algorithm should concentrate on the visible, physical properties of any object held in the hand (Darker, Gale, & Blechko, 2008). This also reflects the principal means of detecting firearms during real-life CCTV surveillance (Darker, Gale, Ward, & Blechko, 2007). Subsequently, a preliminary algorithm has been designed which, in the present staged CCTV footage, has sensitivity to unconcealed firearms that is equivalent to the high levels exhibited by the CCTV operators who participated in this study (Darker, et al., 2009, in press). Future work involving real-life CCTV footage may help refine the algorithm and facilitate its use in complex imagery associated with public space surveillance.

The present study revealed that CCTV operators might possess an element of expertise in the detection of illicit items such as firearms, and that this expertise might relate to the interpretation of a surveillance target's demeanour or overall behaviour. Future work will combine the use of real-life CCTV footage of firearms with eye-tracking technology, in a firearms detection task. Such a study might further elucidate the nature of this expertise and better illuminate any elements involved in the use of demeanour or overall behaviour in firearms detection. This information could then inform further development of the algorithm designed to automate the detection of firearms via CCTV.

References

Anderson, C. A., Benjamin, A. J. J., & Bartholow, B. D. (1998). Does the gun pull the trigger? Automatic priming effects of weapon pictures and weapon names. *Psychological Science, 9*, 308-314.

Berkowitz, L., & Lepage, A. (1967). Weapons as aggression-eliciting stimuli. *Journal of Personality and Social Psychology, 7*, 202-207.

Darker, I. T., Gale, A., Ward, L., & Blechko, A. (2007). Can CCTV reliably detect gun crime? In L.D.Sanson (Ed.) *Proceedings of the 41st IEEE International Carnahan Conference on Security Technology* (pp. 264-271). Piscataway, NJ: IEEE.

Darker, I. T., Gale, A., Ward, L., Blechko, A., & Purdy, K. (2007). Light, camera, action, and arrest. In P. Bust (Ed.) *Contemporary Ergonomics* (pp. 171-177). Loughborough, UK: Ergonomics Society.

Darker, I.T., Gale, A.G., and Blechko, A. (2008). CCTV as an automated sensor for firearms detection: human-derived performance as a precursor to automatic recognition. In E.M. Carapezza (Ed.) *Proceedings of SPIE Europe Security and Defence, Volume 7112* (p. 71120V). Bellingham, WA: SPIE.

Darker, I. T., Kuo, P., Yuan Yang, M., Blechko, A., Grecos, C., Makris, D., et al. (in press 2009). Automation of the CCTV-mediated detection of individuals illegally carrying firearms: combining psychological and technological approaches. *Proceedings of SPIE Defense, Security, and Sensing, Volume 7341*.

Gill, M., Spriggs, A., Allen, J., Hemming, M., Jessiman, P., Kara, D., et al. (2005). *Control room operation: Findings from control room observation.* London, UK: Home Office.

Hales, G., Lewis, C., & Silverstone, D. (2006). *Gun crime: the market in and use of illegal firearms.* London, UK: Home Office.

Harvey Jr, L.O. (2003). Living with uncertainty in an uncertain world: Signal detection theory in the real world. In J. Andre, D.A. Owens and L.O.J. Harvey (Eds.), *Visual perception: The influence of H.W. Leibowitz.* (pp. 23-41). Washington DC, USA: American Psychological Association.

Harvey Jr, L.O. (2007). *Parameter Estimation of Signal Detection Models: RscorePlus User's Manual* (Version 5.5.7. ed.). Boulder, Colorado, USA: University of Colorado.

Klinesmith, J., Kasser, T., & McAndrew, F.T. (2006). Guns, testosterone, and aggression: An experimental test of a mediational hypothesis. *Psychological Science, 17*, 568-571.

Siegel, S., & Castellen, N.J. (1988). *Nonparametric statistics for the behavioural sciences* (2nd ed.). New York, USA: McGraw-Hill.

Troscianko, T., Holmes, A., & Stillman, J. (2004). What happens next? The predictability of natural behaviour viewed through CCTV cameras. *Perception, 33*, 87-101.

Improving Situational Awareness in camera surveillance by combining top-view maps with camera images

Frank L. Kooi & René Zeeders
TNO Human Factors,
The Netherlands

Abstract

The goal of the experiment described is to improve today's camera surveillance in public spaces. Three designs with the camera images combined on a top-view map were compared to each other and to the current situation in camera surveillance. The goal was to test which design makes spatial relationships between camera images most intuitively apparent to the viewer. The results showed that all three designs create a better understanding of these spatial relationships as indicated by faster response times.

Introduction

Current technology

Contemporary camera surveillance in public spaces relies on a video wall with various video screens. The simultaneous overview of many cameras improves surveillance. The situation awareness of the operator is now limited by the ability to correctly *integrate* the individual images to a consistent whole. Various approaches try to aid the (mental) integration of numerous images. The "Flashlight" system projects camera imagery over a 3D model. It is related to the projection of video image in 3D models (Sawhney et al., 2002; Sidenbladh & Ahlberg, 2005). The camera images appear as flashlights, but instead of light, the images are projected; people and other objects appear as textured shadows. The main disadvantage of this system is the increasing distortion of the images as the camera is mounted lower to the ground; analogous to the long shadows at sunset. Fleck et al. (2006) have developed a distributed network of cameras that allows for tracking and handover of multiple persons in real time. The inter-camera tracking results are embedded as live textures in an integrated 3D world model. This enables the user to fly through a virtual 3D world which is a live representation of the real world being monitored. The disadvantages are that all cameras need to be smart or combined with a PC and the inability to deal with occlusions which cause incomplete textures for projection. This complicates the tracking, handover and 3D positioning of individuals. The system is therefore not (yet) suitable for crowded areas such as a busy city square.

In D. de Waard, J. Godthelp, F.L. Kooi, and K.A. Brookhuis (Eds.) (2009). *Human Factors, Security and Safety* (pp. 355 - 368). Maastricht, the Netherlands: Shaker Publishing.

We therefore explored another way to improve the spatial situation awareness by supporting the mental integration of multi-camera "video-wall" surveillance. The present study examines a method, in three variants, to support the spatial integration task. The combined imagery shown in Figure 6 aims to minimize the need for 'mental rotations' (Wickens and Hollands, 2000) by locating the video images on the map such that they match in location and orientation. The three conceptual designs described in this paper are designed to enhance the ability of the operator to intuitively understand the spatial relationship between the surveillance camera image and a map of the area being monitored. It aims to minimize the time needed to re-orient when switching to another camera. Punte and Post (2002) analogously have previously integrated camera observation processes with the same purpose: efficient employment of operators. Punte and Post (2002) proposed an integration of these observation processes by creating a single "Integrated Monitor Centre" to optimize detection of incidents and assignment to the responsible Police department.

Purpose

The goal of this project is to significantly improve the efficiency of camera surveillance. This means that either less personnel would be needed for the same task, or the same amount of personnel will be able to camera-survey a larger area.

We compared three different designs with a slightly enhanced representation of the original video wall situation to see if situational awareness of surveillance personnel could be improved. The three designs combine camera image with a top-view map by placing the camera images on a map of the area being monitored. The way the camera images are placed on the map is explained in detail in the next paragraph.

The design process
This design attempts to fit in with the current camera surveillance workstations, requiring minimal adjustments. We compared various possibilities to visualize the camera imagery in such a way to make the spatial relationships between the cameras intuitively apparent. The design process converged on three variants having the following in common:

- Camera images are placed on a map of the area such that the map area being monitored is apparent to the viewer
- The images are tilted from the perpendicular position to allow the viewer to perceive the map as well as the images

The difference between the designs is the direction of tilt from the perpendicular position. When the image is tilted back, we use the metaphor *prism*, when the image is tilted forward we use the metaphor *mirror*. Thirdly the two can be *mixed* in order to avoid the image to appear upside-down.

Three variables arise when placing a camera image on a (top-view) map: location, size and orientation. We have chosen the following procedure to fix these variables, making the appearance of map and camera images predictable.

improving Situational Awareness in camera surveillance 357

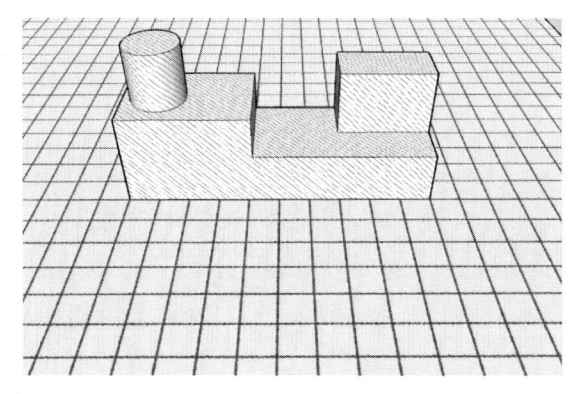

Figure 1. Simple representative image as seen from the camera viewpoint

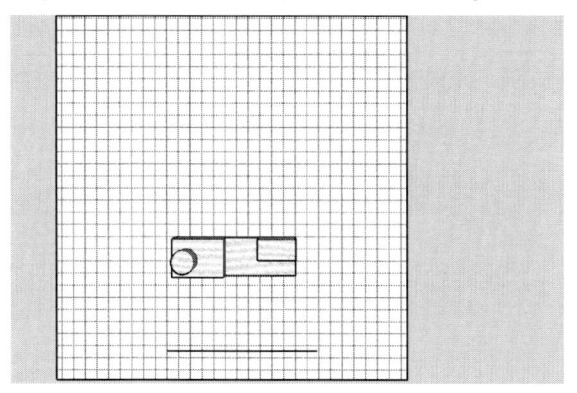

Figure 2. Top-view of the same situation, the horizontal line shows where the ground surface enters the camera view

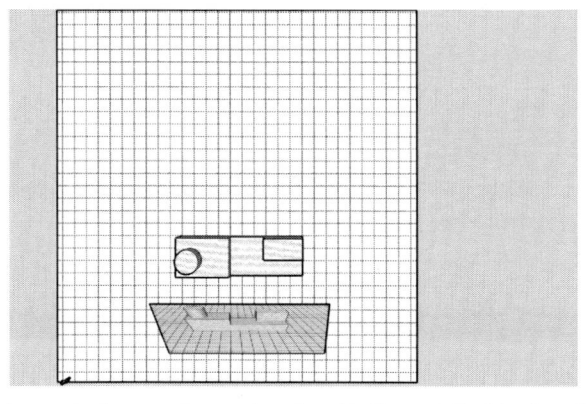

Figure 3. Miniature sized camera image placed on this line and tilted backwards

The *location* of the image is determined by the line that marks the bottom of the image. In effect the map and camera image meet there. We illustrate this in Figures 1 to 3. In Figure 1 the camera image shows an object on a grid, visible out to 6½ squares from the object. Figure 2 marks this line as shown from above (top-view). In Figure 3 the tilted camera image is placed on this line, automatically determining the width and therefore the size of the image.

As shown in the example (Figure 1-3), the line orientation is orthogonal to the viewing direction of the camera. In order to make the miniature image visible from above, it must be tilted. The two design parameters are the direction of tilt – forward or backward – and the amount of tilt. The advantages and disadvantages of the direction and amount of tilt are summarized in Table 1 and Table 2.

For the amount of tilt we chose a midway solution, in-between flat on the ground and perpendicular (48deg tilt). If we avoid making the angle too small, perspective distortion is kept to a minimum. If we tilt the image, the perspective distortion will influence the vertical extent of the image (Tyler, 2005). Therefore we tilt the image such that the image height is ¾ of the original height.

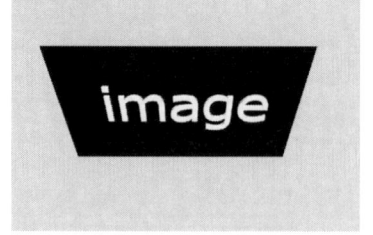

Forward tilt seen from above   *Backward tilt seen from above*

Figure 4. Forward and backward tilt

The tilting direction, forward or backward, distinguishes between the prism and mirror variants. These metaphorical names are chosen, because of their characteristics:

- Prism: a prism bends the light, changing its direction
- Mirror: a mirror reflects the light, changing its direction

The mirror metaphor is used for the tilt towards the camera, and the prism metaphor is used for the tilt towards the scene (Figure 5). These metaphors as such are not necessary to understand the designs, but are used throughout the paper to distinguish between the concepts.

Table 1. Advantages and disadvantages of direction of tilt

Direction of tilt	Forward ('mirror' image)	Backward ('prism' image)
Advantage	• No occlusion of a part of the map that is shown in the image • Mirrored could be an advantage, as it is something we are very familiar with (e.g. shop windows, car mirrors, etc.)	• The image is displayed as it is recorded
Disadvantage	• The image is mirrored. Top becomes bottom and vice versa	• Occlusion of a part of the map that is shown in the image

Table 2. Advantages and disadvantages of amount of tilt

Amount of tilt	Small angle	Large angle (up to flat on the ground)
Advantage	• Makes more sense about how the image is related to the situation; normally the situation is almost perpendicular to the image	• Little information is lost because viewing angle is (relatively) large and perspective distortion is little
Disadvantage	• information is lost because viewing angle is small and there is much perspective distortion	• makes less sense as the situation normally is almost perpendicular to the image

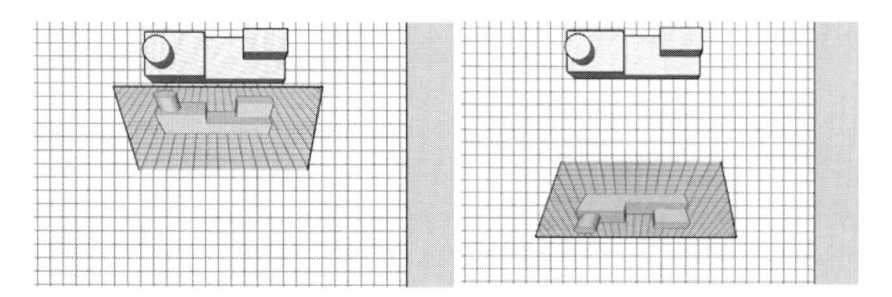

Figure 5. Left: example of the prism metaphor. Right: example of the mirror metaphor

Besides the reference condition (Condition 1), the mirror and prism designs are labeled Condition 2 and 3. The 4th design is a mixture of the prism and mirror designs, avoiding camera images to appear upside-down. Our hypothesis was that the mirror design would be most intuitive and therefore give the best situation awareness; as humans we are for example very accustomed to viewing in (car) mirrors at odd orientations. To test whether upside-down images indeed are harder to interpret, we mixed the mirror and prism designs. All images are placed on the top-view map as if it were mirrors; if the image would be perceived upside down, then it would be placed as a prism.

Research question and hypotheses

The goal of this experiment was to measure the effect of combining camera image with a map of the environment on situational awareness, as measured with a search task. The main research question was: *do the mirror, prism and mixed designs improve situational awareness more than the current video-wall design?*

Sub questions were:

1. Do the mirror, prism and mixed designs differ in their support of spatial situational awareness?
2. Do the new designs also help the operator to make fewer mistakes in locating the target persons?

We hypothesized that all three designs would improve situational awareness over the current video-wall situation and secondly, that of the three variants the mirror design would perform best. We did not have explicit expectations towards the relationship between speed and correctness, as we noticed it is hard to predict given the information we had.

Methods

Subjects

Participants were selected from students working on projects at TNO Human Factors, paid subjects, and TNO employees. All in approximately the same age group. In subsequent studies we plan to incorporate actual operators.

Seven females and five males participated in the study. Their mean age was 24.4 years (range 18-32 years). All participants were in or completed a medium to high level of education (medium: HBO, high: WO). Ten participants were trainees or employees at TNO Human Factors and two were students at high school.

Stimuli

The computer generated stimuli gave complete control over all aspects of the (simulated) camera images. The assumption is that we can extrapolate the results from the stylistic experiment to more complicated real world scenes. The experiment is based on static instead of dynamic images, further simplifying the experiment.

We created a 3D street model consisting of alleys in Google Sketchup[*]. To create some clutter, the walls were painted yellow and white. To resemble a real world situation, the ground was tiled with grey tiles and the sky was painted blue. In the

---

[*] See http://sketchup.google.com/

model five alleys were being monitored by one (virtual) camera each, amounting to a total of five not overlapping cameras. The static (non-dynamic) camera images were shown to the subjects in the four variants: 1) Control condition, 2) Mirror condition, 3) Prism condition, 4) Mixed condition.

Within the field-of-view of each camera a uniquely coloured tile was placed on the second street crossing. These tiles functioned as the only distinguishable factor between the camera images. As the top-view map is rotated and flipped from trial to trial, the coloured tiles end up at (semi) random locations. Because the top-view is a parallel projection, the walls were somewhat slanted to make their colours visible to the viewer.

The cameras were indicated by white squares with their fields-of-view represented by white, transparent cones. The cones did not reach as far as the real field-of-view; this was a design choice to avoid overlap of the fields-of-view. The contrast of the top-view map is lowered, to make the camera image more apparent to the viewer. The part of the map that is not visible on the cameras is therefore darker, to indicate non-information.

This same design of the top-view map was used for all four conditions. By using this design in the control condition we believe we created a slight enhancement compared to the current video-wall situation, as the locations and orientations of the cameras are more obvious. Because of this enhancement and its function in the experiment we call it the control design.

Four top-view geometries of each variant were created in Adobe Photoshop[*]: The original, the original rotated 180 degrees, the original flipped horizontally, and the original flipped vertically.

Three 'camera images' were created by inserting two persons (one red and one blue) in an alley as shown in Figure 7:

1. Red person in the first part of the alley, blue in the second. Alley contains green tile
2. Red person in the second part of the alley, blue in the first. Alley contains yellow tile
3. Red person in the third part of the alley, blue in the first. Alley contains purple tile

These three parts of the alley are:

1. From the camera to the middle of the first crossing
2. From the middle of the first crossing to the middle of the second crossing
3. From the middle of the second crossing to the middle of the third crossing

---

[*] See http://www.adobe.com/nl/products/photoshop/photoshop/

362   Kooi & Zeeders

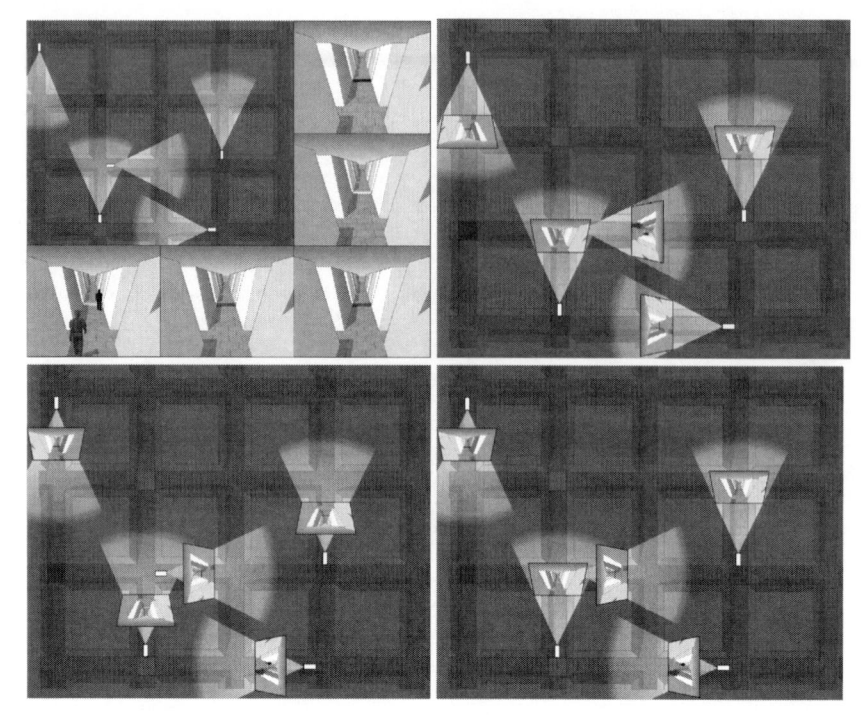

Figure 6. The four top-view maps: the Control condition (left-above), the Prism or 'Window' condition (left-below), the Mirror condition (right-above), and the mixed condition (right-below). Colour images of all four stimuli can be viewed at http://extras.hfes-europe.org

Combining the four top-view maps with the three 'camera images' in Photoshop results in twelve (4x3) stimuli per condition. In these combinations three constraints were applied. Each top-view map contains five positions for a 'camera footage' image. Each top-view map contains different camera positions and different camera orientations (north, east, south, west), balanced over the twelve images.

In the experiment each image was followed by an 'answer image': identical except without the camera images, as shown in the top-left of Figure 6.

Because of the designs two confounding variables arose:

1. Difference in size of the camera images between the control conceptual design and the other designs.
2. Difference in size of the top-view map between the control design and the other designs.

These differences are a result of the designs, and we concluded that the negative effect of having smaller images in the other three designs should be accepted. Making the images smaller in the control condition would give cause to two other confounding variables: unequal screen usage and unfair comparison with today's manner of camera surveillance. The fact that the top-view map is smaller in the control condition is also something that had to be accepted, though we do not expect that this would influence the results.

The experimental software was written by TNO. The time elapsed between the stimulus appearance and the participant response was recorded in milliseconds, the delay caused by the mouse being minimal compared to the length of the search tasks.

Figure 7. Example of a 'camera image' image. The person close to the camera is 'red' in the coloured version (see http://extras.hfes-europe.org).

Experimental design and procedure

The experiment was based on a repeated measures 1-factor design. The independent variable is the manner in which the camera imagery is combined with the map. The four variants are illustrated in Figure 6. The dependent variable is the situational awareness support as measured by the reaction time and error rate. The four conditions were presented twice (i.e. repeated once) in blocks containing twelve stimuli. Partial counterbalancing over all participants was performed through the Latin square and reverse counter balancing was used to balance over a single participant. In total each participant went through eight blocks, each consisting of twelve stimuli, resulting in a total of 96 stimuli.

The task of the participants was to locate the red person on the top-view map given the camera images. From the search time and correctness we inferred how easily the participant perceives the spatial relationships between the cameras and the top-view map.

The task involved two steps:

1. Based on the camera image, determine the location of the red person on the top-view map as quickly as possible, then press the button
2. After the camera image has disappeared, indicate the location of the red person on the top-view map

In step 1 the participant was instructed to press the left mouse button as soon as they thought they knew the location of the red person on the top-view map. The goal was to quickly/intuitively locate the red person, and secondly to do it correct (without double-checking). The time needed to locate the red person was recorded in milliseconds. After the mouse click, an answer screen appeared (step 2) on which they could pin-point the exact location. The answer was correct when the location pinpointed by the participant and the actual location of the red person was in the same street block (between two adjacent crossings). On this answer screen the location could be given precisely, and the participant could take his/her time to ensure that the location the participant had in mind and the location pinpointed coincided. The answer was correct when the location pinpointed by the participant and the actual location of the red person was in the same street block. The participant was told (and from time to time reminded) not to pinpoint the location on the first screen. Otherwise differences in mouse experience would introduce apparent differences in reaction times.

After pinpointing the location, a blank screen appeared for 600 ms followed by the next stimulus. The geometric center of the blank screen displayed the mouse pointer as a text cursor, drawing the attention of the participant. Each block was followed by a brief pause (about one minute).

Afterwards a questionnaire collected personal data (Age, Sex, Educational Level, Hours per week spent on computer games) and inventoried the subject's design preferences as an order from most to least preferred.

The four designs were described by the explanatory texts shown below, accompanied by one example corresponding image and a verbal explanation:

- Control: "Five cameras are displayed on the map symbolized by white squares. Their fields of view are visualized by white, transparent cones. The images around the top-view map are the images of the cameras. You can distinguish between them by using the coloured squares."
- Mirror: "The top-view map is represented in the same way as the top-view map in the control design, but in this design the images of the cameras are placed on the map near the cameras they belong to. In this design the images are tilted toward the cameras."
- Prism: "This design resembles the mirror design. The difference is that the images are tilted away from the cameras."
- Mixed: "This design is almost the same as the mirror design. The only difference is that when an image would be placed upside down for the viewer by tilting it toward the cameras, it instead is tilted away from the camera. This way an image will never be placed upside down."

In the experiment the words control, mirror, prism and mixed were never mentioned, but stated here for convenience for the reader. The reason why the names of the conditions were not mentioned is the possible positive or negative influence it could have on the performance of the participant, whether by participant bias, or by the positive or negative effect of intentionally using a metaphor.

Before each block the subject viewed some practice trails to familiarize with the design. Before each of the first four blocks six familiarization trials took place, four accompanied by the experimenter to make sure the participant understood the visual design. The answer screens included the camera images to allow the experimenter to explain where and why the red person was located when the participant's answer was incorrect. The last two of these six practice trials were identical to the real task to prepare the participant for the actual experiment. After each design had been presented once (the first four blocks), the second set were preceded by just two practice trials. This was sufficient to re-focus on the design condition ahead.

After all eight blocks were completed the subjects completed a questionnaire.

Results

The results are presented separately for the first and second runs to also show the training effect.

Search times

In Figure 8 the means and standard errors are given for the four design conditions and the 1st / 2nd run. Search times from incorrect answers are ignored.

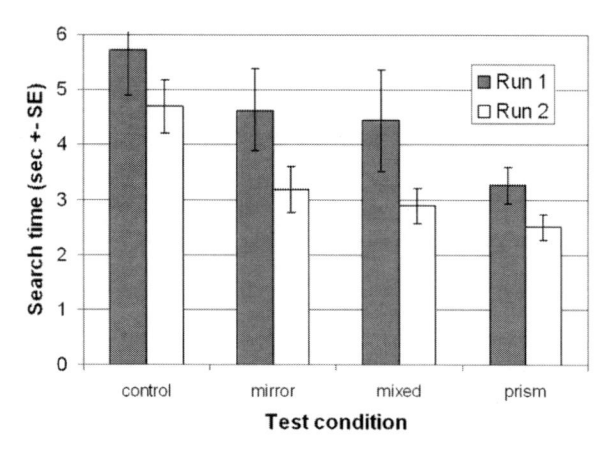

Figure 8. Means and standard errors in reaction time. Subjects are faster in the 2nd run, in particular in the three conditions with the video image embedded in the map

There was a significant effect for type of condition, $F(3,33) = 18.493$, $p < .0005$, partial $\eta^2 = .63$.

Figure 8 shows that participants take the longest time to locate the red person in the control condition and the shortest time in the prism condition. An ANOVA statistical test was performed to do pair wise comparisons. The results are summarized in Table 4. The significant differences between means are shown in seconds.

Table 4. Grey areas are significant differences between conditions over all stimuli in seconds, $p < 0.01$

	Control	Mirror	Prism	Mixed
Control		1.3	2.3	1.6
Mirror	-1.3		1.0	
Prism	-2.3	-1.0		
Mixed	-1.6			

Percentages correct

Figure 9 shows the means and standard deviations of the errors. Only the mirror-mixed and the mixed-prism conditions do not differ significantly within $p<0.05$.

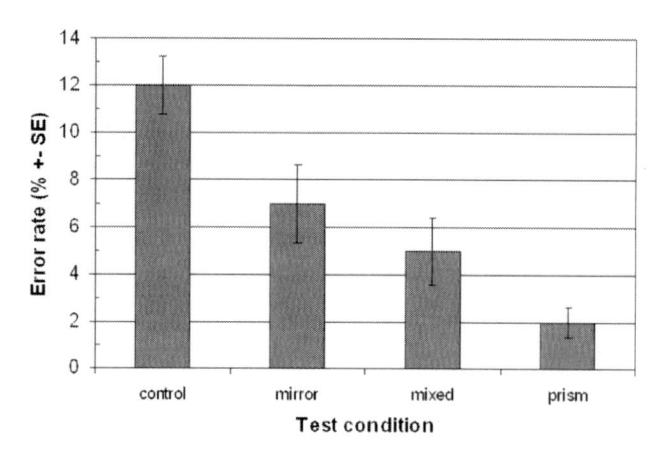

Figure 9. Means and standard errors in % errors, averaged over the two runs. The control condition is far inferior.

These results show the same ordering as seen with the search time data: as search time goes up, correctness goes down. The conditions that significantly differ in search time, also significantly differ in correctness. Contrary to our expectation, performance in the prism condition is significantly faster and better (less errors) than in the mirror and control conditions.

Subjective questionnaire

The participants rated the four designs. The subjective ratings are summarized in Figure 10. The control condition is practically always judged as least favorite (ranked 4th). Prism and Mixed are rated as best and differ marginally. No significant results could be derived from the experience with computer games or any of the other questionnaire data.

Questionnaire ratings

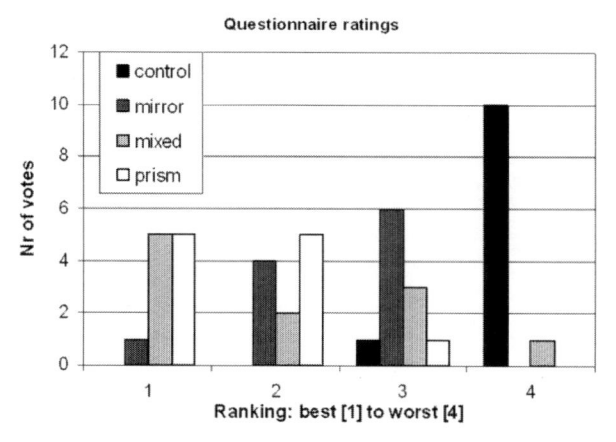

Figure 10. The subjective ranking of each design

Discussion

The results convincingly show the superior situation awareness performance in the mirror, prism and mixed designs over the control design, both in terms of speed and accuracy. The prism and mixed designs do best, showing no significant differences. The Mirror design leads to more errors and longer reaction times and correspondingly is judged as less favourite. The ordering in situation awareness performance (first being best) is therefore: Prism and Mixed (best), Mirror (2nd best), and Control (poorest).

Reaction time in the prism condition is virtually 2 times faster than in the control condition for participants without directly relevant experience. Preliminary data on trained subjects show that the benefit further increases to the point that the location of the person-in-red is virtually instantaneously clear. The data confirm our hypothesis that all three designs lead to a higher level of situational awareness compared to the current situation in camera surveillance. Our expectation about mirror being the best is incorrect, as performance in the prism condition was significantly higher than performance in the mirror condition. Apparently the mirror design can give cause to confusion. If the red person was located further away from the camera, the object was positioned higher on the camera image (closer to the horizon due to perspective), and therefore closer to the camera, because the image was tilted towards the camera. This was confusing for the participant. With the prism condition this did not occur.

During the experiment some participants continuously tilted their heads in the prism, mixed and mirror conditions to correct for the rotation of the images. The question that arises is whether any of these three concepts are justified from the perspective of ergonomics. A logical alternative is to show the camera image upright and thereby not adhere to the principle to match the camera image to the map as closely as possible. We will experimentally compare these conditions next.

Afterwards we realized the design would lead to more training on the three similar conditions: prism, mirror and mixed than on the 'oddball' control condition. Nevertheless we believe that the objective results and the questionnaires indicate that the mirror, prism and mixed designs improve performance.

Our expectation is that when the prism design is used regularly and the operator is trained in using it, the increase in reaction speed over the control design will further increase. This is the second topic of research we foresee.

References

Fleck, S., Busch, F., Biber, P., & Straßer, W. (2006). 3D Surveillance – A Distributed Network of Smart Cameras for Real-Time Tracking and its Visualization in 3D. In *Proceedings of the 2006 Conference on Computer Vision and Pattern Recognition Workshop* (pp 118). Tübingen, Germany: University of Tübingen.

Hedvig Sidenbladh, J., & Ahlberg, L.K. (2005). *New Systems for Urban Surveillance*. Defense Research Agency, Sensor Technology, Linköping, Sweden. Retrieved April 14, 2009, from http://www2.foi.se/rapp/foir1668.pdf

Punte, P.A.J. & Post, W.M. (2002). Werkwijze Geïntegreerde Monitor Centrale Politie Haaglanden [Way-of-work Integrated Monitor Centre Police department Haaglanden] (TNO Technische Menskunde Report TM - 02 - C062). Soesterberg, the Netherlands: TNO Human Factors.

Sawhney, H.S., Arpa, A., Kumar, R., Samarasekera, S., Aggerwal, M., Hsu, S., Nister, D., & Hanna, K. (2002). Video Flashlights. – Real Time Rendering of Multiple Videos for Immersive Model Visualization. In P. Debevec and S. Gibson (Eds). Thirteenth Eurographics Workshop on Rendering (pp 157-168). Aire-la-Ville, Switzerland: Eurographics Organization.

Tyler, C.W. (2005). A Horopter for Two-Point Perspective. In Rogowitz, B. E., Pappas, Thrasyvoulos N., Daly, S.J. *Proceedings of the SPIE, Volume 5666: Human Vision and Electronic Imaging X*, (pp. 306-315). SPIE--The International Society for Optical Engineering.

Vandenberg, S.G. & Kuse, A.R. (1978). Mental rotations, group test of three-dimensional spatial visualization. *Perceptual and Motor Skills, 47*, 599-601.

Wickens, C.D., & Hollands, J.G. (2000). *Engineering psychology performance and human performance* (3rd Ed). New Jersey, USA : Prentice-Hall Inc.

'Sounds like trouble'

Peter W.J. van Hengel[1], M. Huisman[2], & J.E. Appel[1]
[1]Fraunhofer Institut Digitale Medientechnologie
Oldenburg, Germany
[2]Sound Intelligence BV
Groningen, the Netherlands

Abstract

The newly founded project group Hearing, Speech and Audio Technology of the Fraunhofer Institute for Digital Media Technology IDMT focuses on applications based on models of the human hearing. One of the domains for application of these models is the detection of incidents in the surveillance domain based on audio information. For this application a system designed to detect human verbal aggression was developed by Sound Intelligence. A case study with this system will be described here as an example of what can be achieved using audio technology in a supportive role for surveillance .Models of human sound processing incorporating aspects of neural processing are now further developed to indicate a surveillance operator possible incidents or unexpected events.

Introduction

To improve the overall usefulness of camera-based surveillance systems it is important that situations with a high risk of injury and a relatively fast development, such as street-fights, collapses or accidents, are detected as quickly and as reliably as possible. Only then, appropriate action can be initiated. A system which can prioritize potentially dangerous situations autonomously and presents high priority events to a human observer for further analysis of the situation, would greatly reduce the chances of incidents being missed.

An evaluation of CCTV in one of the London boroughs in 2004 revealed that only about 30% of incidents such as criminal damage and emergency incidents happening in view of a surveillance camera, were detected by CCTV (Gill & Hemming, 2004). This rather low percentage is in contrast with the expectation of over half the people in residential areas believing that the police will respond quickly to incidents when CCTV cameras are installed (Spriggs et al., 2005).

With each operator handling about 80 cameras on average, the chance of looking at the right place at the right time is low. As one of the CCTV operators is quoted in (Gill & Hemming, 2004): "we are the eyes of the police but we don't know where to look".

In D. de Waard, J. Godthelp, F.L. Kooi, and K.A. Brookhuis (Eds.) (2009). *Human Factors, Security and Safety* (pp. 369 - 375). Maastricht, the Netherlands: Shaker Publishing.

Where Gavin J.D. Smith in his 2004 paper said that: "Most writers seem almost to forget that, by and large, CCTV cameras are neither conscious, nor autonomous, and require, in order to be effective, constant monitoring and control by human beings in a work-like situation, so that the millions of images produced can be watched, interpreted and acted upon" (Smith, 2004), the human factor in the use of camera surveillance is getting more attention in recent years,

Many technologies are being developed to assist the camera operator with this task. Most of these technologies focus on detecting incidents in the video signal picked up by the cameras. Although exciting work is being done and results of new developments are promising, it is clear that video analysis alone has its limitations.

Of course, humans on the street do not only see but listen as well. In fact we typically use our hearing to raise our alertness and to focus visual attention in the correct direction; i.e. audition guides vision.

Research on the use of audio in a supporting role for surveillance was started at the university of Groningen and the company Sound Intelligence, in cooperation with TNO departments FEL and TPD, the Dutch Railways (NS) and ProRail in 2003 (Van Hengel & Huisman, 2003). This resulted in the development of a technology for the detection of verbal aggression (Huisman, 2004), which was extensively tested e.g. in the inner city of Groningen.

In contrast with commonly used classification systems based on neural network techniques, the verbal aggression detection system was designed to operate independent from the acoustic environment and with a minimum of training material. However, a shortcoming of the system is that it looks only for a specific type of signal. In practice a system which could respond to anything 'out of the ordinary' would be preferable.

Verbal aggression as a cue for incidents

The idea to use verbal aggression as the most important (acoustic) cue for the detection of incidents is given by the fact that approximately half of the incidents are reported to be linked to some form of aggression or violence. This proportion is substantially higher during the weekend nights. Acts of aggression are generally accompanied by aggressive shouting. This aggressive shouting in most cases precedes any visual cue of incident developing. Furthermore aggressive shouting triggers human attention and must therefore possess acoustic properties which can easily be identified by human listeners. A model of the human hearing which produces quantities such as audibility, tonality and pitch could therefore be expected to give good results in identifying aggressive shouting, and thus detecting incidents.

The investigation by Huisman 0 showed that aggressive shouting can be described in terms of changes to the anatomy and use of the vocal tract, and therefore by specific acoustic cues in the produced voice signal. He showed that a number of acoustic cues are connected to aggressiveness in the human voice. A shortcoming of this study was, however, the use of acted aggression. The outcomes showed an ability of human listeners to identify most of the aggression audio samples as 'realistic but not real'. For this reason the use of the cues based on this data set was treated with great caution. After all, the final system should respond to real aggression, not to acted aggression.

As a first approach it was decided to use what is known of the Acoustic Startle Reflex (ASR). The ASR is the well-studied reflex that makes humans (and animals) look up and/or turn their head at an unexpected loud sound. Since the human behaviour when hearing aggressive shouting is similar to the ASR it seemed sensible to start off by trying to model the ASR. It is known that only the lowest nerve centres in the acoustic pathway (up to the cochlear nucleus) are involved in eliciting the ASR (Davis et al., 1982). From what is known about the function of these centres it seems that only such properties as the overall loudness of the signal, its audibility above the background sound and its location could be of significance. Since localisation cues are only relevant in a binaural system, which would imply solving several technical difficulties in the development and placement of such a system, the decision was made to focus on monaural cues only.

Experiences with the aggression detection system

A system using cues for audibility of signals above background was implemented and installed in the inner city of Groningen. After solving several technical difficulties, such as finding microphone types that could cope with the expected atmospheric conditions, finding hardware to perform the audio analysis that could be installed on-site and finding ways to transport the detection signals to the surveillance room, a system comprised of 11 sensors started operation in the pub district in the inner city of Groningen in 2005. The primary aim of the installation was to test the systems performance against the incidents observed by and reported to the camera operators. For this reason an area was chosen where the number of incidents was likely to be high and the chance of incidents going unnoticed by means of normal surveillance operation was low. The primary aim of the system therefore was to collect data for further investigation and development of the aggression detection algorithms. A database of some 15000 sound 'events' was collected over a period of 3months. These events were collected by a system using only an audibility cue set at a low level and making recordings of all detected events. These event recordings were all annotated by at least two listeners and in case of disagreement the opinion of a surveillance operator was used to classify the event. Five different classes were used:

1. non-human: this class included sounds such as car horns, jack hammers and angle grinders
2. unclear: this class included mostly sound events from multiple sources sounding simultaneously
3. shouting: this class contained all human non-aggressive shouting
4. irritation: this class contained all human shouting where a first indication of aggression could be heard, but no incident followed
5. aggression: this class contained all human aggressive shouting that should result in a warning being given to the surveillance operator

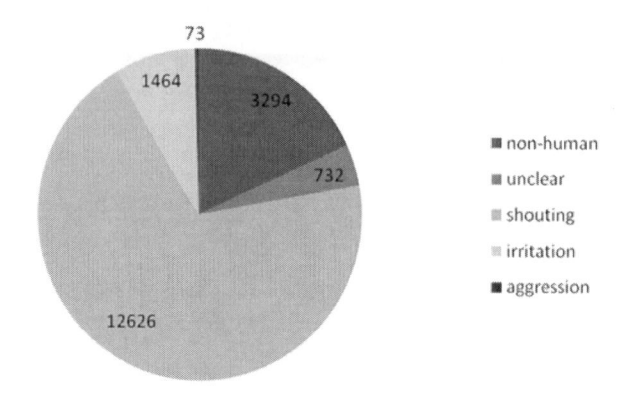

Figure 1. Distribution of original sound event database over the classes. Shouting by far is most common (12626 events or 70% of all events) and agression least common (73 events or 0.4% of all events).

Figure 1 clearly shows that the majority of sound events picked up by the system consist of non-aggressive human shouting. This is what could be expected given the location of the system in the busiest pub-district of Groningen. The number of only 73 aggressive events (0.4%), reflects the rarity of incidents in the inner city of Groningen when compared to the total number of sound events that could or would attract the attention of a listener.

Monthly evaluations and discussions with the surveillance operators made clear that a system which would reproduce the human auditory attention grabbing process, and give a useful alert in only 0.4 % of the cases (or 8.5 % if 'irritation' events are also considered useful) would not be of any additional value. The distraction caused by the huge amount of false alarms would outweigh the potential benefit of being alerted to an incident in an earlier stage.

As a first attempt to improve the performance of the system one might try to make the aggression detection less sensitive, that is to make the threshold of audibility for an event to be passed on higher. An increase of the audibility threshold from 6 dB (the standard) to 18 dB significantly reduced the total number of incidents, but unfortunately did not improve performance. Now 96 % of the detected events consisted of non-human sounds, with the remaining 4 % being non-aggressive shouting.

It was apparent already from the study by Huisman (2004) that human listeners do not just respond to loudness or audibility of an event. Studies by e.g. Young et al. (1996) have shown the neurons on the level of the cochlear nucleus to be sensitive to spectral shape. Other studies indicated a feedback from higher neural levels may influence the operation of the cochlear nucleus (Malmierca et al., 1996), (Schofield, 2000).

Therefore, it was reasonable to assume that not only audibility, but also spectral cues have to be considered in the steering of attention upon hearing aggressive shouting.

Including such cues as pitch presence, pitch frequency and spectral distortion gave a marked improvement of the performance, as can be seen in Figure 2.

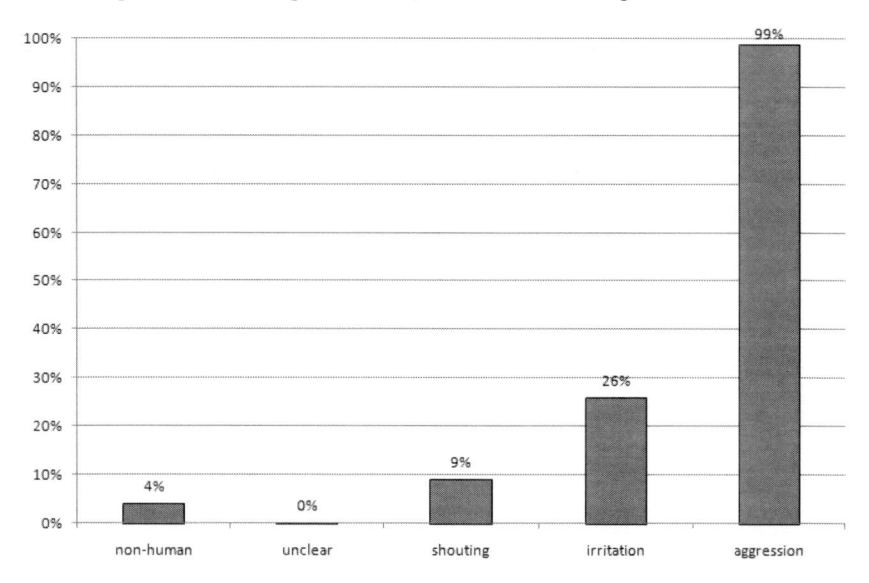

Figure 2. Percentage of original database events identified as aggressive by the recognition system, as a function of the event class. Aggression is virtually always identified as such, there are no 'misses'. Shouting is so common (12626 events or 70% of all events) that the 9% erroneously identified as aggression causes 16 times more aggression false alarms than the true aggression hits.

Based on this representation of the results it is fair to say that the aggression detection system performs well. When taking into account the fact that this performance has been achieved under difficult real-life acoustic conditions, it may even be described as remarkably well. However, when these numbers are translated into the distribution of alarms given by the system, the picture as seen by the surveillance operator is somewhat different.

Due the hugely skewed distribution of events, even a score of only 9 % false alarms on non-aggressive shouting versus a detection score of 99% for the aggressive shouting, implies that the alarms on non-aggressive shouting outnumber the genuine incidents by a relation of 16 to 1. Even the alarms on car horns and angle grinders are twice the ones produced by incidents.

In part this may be related to the choice of locations for this test, i.e. locations close to bars where the proportion of non-aggressive shouting by boozed pedestrians can be expected to be much higher than at other places. By choosing a location where the number of non-aggressive shouting is substantially lower, the fraction of events requiring an alert higher, or the chance of an incident otherwise going unnoticed is substantially higher, may greatly enhance the added value of the acoustic aggression detection. Nevertheless, the technology must be properly explained to the surveillance operator, with its shortcomings, and integrated into the workplace so as not to become a hindrance. It is especially important to note that false alarms will never be fully eliminated.

Future developments

The experience with the aforementioned detection technology has brought to light two fundamental problems. First of all, the so far developed system makes errors on sounds that can quite easily be distinguished from aggression by human listeners. Secondly, the system only responds to aggressive shouting, which is not the only possible (acoustic) indication of an incident. This implies that the system on the one hand creates false alarms that are difficult to understand for the operator, and on the other hand does not respond to all incidents. It is clear that the combination of these effects may rapidly decrease operator acceptance of this technology.

The fact that the system responds to sounds that can easily be identified by human listeners as something completely different than an aggressive shout, is caused by the difference in which human listeners interpret the sound. Where the system treats the incoming sound as a continuous signal with differing properties, the human listener perceives a stream of discrete sound events, each with more or less stable properties. Certain properties of the sound may vary, reflecting changes in the sound source, but other properties remain stable. E.g. for a passing car the speed may vary and the sound may even change by a shift in gear. But the sound remains that of a car of a certain model and/or size. A human voice may produce different words at different intonations with different emotions, but it remains the voice of the same person.

The process by which the human auditory system divides up the incoming continuous sound signal into these discrete events, assigns the proper parts of the signal energy (in time and frequency) to each, and derives both constant and varying properties from this, is called auditory streaming. It has been extensively studied since the ground-breaking work of Bregman (1990), but unfortunately many aspects of this amazing ability still elude scientists. Nevertheless, this is one of the subjects for which further research and algorithm development is needed.

The problem that other sounds, besides aggressive shouting, may indicate an incident, could be solved by introducing more detection systems, each tuned to its own incident sound. However, this would not only increase the number of false alarms with each additional detection sound class, it is also impossible to describe every possible sound indicative of an incident.

Actually it is in most cases much easier to describe the normal situation, and consider any deviation from that as an indication that there may be something wrong. A better approach to the use of audio in such situations would therefore be to 'train' a system to recognise all normal sound events and only raises an alarm when it encounters unknown events. It is this approach that is followed in the European project DIRAC, which is a cooperation of the Fraunhofer HSA group with the Carl-von-Ossietzky Universität Oldenburg and many other partners on detecting the unexpected (www.diracproject.org). We expect this approach to find applications in many areas, but most certainly in surveillance. Not only the video surveillance in inner city areas, as we are familiar with today, but also indoor surveillance of e.g. elderly people suffering from dementia.

Conclusions

Although the case study with the system for the detection of human verbal aggression has provided us with promising results and valuable insights, it has also proven that the value of a system which is focussed on the detection of a single type of signal is limited. Integration of this technology with other sensor systems, but especially the development of new strategies for auditory streaming and the interpretation of sound will however, be of great value for surveillance applications in many domains.

References

Bregman, A.S. (1990) *Auditory Scene Analysis: The Perceptual Organization of Sound*. Cambridge, Mass: MIT press.

Davis, M., Gendelman, D.S., Tischler, M.D., & Gendelman, P.M. (1982) A primary acoustic startle circuit: lesion and stimulation studies. *Journal of Neuroscience, 2*, 791-805.

Gill, M. & Hemming, M. (2004) *Evaluation of CCTV in the London Borough of Lewisham*. Perpetuity Research & Consultancy International

Huisman, M. (2004) Akoestische Effecten van Emoties in Spraak: De Waarneming van Verbale Agressie" ("Acoustic effects of Emotions in Speech: the Perception of Verbal Aggression). Master thesis. Groningen: Department of AI, University of Groningen.

Malmierca, M.S., Le Beau, F.E.N., & Rees, A. (1996) The topographical organization of descending projections from the central nucleus of the inferior colliculus in guinea pig. *Hearing Research, 93*, 167-180.

Schofield, B.R. (2000) Origins of projections from the inferior colliculus to the cochlear nucleus in guinea pigs. *Journal of Comparative Neurology, 429*, 206-220.

Smith, G.J.D. (2004) Behind the Screens: Examining Constructions of Deviance and Informal Practices among CCTV Control Room Operators in the UK. *Surveillance & Society CCTV Special 2 (2/3)*, 376-395

Spriggs, A., Argomaniz, J., Gill, M., & Bryan, J. (2005) *Public attitudes towards CCTV: results from the Pre-intervention Public Attitude Survey carried out in areas implementing CCTV,* Home Office Online Report 10/05. Retrieved 17.04.2009 from http://www.homeoffice.gov.uk/rds/pdfs05/rdsolr1005.pdf

Van Hengel, P.W.J. & Huisman, M. (2003) Rapportage haalbaarheidsstudie audioclassificatie voor cameratoezicht. internal report

Young, E.D., Spirou, G.A., Rice, J.J., & Voigt, H.F. (1992) Neural organisation and responses to complex stimuli in the dorsal cochlear nucleus. *Philosophical Transactions of the Royal Society* B: *Biological Sciences, 336*(1278), 407-413

Detecting deviants within flocks

Robert J. Houghton & Chris Baber
The University of Birmingham,
Birmingham, UK

Abstract

Flocking refers to a family of grouping behaviours commonly seen in nature (fish, birds, moths etc.) Whilst flocking is a complex pattern of behaviour, Reynolds (1987) and others have demonstrated that it is underpinned by the application of three simple rules –separation, alignment and cohesion– that expressed as vectors can be summed together to give the eventual heading of any one agent and thus produce an emergent flock from multiple independent agents. However, despite the ubiquity of flocking to the natural world and its recent use as a basis for the simulation of human crowds, relatively few studies have examined the perception of flocking. In the present study we examined the ability of observers to pick out 'deviants' (that is, agents that only obey a subset of the rules described by Reynolds) within flocks under cued and un-cued conditions. The results show that cuing causes differences in deviancy detection suggesting that whilst some violations are easier to detect by directing attention to individual agents, others are more reliably detected by considering the flock as a whole. This finding has resonance with accounts of real-world crowd monitoring and suggests that both the global observation of a crowd as a whole and the monitoring of individuals is required to detect a full range of potentially noteworthy deviant behaviours. It is speculated that this requirement arises as a result of the underlying processes that cause crowd behaviour to emerge from the interactions of individuals.

Introduction

Flocking refers to a family of group aggregating behaviours commonly seen in nature, perhaps most visibly in the case of birds and fish but found across a wide range of taxa (see Parrish & Hamner, 1997). Despite being an apparently complex activity at the group level, particularly in terms of how flocks can split and reform to navigate around obstructions, it was demonstrated by Reynolds (1987) through his "Boids" (bird androids) model that flocking could be an emergent group behaviour that might arise from the interaction of agents individually adhering to a small set of simple steering rules. Since this initial demonstration of emergent flocking, similar approaches have been taken to modelling crowds of humans (Reynolds, 1999) and flocking rules have also been used in data visualisation and within architected systems (e.g., Unmanned Aerial Vehicles and reconnaissance satellites). However, despite this wide-spread interest in the technique and apparent ubiquity of flocking in nature, there have been relatively few studies of the human perception of flocking. In the present paper we address this lacuna by examining how easily so-called *deviants*, that is agents that do not behave as others do, can be detected within a flock and whether cuing which focuses attention on a subset of agents within the flock modulates this process.

Possible applications include detecting trouble-makers and alternatively, those in distress within crowds based on the hypothesis that they will show enough commonality in behaviour to be part of the crowd but that their behaviour may show small differences. The question thus becomes whether these individuals are better noticed through viewing the crowd at large or whether an intelligence-led focus on specific individuals results in better detection. Given the use of flocking in visualisation and control systems, we might also wish to know how sensitive observers are to detecting deviants to add appropriate scaling factors in visualisations. In the case of, for example UAVs, we might want to know how easy it would be to detect either a malfunctioning unit (if the UAVs are our own) or to spot a deviant unit (perhaps carrying weapons) amongst reconnaissance UAVs directed by an opposing force. A general aim of this research is to investigate the possibility of using a form of stimuli that are perhaps more complex than those usually used in the laboratory that through the use of relatively simple algorithms possess some of the qualities that flocks (and thus other aggregations of individuals) typically have.

Background

The boids model (Reynolds, 1987)

The first rule, Cohesion leads individual entities (termed agents) to seek to move towards the flock's centre of gravity. A second rule is Avoidance wherein agents seek to avoid collision with other flock members. A third rule is Velocity matching (or Alignment) wherein agents seek to follow the average heading of other flock members. Each steering rule generates a velocity (i.e., a heading and a speed) which are, in the simplest implementations of the approach, simply summed together to give a final heading and speed for each agent in a given epoch ('tick') within a simulation. It is possible to extend the basic boids model by adding additional rules. One might for example modulate the weightings given to different rules to give agents a greater tendency to seek cohesion in certain circumstances such as the presence of a threat or to relax the tendency toward avoidance to model a relaxation of a sense of personal space. Alternatively, one may wish to encourage attraction to, or avoidance of, a given spatial location or to impose a tendency toward following a certain path or goal. Further fidelity may be added by modelling the perceptual limitations of individual agents and/or constraints that emerge from the mechanics of movement in agents (Tu & Terzopoulous, 1994; Brogan, Metoyer, & Hodgins, 1998).

Emergent systems for crowd modelling

Since Reynold's initial report, emergent flocking has been used as a basis for crowd modelling and animation alongside related approaches such as particle systems

(Bouvier, Cohen, & Najman, 1997; Moore et al., 2008) which together with flocking, fall under the general heading of 'swarming' models. Additional to these are approaches in which agents are given greater intelligence through more complex sets of rules and internal states that are updated to provide motivations toward certain behaviours depending on the results of prior behaviours (e.g., Musse, Garat & Thalmann, 1999). Reynolds has suggested some rule extensions to his own initial design that include seek, flee, containment, and leader following that would be sufficient to produce autonomous character behaviour in animation (Reynolds, 1999). While it may be argued that flocking rules alone may be insufficient to model the complexities of crowds of thinking and perceiving individuals (see Sime, 1995 for a critique of engineering models of crowd movement), emergent systems based upon simple rules have been able to spontaneously generate various kinds of recognisable social behaviour (McPhail, Powers, & Tucker, 1992). Arguably, whilst difficult to validate given a lack of real-world data, individual-based models of crowd dynamics have an explanatory power that previous flow-based models which treated crowds as if they were continuously flowing liquids lack (Low, 2000). The general thrust of this research is that groups are best modelled not as single entities but rather as conglomerations of individuals whose behaviours interact to produce grouping phenomena.

Uses of flocking algorithms in artificial systems

Quite apart from its use to simulate or model human and animal behaviour, flocking algorithms have also been used to produce flocking within artificial systems. There has been considerable interest in the use of flocking algorithms as a basis for the automated control of robots (e.g., Mataric, 1992; Balch & Arkin, 1998) and Unmanned Aerial Vehicles (e.g., Crowther, 2003, 2004). Advocates of flocking have generally favoured it as a heuristic approach to robotic swarm control over more tuned so-called optimal algorithms on the basis that what it may lack in efficiency or effectiveness in specific situation is offset by its robustness (see Gurfil, 2005 for an evaluation). However, it is possible this very robustness may be a problem for operators, in so far as a damaged or malfunctioning UAV may not be easily detectable as the flocking dynamic lends it a form of graceful degradation. For example, if one UAV suffers engine trouble that causes it to slow down, rather than dropping out of the group, the rest of the group may slow down to compensate for it; it may not be immediately obvious to an operator which drone is suffering from engine problems.

Leaders and deviants

In the present paper we describe agents that do not follow the same rules as the majority as deviants. Within sociology, the term deviance relates to behaviour that violates *norms*; a set of informal rules that regulate behaviour. In many instances these norms are socially constructed and limited to specific contexts, for example behaviour appropriate to being in the crowd at a football match would be inconspicuous on the terrace but one would however clearly stand out if one behaved the same way at the opera. In broad terms, detection of a deviant presumably requires the observer to build up some form of representation of what constitutes normative flocking behaviour and to observe violations of these norms. We emphasise this reliance on shifting norms because in the absence of the other members of a flock, the motion of an individual boid just appears to be noise as individual behaviour is stochastic (White, 2005). An additional difficulty in deviant detection is that within flocks whilst a deviant may be guided differently from their flockmates, the remainder of the flock still include the position and vector heading of the deviant in their own computations (for example, they will avoid collision with it even if the deviant takes no avoiding action) with the result that the deviant has some influence over the flock itself. This interactivity within a flocking stimulus means it differs somewhat from the stimuli commonly used in the laboratory to investigate sensitivity to differences in the detection of differing motion paths (e.g., Horowtiz et al., 2007). There is support for the notion that individuals can influence groups in mathematical studies of animal movement (Couzin et al., 2005), escape panic (Helbing, Farkas, & Viscek, 2000) and from a human study which showed that in a walking task individuals with knowledge of target locations exerted significant influence over the paths of naïve individuals without any explicit communication or signalling taking place (Dyer et al., 2008). More prosaically, it is common to notice that it does not take many 'leaders' to start pushing or running for the rest of the crowd to take up this form of behaviour.

In terms of the specific stimulus manipulation carried out here – the suspension of selected steering rule(s) to create deviance – this general approach has been used as a basis for the modelling of alcohol-related violence (Moore et al., 2008). In Moore et al.'s simulation while all particle-modelled agents employed their usual repertoire of behaviours, 'emergent affiliative behaviours' (which is to say, grouping behaviours) minimised collisions between sets of individuals. When these behaviours were inhibited (modelling the effects of alcohol) it gave rise to the kinds of interactions sometimes seen between "drunk" individuals that result in violations of personal space.

Experiment

Participants

Forty students (34 male, 6 female) at the University of Birmingham participated in the experiment for course credit or monetary compensation. None of them were aware of the purpose of the experiment and all reported having normal or corrected vision.

Apparatus and stimuli

The software controlling the generation of stimuli and the running of the experiment was written using the Python programming language. Stimuli were presented on a standard PC using a 22" LCD monitor refreshing at a rate of 60 Hz with a resolution of 1680 x 1050 pixels. This was observed from a view distance of 57 cm. Participants recorded responses using pen and paper.

Boids

Stimuli consisted in every trial of nine circular "boids" which began in randomly generated positions. The motion of each boid was calculated using the three steering rules described Reynolds (1987, for guidance on implementation see Parker, 1995, and see Davison, 2005, for further elaboration). The steering rules are depicted in Figure 1 and are:

1. Cohesion (steers toward the average position of local flockmates)
2. Separation (steering to avoid local flockmates)
3. Alignment (steering towards the average heading of local flockmates)

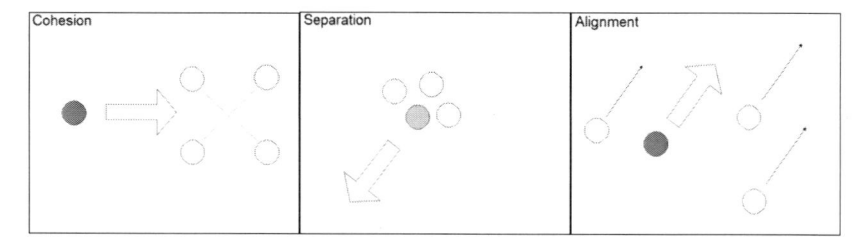

Figure 1. The three boid steering rules after Reynolds (1987). The left-hand panel shows the cohesion rule, the centre panel the separation rule and the right-hand panel the alignment rule. In each case the resultant steering vector for the shaded circle is shown as a large arrow. The wall rebound rule is an adaptation of the Separation rule applied to the sides of the screen

When the results of the three algorithms are added together, this gives a final vector heading that the boid then follows in the next epoch. In the case of the single deviant boid, only a subset of the three steering rules were calculated, which gave raise to six variants of deviant:

1. Cohesion only (suppression of Separation and Alignment)
2. Separation only (suppression of Cohesion and Alignment)
3. Alignment only (suppression of Cohesion and Separation)
4. Cohesion & Separation (suppression of Alignment)
5. Cohesion & Alignment (suppression of Separation)
6. Separation & Alignment (suppression of Cohesion)

In addition to these rules, a further rule was added, which was activated when a boid reached the edge of the screen in order to avoid colliding with it. This "wall repulsion rule" can be viewed as a version of the separation rule that applies to non-boid objects in the environment. The positions of each agent were calculated and updated 40 times per second (40 Hz) and each circular agent had a diameter of approximately 1.5 cm. In the uncued condition, all boids were coloured white and labelled with black numbers from 1 to 9 inclusive. In the cued condition boids were not numbered, but the two cued boids (one of which was the deviant) were coloured red and blue.

Design and procedure

The experiment had a mixed design, with participants allocated to either the cued (boids numbered from 1 to 9 inclusive, see Figure 2a) or uncued group (boids unnumbered but two coloured red and blue, see Figure 2b).

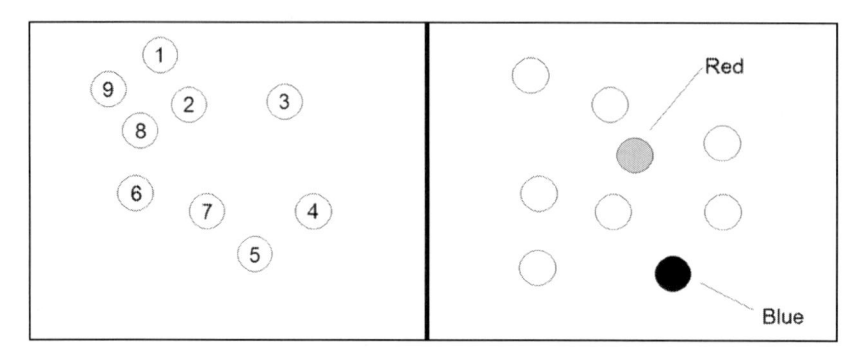

Figure 2. Apperance of stimuli. The left hand panel shows the uncued (numbered) stimuli, the right hand panel the cued (coloured) stimuli

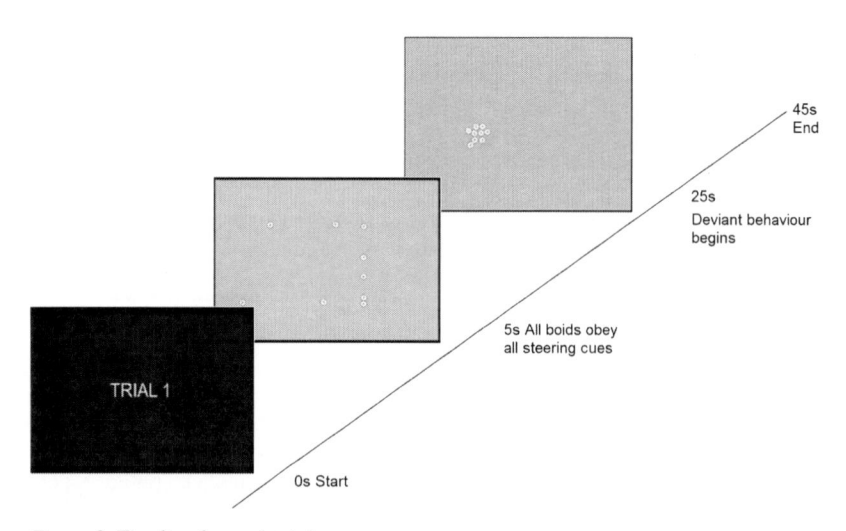

Figure 3. Timeline for each trial

Participants were asked to decide which boid appeared to behave differently from the others and responded at the end of each trial with either a number in the uncued group or a colour (red or blue) in the cued group. There were six deviance conditions (all single and paired steering rules) and each participant saw each form of deviance five times for a total of 30 trials in all. Each trial began with a blank screen and 25 point text stating the trial number. After 5 s, 9 circular boids appeared in randomly generated locations. The screen background was grey in colour which indicated to the participant that all boids would be obeying all steering cues. The boids then began movement. After 20 s the background changed to light blue which indicated one of the boids was now deviant and the trial continued for a further 20 s (see Figure 3).

Results

Splitting the results by cueing condition we find the results as per Figure 4 (uncued) and Figure 5 (cued). It is notable that participants were able to detect deviants but that performance varied considerably with both the steering rules used by the deviant and cueing condition. In both cueing conditions the detection of deviant using only coherence or separation rules was highly successful. Analysis of the stimuli revealed that in these conditions deviants frequently broke away from the flock suggesting detecting this form of deviance was relatively trivial. However, when a deviant used only the alignment rule they were in both cued and uncued conditions very difficult to detect with results at or around chance in both cases. Indeed, there is a general trend in both sets of data to suggest that adding an additional steering rule to alignment actually made deviants easier to detect.

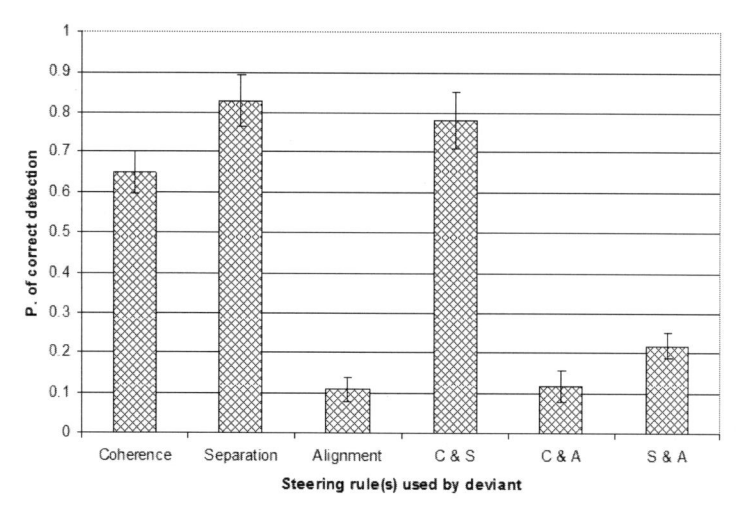

Figure 4. Probability of deviant detection in the uncued condition. Error bars show Standard Error

In order to normalise the results for comparison, we adjusted the probability of detection with reference to the guess rates in each condition (1:9 for the uncued condition, 1:2 for the cued condition). The results, where 0 on the Y-axis constitutes chance performance, is show in Figure 6. A Mixed ANOVA was then carried out on the normalised data. There was not a main effect of condition ($F(1,38) = 1.82$, $MSE = .419$, $p = .185$.), however there was a main effect of steering rules used by the deviant ($F(5,38) = 33.75$, $MSE = 4.07$, $p < .001$). In Figure 6 we can see that cueing (which in an applied CCTV situation would be the equivalent of an operator having prior intelligence about which individuals to concentrate on) clearly improved performance as one might expect, however there is a notable exception when the deviant employed the coherence and separation steering rules where uncued performance was actually better. This interpretation is supported by an interaction between cueing condition and steering rule (F(1,5) = 2.95, MSE = .356, p = 0.014).

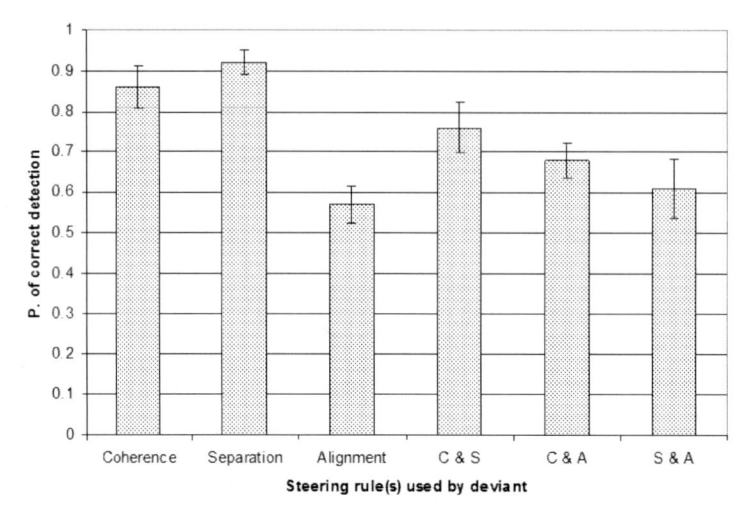

Figure 5. Probability of deviant detection in the cued condition. Error bars show Standard Error

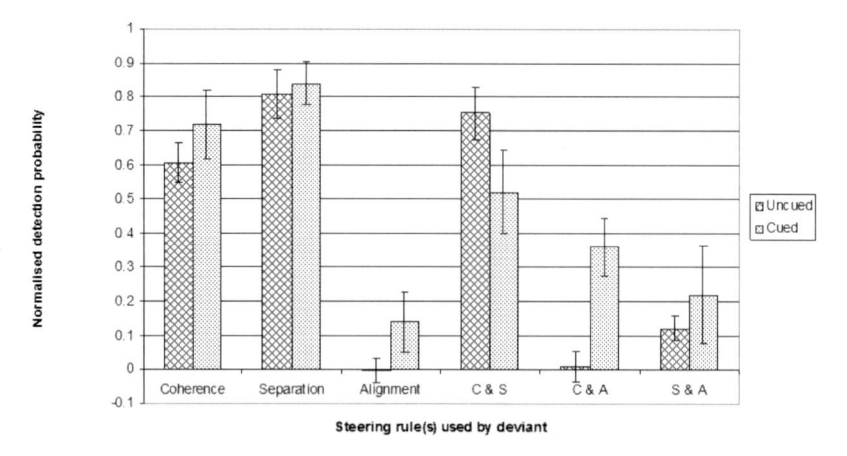

Figure 6. Probability of deviant detection in both the cued and uncued conditions normalised for the different guess rates across conditions. Error bars show Standard Error

Conclusions

As an experimental stimulus, a flock has several interesting properties that make it more complex than those commonly used in the laboratory such as Random Dot

Kinematograms. First, the agents do not necessarily follow simple linear paths of motion and these paths are determined in real time. Second, these paths of motion are a result of reciprocal interactions between agents. As a result of this complexity, defining what is normal and what is unusual is presumably a relative complex process for the observer that would presumably require some ongoing build up of a representation of what constitutes normality and related expectations, and it is in the violation of these expectations that the identification of deviants takes place.

A first finding from this experiment is that naïve observers can indeed detect deviants, but only when this deviance expresses itself in the suppression of specific behaviours (steering rules). Suppressing all steering rules except alignment, (sharing approximately the same heading but neither avoiding others or attempting to cohere with them) proved to be the hardest for observers to detect. In turn this suggests that when alignment behaviour was not present, violation of what has been termed in Gestalt theory 'Common Fate' was easiest to detect compared to either collisions between agents (suppression of separation) and the tendency to drift away from the flock (suppression of cohesion). This is supported by the finding that in the specific condition of deviants using "coherence and separation" (but alignment) this change was actually better detected in the uncued condition which was likely due to observers examining the entirety of the flock and thus viewing the display in a more global manner. Contrastingly, drift away from the flock and collisions were much better detected in the cued condition, suggesting that these forms of deviancy required more observation time (or more frequent sampling) to become apparent.

There is an additional, perhaps more subtle hypothesis that may also apply which concerns the actions not of the deviant, but rather the perturbations it causes amongst other agents that causes the deviant to be not so much missed but rather non-deviants to themselves appear 'suspicious' in that the normal pattern of behaviour has been disrupted for all. This is particularly evident in conditions where separation is suppressed in the deviant and therefore causes it to stray into the paths of others. The remaining agents have then to be relied upon to take evasive action with the effect that in certain circumstances this can cause a general scattering of the flock that effectively hides the deviant. Determining between these two hypotheses is an interesting question for future study and possibly post-hoc analysis of flocking dynamics.

Discussion

The present findings, although drawn from a very simple laboratory paradigm appear to have resonance with the content of informal discussions we have had with CCTV and surveillance system operators. Particularly for purposes of crowd observation, aerial surveillance is employed that is limited in terms of resolution and therefore, for the operator viewing the video feed, detail. Thus, instead of looking at the behaviour of individuals it is more common to try to identify unusual crowd behaviour through this 'global' view and then to use a 'local' view – either other fixed CCTV cameras or individuals on the ground – to follow up initial observations. For example, if an individual produces a weapon this may not be directly observable but the scattering of individuals who have themselves seen the weapon and wish to move away from it may well be. Furthermore, some crimes are only practically observable from a global view and not a local view. A specific example of this would be a form of robbery known as steaming. The so-called steamer gang consists of a small group of individuals who push through a crowd usually in the opposite direction to its general flow grabbing bags, jewellery and mobile phones from individuals. The individual making the theft is preceded and followed by accomplices who prevent victims from stopping the thief. On the ground this is very hard for individuals to observe, indeed the crime relies upon surprise, confusion and the lack of visibility within a packed space, but often very clear in aerial surveillance video as a spreading perturbation that travels across a crowd. Other crimes are of course only apparent through the close observation of individuals about whom suspicion may already exist.

The over-arching finding of the present study is that different forms of deviant behaviour are best detected in different ways. Some forms of deviancy are better observed when taking a 'global' view of the crowd and pop-out for the observer while other forms of deviancy are best detected through careful observation of a couple of individuals. This would appear to be at least analogous to real-world surveillance and would appear to suggest that both global and local views are required for crowd monitoring applications. Whilst in the present experiment this global/local split was produced through the manipulation of the display and instructions to participants, in an applied setting it would be created by the viewing distance and resolution of cameras. Neither view is sufficient in itself owing to the emergent processes that we hypothesise produce crowd phenomena and how these processes function when an individual violates the norms of crowd behaviour. This may then require two different systems to work together (e.g., air and fixed CCTV) or else for zoom functionality to be provided (zooming out being as important as zooming in). Furthermore, this suggests that a future way to do crowd observation research may be to develop models of crowd behaviour in tandem with perceptual studies concerning observation of those models so that the potential drivers of crowd behaviour can be assessed as moderators of performance in crowd perception. Thus far crowd modelling and perceptual studies of crowd monitoring have tended to belong to separate literatures.

A topic for further research would be the graphical fidelity of the displays used. In the present experiment we used simple shapes moving against an empty background. The literature concerning perceptual causality and animacy (the interpretation of meaning in moving displays and the attribution of causes for what is observed) suggests that observers have a strong tendency to attribute motivations, intentional behaviours and even emotions to very simple displays (for a review see Scholl & Tremoulet, 2000). However, it would be interesting to see how the perception of crowd behaviour is affected by manipulations in the complexity of the graphical rendering of individuals; it may even be the case that incidental details distract from the perception of behaviour. This in turn would further inform the resolution requirements for CCTV and surveillance systems.

References

Balch, T., & Arkin, R.C. (1998). Behavior-based formation control for multirobot teams. *IEEE Transactions on Robotics and Automation, 14*, 926-939.

Bouvier, E., Cohen, E., & Najman, L. (1997). From crowd simulation to airbag deployment: particle systems, a new paradigm of simulation. *Journal of Electronic Imaging, 6,* 94-107.

Brogan, D., Metoyer, R., & Hodgins, J. (1998). Dynamically simulated characters in virtual environments. *IEEE Computer Graphics and Applications, 18* (5), 58-69.

Couzin, I.D., Krause, J., Franks, N.R., & Levin, S.A. (2005). Effective leadership and decision-making in animal groups on the move. *Nature, 433,* 513-516.

Crowther, B. (2003). Flocking of autonomous unmanned air vehicles. *Aeronautical Journal, 107,* 99-110.

Crowther, B. (2004). Rule-based mechanisms for flight vehicle flocking. *Proceedings of the Institute of Mechanical Engineers, Part G: Journal of Aerospace Engineering, 218* (2), 111-124.

Davison, A. (2005). *Killer game programming in Java.* Sebastopol, CA: O'Reilly.

Dyer, R.D., Ioannou, C.C., Mitchell, L.J., Croft, D.P., Couzin, I.D., Waters, D.A., & Krause, J. (2008). Consensus decision making in human crowds. *Animal Behaviour, 75,* 461-470.

Gurfil, P. (2005). Evaluating UAV flock mission performance using Dudek's taxonomy. *American Control Conference, FrC04.4* (pp. 4679-4684). Portland, OR: IEEE.

Helbing, D., Farkas, I., & Vicsek, T. (2000). Simulating dynamical features of escape panic. *Nature, 407,* 487.

Horowitz, T.S., Wolfe, J.M., DiMase, J.S., & Klieger, S.B. (2007). Visual search for type of motion is based on simple motion primitives. *Perception, 36,* 1624-1634.

Jian, J., Matsuka, T., & Vickerson, J.V. (2006). Towards deceptive intention: Finding trajectories and its analysis. *Proceedings of the Human Factors and Ergonomics Society 50th Annual Meeting 2006,* (pp. 324-328). Santa Monica, CA: HFES.

Low, D.J. (2000). Statistical physics: Following the crowd. *Nature, 407,* 465-466.

Mataric, M. (1992). Designing emergent behaviors: From local interactions to collective intelligence. *Proceedings of the International Conference on Simulation of Adaptive Behavior: From Animals to Animats 2* (pp. 432-441). Cambridge, MA: MIT Press.

McPhail, C., Powers, W.T., & Tucker, C.W. (1992). Simulating individual and collective action in temporary gatherings. *Social Science Computer Review, 10,* 1–28.

Moore, S.C., Flajslik, M., Rosin, P.L., & Marshall, D. (2008). A particle model of crowd behavior: Exploring the relationship between alcohol, crowd dynamics and violence. *Aggression and Violent Behavior, 13,* 413-422.

Musse, S.R., Garat, F., & Thalmann, D. (1999). Guiding and interacting with virtual crowds in real time. In N. Correia, T. chambel, and G. Davenport (Eds.) *Multimedia '99: Proceedings of the Eurographics Workshop in Milano, Italy, September 7-8, 199.* Wien: Springer.

Parker, C. (1995). Boids pseudocode. Retrieved 09.06.2008 from http://www.vergenet.net/~conrad/boids/pseudocode.html.

Parrish, J.K., & Hamner, W.M. (1997). *Animal groups in three dimensions: How species aggregate.* Cambridge, UK: Cambridge University Press.

Reynolds, C.W. (1987). Flocks, herds and schools: A distributed model. *SIGGRAPH'87: Proceedings of the 14th annual conference on computer graphics and interactive techniques*, (pp. 25-34). New York: ACM Press.

Reynolds, C.W. (1999). Steering behaviors for autonomous characters. *Proceedings of the Game Developers Conference 1999, San Francisco, California.*

Sime, J.D. (1995). Crowd psychology and engineering. *Safety Science, 21*, 1-14.

Tu, X., & Terzopoulos, D. (1994). Artificial fishes: Physics, locomotion, perception, behaviour. *Proceedings of SIGGRAPH'94* (pp. 43-50). New York: ACM Press.

White, T. (2005). Expert assessment of stigmergy: A report for the Department of National Defence. Defence Research and Development Canada.

Theories and miscellaneous

WORKING TOGETHER IN THE OPERATING ROOM

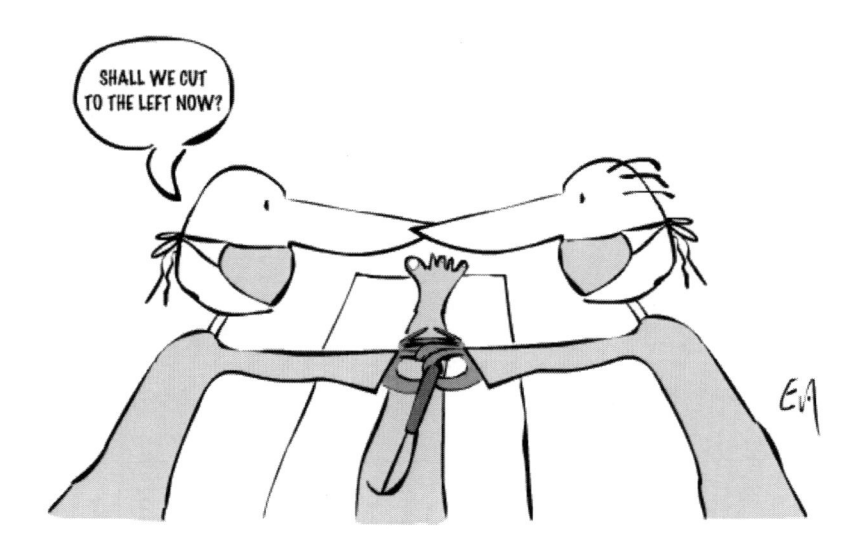

Shall we cut to the left now?

In D. de Waard, J. Godthelp, F.L. Kooi, and K.A. Brookhuis (Eds.) (2009). *Human Factors, Security and Safety* (p. 389). Maastricht, the Netherlands: Shaker Publishing.

Using social network analysis to identify sub-groups in the operating room

Tita A. Listyowardojo[1], Christian E.G. Steglich[1],
Stephen Peuchen[2], & Addie Johnson[1]
[1]University of Groningen
[2]University Medical Centre of Groningen
The Netherlands

Abstract

The frequency with which operating room (OR) staff work together can impact patient safety because staff who often work together share a set of experiences which may enable them to anticipate each other's actions and reactions in the future. Identifying sub-groups of staff who frequently work together is thus a significant step in investigating team skills and the knowledge needed to prevent mishaps. Here, social network analysis (a set of statistical techniques for analyzing networks of interactions) is used to quantify the frequency with which individual OR staff members in a large university hospital worked together on three types of operation over a period of 2.6 years. Details of the specific techniques used are given. It is concluded that social network analysis is a viable method to identify sub-groups in the OR.

Introduction

Operating room (OR) staff who frequently work together may be more effective in preventing mishaps because these staff share a set of experiences which enables them to anticipate each other's actions and reactions, especially in emergency situations where communication may otherwise be ineffective. An example of such *tacit communication* is described in a study by Friedman and Bernell (2006): A perfusionist was being interviewed when she suddenly excused herself and walked into the OR where her team was performing a coronary artery bypass graft surgery. She immediately prepared the heart-lung machine before hooking the patient's heart up into the machine without being told to do so by anybody. When asked about this afterwards, she told the interviewer that when she saw the monitor which provided the patient's vital signs, she knew that the procedures to isolate a small part of the patient's heart without using the heart-lung machine (an "off-pump" procedure) had gone wrong and that the surgeon, whom she had worked together with for many years, would want to put the patient on the heart-lung machine (an "on-pump" procedure) to circulate and oxygenate the blood while the heart was stopped and being operated. As a result, without having any explicit communication between the perfusionist and the surgeon, a potential mishap was prevented.

Ineffective communication has been identified as the cause of many medical errors (Awad et al., 2005; Lingard et al., 2004), including wrong-site operations (Makary et al., 2007; Michaels et al., 2007), and has resulted in "near-misses" (Brennan et al., 1991; Leape et al., 1991) and deaths (Calland et al., 2002; Kawashima et al., 2003). The aforementioned example suggests that staff who frequently work together may develop unique work styles and habitual routines and become better able to use the information conveyed by body language and facial expressions as a result of repeated exposure to each other (Gersick & Hackman, 1990). The tacit communication which may thus be enabled may improve performance when explicit communication breaks down.

An investigation of the ability to anticipate actions and reactions in the OR requires the identification of staff who frequently work together. Once these staff members are identified, an interview study can be conducted with, for example, the critical incident technique. In this method of probing interaction styles, participants are asked to recall a potential mishap which they successfully anticipated and prevented in the past. The details on how they anticipated the potential mishaps are then probed to uncover the roles of different types of communication.

Identifying OR staff who frequently work together is a problem in itself because of the complexity of OR environments. The density of the data (high number of operations and OR staff) and *ad hoc* scheduling practices in which any available OR staff are assigned to operations are some of the barriers to identifying sub-groups by "hand". Specific analysis techniques are thus needed to identify these sub-groups. A general method with specific techniques that may be able to capture sub-groups in the OR environment is *social network analysis*. Social network analysis (SNA) can be used to analyze various types of relational data (Scott, 1991) and has been extensively utilized. For example, SNA has been used as an analytic tool to capture decision-making patterns in health care practices (Scott et al., 2005). In this case, participants were asked to identify whom they referred to when they had to make important decisions in the work setting. Networks of decision making patterns were captured that allowed the identification of significant differences between health care workers and which suggested interventions to promote practice organizational change.

Another example of the use of SNA is in describing the structure of multidisciplinary long-term care teams by identifying the relationships that develop among staff as they do their work (Cott, 1997). Keating et al. (2007) showed that SNA can also be used to capture factors that affect influential discussions among physicians in a primary care practice, such as the accessibility of colleagues based on location and schedule. This information can be used to organize practices to promote more rapid dissemination of medical information.

Social network analysis differs from other analysis techniques by focusing on patterns of relationships and by emphasizing empirical observation of these relationships. The current study aims to identify sub-groups in the OR whose staff members work together the most frequently. This data will be used in subsequent research to select groups for an interview study to investigate how staff who frequently work together prevent mishaps in the OR.

Method

Data

The data were drawn from the OR staffing records of a large university hospital in the Netherlands. To reduce noise in the data, medical students in training were excluded from the data analysis, as were staff who were involved only at the beginning (e.g., less than 1.5 years) or at the end (e.g., started to be involved in the operations at the second year) of the 2.6 year period, or who were involved in only a few operations.

All appendectomies, esophageal resections, and liver resections that took place between June 2, 2005 and December 31, 2007 were included, for a total of 1,423 operations with 63 staff. Of the 63 staff, 33 were female and 25 were male; 5 were not identified. The average age was 44 years old (SD=9.7; range from 25 to 62 years old). In the period of approximately 2.6 years, each staff member performed a minimum of 32 operations and a maximum of 238 operations (M=78 operations; SD=40).

Three types of operation were compared because the fact that they differ in difficulty level may affect staff interactions. Liver resections were the most difficult of these operations and appendectomies the simplest. Liver resections were the most common operations in the dataset (N=627; 44.1%) followed by esophageal resections (N=413; 29%) and appendectomies (N=383; 26.9%).

Almost all of the staff took part in all types of operation: appendectomies (N=60), esophageal resections (N=62), and liver resections (N=63). The roles in the OR were assistant surgeon/surgeon, scrub nurse, nurse anesthetist, and anesthetist (see Table 1). Six staff performed both as anesthetist and assistant surgeon/surgeon and one as both scrub nurse and nurse anesthetist (see Table 1).

Table 1. Number of staff per role

Staff Roles	Value
Anesthetist	4
Assistant surgeon/surgeon	10
Anesthetist and Assistant surgeon/surgeon	6
Scrub nurse	30
Nurse anesthetist	12
Scrub nurse and nurse anesthetist	1
Total	63

Procedure

The data were first put in the form of an adjacency matrix (see Figure 1) for analysis in UCINET VI (2002), a computer programme developed for performing SNA. For confidentiality purposes, staff identities were converted into individual IDs. The following codes were used: "A" for anesthetist, "S" for surgeon or assistant surgeon, "SN" for scrub nurse, "NA" for nurse anesthetist, "A/S" for anesthetist/assistant surgeon/surgeon, and "NA/SN" for nurse anesthetist/scrub nurse. Each OR staff member was assigned to a column and a row in the matrix. The number of operations performed together for each pair of staff was entered in each matrix cell. For example, in Figure 1, surgeon 1 (S1) worked together for 5 operations with scrub nurse 1 (SN1), therefore, 5 is entered in the matrix cell indexed by S1 and SN1. The matrix cells below the diagonal were identical to those above the diagonal.

	S1	SN1	A1
S1	--	5	10
SN1	5	--	0
A1	10	0	--

Figure 1. An example of an adjacency matrix

A *clique at level C* analysis was then used to identify sub-groups whose members worked together frequently. A *clique* is defined as the maximum number of actors who share all the possible ties among themselves (Wassermann & Faust, 1994). That is, it is a group within which all members interact with each other. 'A clique at level C' can be defined as a sub-graph in which the relationships between all pairs of staff have values of C or greater. In other words, the frequency with which staff within a clique work together is given by the parameter C. For example, if C is a set at 10, only staff who work together at least 10 times will be included in a clique. Although Wassermann and Faust define a clique as having a minimum of three members, in this study, cliques of only two members were allowed because the minimum number of staff to perform an operation is two.

One person can belong to many cliques. However, because clique analysis in its most general form (C = 1) simply connects all people who have worked together at least once, staff who have performed ten operations together cannot be distinguished from those staff who have performed only one operation. To identify sub-groups whose members have worked together at a certain level of frequency, C, a value for C must first be determined.

In order to define a threshold level for C, hierarchical clustering is performed to determine the frequency with which pairs of staff work together (Reffay & Chanier, 2003). Hierarchical clustering is based on the strength of the relationship (i.e., similarity) between actors (here, the number of operations performed together). Clusters are produced by joining the most similar pairs of staff based on their proximity. For example, if actor A and actor B work together 10 times, and not more than ten times with anyone else, A and B will be joined in a single cluster (A/B at level 10). If both A and B work together 8 times with actor C, then cluster A/B is identifying sub-groups in the operating room 395

joined with C to create cluster A/B/C at level 8. However, if A only works together with D for 7 times and 7 times is the highest level of interaction for D, then A and D are joined into cluster A/D at level 7 (see Borgatti, 1994).

Results

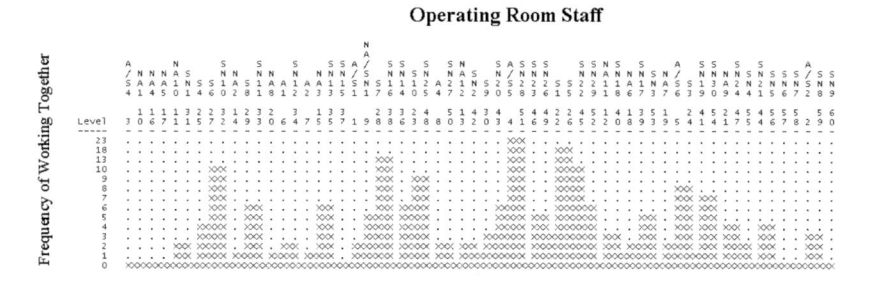

Figure 2. Hierarchical clustering of appendectomies

Figure 3. Hierarchical clustering of esophageal resections

Figure 4. Hierarchical clustering of liver resections

Figure 2, 3, and 4 show the results of the hierarchical clustering analysis. Each group of three or more Xs represented the clusters involving the staff of the corresponding columns. For example, for appendectomies (see Figure 2), staff member A/S5 and SN28 performed 23 operations together. They performed six operations with SN20, and three with SN20 and S9. The highest level of interaction was reached for liver resections, with 37 operations performed together by S3 and SN28 (see Figure 4). The frequency of working together by any one pair was the lowest for appendectomies, with 23 operations performed together by A/S5 and SN28 (see Figure 2). On the basis of frequency of working together, a threshold level of C = 10 was selected for the clique at level C analysis. This resulted in the selection of nine staff for appendectomies, fourteen for esophageal resections, and eighteen for liver resections.

Clique at Level C analyses

NetDraw (2002) was used to visualize the figures: five sub-groups were found for appendectomies (see Figure 5), eight for esophageal resections (see Figure 6) and sixteen for liver resections (see Figure 7). The members of each sub-group per type of operation are given in Table 2.

Table 2. Sub-groups Found for Each Type of Operation

Type of Operation	Sub-groups Found
Appendectomies	S1, S5, SN22; S1, S5, SN28; A/S5, SN28; S6, SN10; S7, SN16
Esophageal resections	S3, S5, SN20, SN28; A/S6, S3, S5, SN28; S1, S5, SN11; S5, SN11, SN29; S10, S4, SN10; S10, S4, SN25; A/S6, SN12; SN26, SN29
Liver resections	A/S5, S1, S5, SN28; S1, S5, SN20, SN28; S5, SN11, SN28; A/S6, S5, SN28; A/S5, S5, SN24; S5, SN11, SN24; A/S5, S3, SN28; A/S6, S3, SN28; A/S6, S8, SN28; S5, SN29; S10, SN10; S10, SN25; S4, SN10; S4, SN25; S7, SN16; SN26, SN29

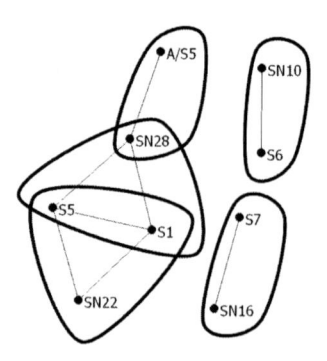

Figure 5. The five sub-groups found for appendectomies

The clique at level C analysis also produces a so-called *hierarchical clustering of overlap matrix*. The hierarchical clustering of overlap matrix shows how many sub-groups were shared between two or more staff (see Figures 8, 9 and 10).

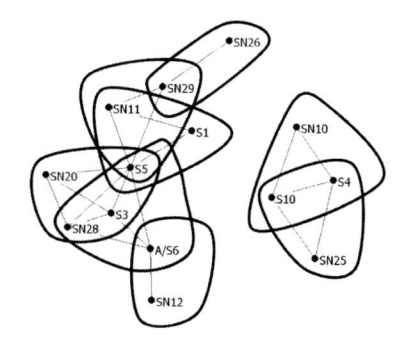

identifying sub-groups in the operating room 397

Figure 6. The eight sub-groups found for esophageal resections

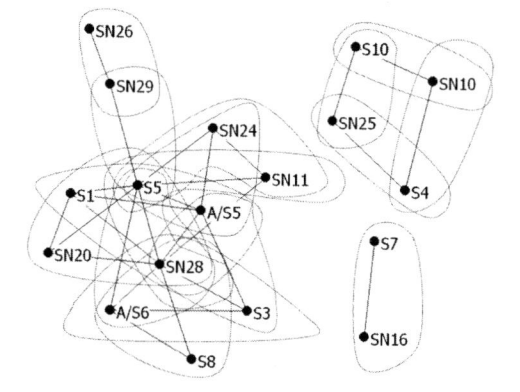

Figure 7. The sixteen sub-groups found for liver resections

Figure 8. The hierarchical clustering of overlap matrix for appendectomies

In the hierarchical clustering of overlap matrix for appendectomies, S5 and S1 were seen to share two sub-groups (see Figure 8). Because no other sub-group shared two or more sub-groups, this sub-group (S5 and S1) was considered as the most central one in appendectomies. For esophageal resections, the hierarchical clustering of overlap matrix revealed two sub-groups as the most central ones, both sharing members of two other subgroups: {S4 and S10} and {S5, S3, and SN28} (see Figure

9). Staff S5 and SN28 shared four sub-groups and were thus considered as the most central sub-group in liver resections (see Figure 10).

Operating Room Staff

Number of cliques shared

```
            S S A S S     S   S S S
          S N N / N N     N   N N N
        S 1 1 2 S 1 2 S S 2 S 1 2 2
        4 0 0 5 6 2 0 5 3 8 1 1 6 9

              1     1     1     1 1
Level   5 3 7 1 1 9 0 6 4 3 2 8 2 4
-----   - - - - - - - - - - - - - -
2.000   XXX . . . . . XXXXX . . . .
1.000   XXXXX . XXX XXXXXXX XXX XXX
0.444   XXXXX . XXX XXXXXXX XXXXXXX
0.400   XXXXXXX XXX XXXXXXX XXXXXXX
0.267   XXXXXXX XXXXXXXXXXX XXXXXXX
0.160   XXXXXXX XXXXXXXXXXXXXXXXXXX
0.000   XXXXXXXXXXXXXXXXXXXXXXXXXXX
```

Figure 9. The hierarchical clustering of overlap matrix for esophageal resections

Operating Room Staff

Number of cliques shared

```
          S   S   S   S S S     A A   S S S
          N S N   N   N N N     / /   N N N
        S 1 1 1 S 2 S 1 2 2 S S S S S 2 2 2
        7 6 0 0 4 5 8 1 4 0 3 1 5 6 5 8 6 9

          1   1   1   1 1 1           1 1 1
Level   8 2 4 0 6 5 9 1 4 3 5 3 1 2 7 7 6 8
-----   - - - - - - - - - - - - - - - - - -
4.000   . . . . . . . . . . . . . . XXX . .
2.333   . . . . . . . . . . . . . XXXXX . .
1.500   . . . . . . . . . . . . XXXXXXX . .
1.400   . . . . . . . . . . . XXXXXXXXX . .
1.000   XXX XXX XXX . XXX . XXXXXXXXXXX XXX
0.571   XXX XXX XXX . XXX XXXXXXXXXXXXX XXX
0.444   XXX XXXXXXX . XXX XXXXXXXXXXXXX XXX
0.417   XXX XXXXXXX . XXXXXXXXXXXXXXXXX XXX
0.307   XXX XXXXXXX XXXXXXXXXXXXXXXXXXX XXX
0.062   XXX XXXXXXX XXXXXXXXXXXXXXXXXXXXXXX
0.000   XXXXXXXXXXXXXXXXXXXXXXXXXXXXXXXXXXX
```

Figure 10. The hierarchical clustering of overlap matrix for liver resections

Discussion

The aim of this preliminary study was to identify sub-groups whose members frequently work together. Techniques from social network analysis – hierarchical clustering and clique at level C analysis – successfully identified these sub-groups. How each technique complements the other and the implications for further study are discussed here.

In hierarchical clustering, two items which are joined in a cluster at an earlier stage cannot belong to different clusters at a later stage: Only the highest frequency of working together between a pair of staff is included. For example, if A and B work together for 10 times, but B also work together with C for 12 times, and A also works together with D for 13 times, hierarchical clustering will only identify the relationships between B and C and between A and D, and the relationship between A and B will not be documented. The relationship between A and B is made visible using clique at level C analysis (C=10). As a result, these two techniques are complementary.

The results showed that most of the sub-groups consist of staff members with diverse roles. This can be considered as particularly beneficial for the planned interview study which will focus on how staff who frequently work together prevent mishaps in the OR. The communication between disciplines (i.e., surgeons, nurses, anesthetists) in the OR has been suggested to be often ineffective which may lead to mishaps (Awad et al., 2005; Lingard et al., 2002). We hope to learn how staff from different disciplines prevent mishaps, for example, by investigating how they improvise and communicate during the crisis, and how these behaviours depend on frequency of working together.

In conclusion, clique at level C analysis is a viable technique to highlight the most central sub-groups whose members frequently work together in the operating room. The techniques should lend themselves to more relational studies to identify sub-groups in complicated environments.

References

Awad, S.S., Fagan, S.P., Bellows, C., Albo, D., Green-Rashad, B., De la Garza, M., & Berger, D.H. (2005). Bridging the communication gap in the operating room with medical team training. *American Journal of Surgery, 190,* 770-774.

Borgatti, S.P. (1994). How to Explain Hierarchical Clustering. Retrieved 01.02.2009 from http://www.analytictech.com/networks/hiclus.htm.

Borgatti, S.P., Everett, M.G., & Freeman, L.C. (2002). UCINET 6 for Windows: Software for Social Network Analysis [Computer software]. Harvard, MA: Analytic Technologies.

Brennan, T.A., Leape, L.L., Laird, N. M., Hebert, L., Localio, A.R., Lawthers, A.G., Newhouse, J.P., Weiler, P.C., & Hiatt, H.H. (1991). Incidence of adverse events and negligence in hospitalized patients. Results of the Harvard Medical Practice Study I. *New England Journal of Medicine, 324,* 370-376.

Calland, J.F., Adams, R.B., Benjamin, D.K., Jr., & O'Connor, M.J., Chandrasekhara, V., Guerlain, S., Jones, R.S. (2002). Thirty-day postoperative death rate at an academic medical center. *Annals of surgery, 235,* 690-696.

Cott, C. (1997). "We decide, you carry it out": A social network analysis of multidisciplinary long-term care teams. *Social Science & Medicine, 45,* 1411-1421.

Friedman, L.H. & Bernell, S.L. (2006). The importance of team level tacit knowledge and related characteristics of high-performing health care teams. *Health Care Management Review, 31,* 223-230.

Gersick, C.J.G. & Hackman, J.R. (1990). Habitual Routines in Task-Performing Groups. *Organizational Behavior and Human Decision Processes, 47,* 65-97.

Kawashima, Y., Takahashi, S., Suzuki, M., Morita, K., Irita, K., Iwao, Y., Seo, N., Tsuzaki, K., Dohi, S., Kobayashi, T., Goto, Y., Suzuki, G., Fujii, A., Suzuki, H., Yokoyama, K., & Kugimiya, T. (2003). Anesthesia-related mortality and morbidity over a 5-year period in 2,363,038 patients in Japan. *Acta Anaesthesiologica Scandinavica, 47,* 809-817.

Keating, N.L., Ayanian, J.Z., Cleary, P.D., & Marsden, P.V. (2007). Factors affecting influential discussions among physicians: A social network analysis of a primary care practice. *Journal of General Internal Medicine, 22,* 794-798.

Leape, L.L., Brennan, T.A., Laird, N., Lawthers, A.G., Localio, A.R., Barnes, B.A., Hebert, L., Newhouse, J.P., Weiler, P.C., Hiatt, H. (1991). The nature of adverse events in hospitalized patients. Results of the Harvard Medical Practice Study II. *New England Journal of Medicine, 324,* 377-384.

Lingard, L., Espin, S., Whyte, S., Regehr, G., Baker, G.R., Reznick, R., Bohnen, J., Orser, B., Doran, D., & Grober, E. (2004). Communication failures in the operating room: an observational classification of recurrent types and effects. *Quality and Safety in Health Care, 13,* 330-334.

Lingard, L., Reznick, R., Devito, I., & Espin, S. (2002). Forming professional identities on the health care team: discursive constructions of the 'other' in the operating room. *Medical Education, 36,* 728-734.

Makary, M.A., Mukherjee, A., Sexton, J.B., Syin, D., Goodrich, E., Hartmann, E., Rowen, L., Behrens, D.C., Marohn, M., & Pronovost, P.J. (2007). Operating room briefings and wrong-site surgery. *Journal of the American College of Surgeons, 204,* 236-243.

Michaels, R.K., Makary, M.A., Dahab, Y., Frassica, F.J., Heitmiller, E., Rowen, L. C., Crotreau, R., Brem, H., & Pronovost, P.J. (2007). Achieving the national quality forum's "never events": prevention of wrong site, wrong procedure, and wrong patient operations. *Annals of surgery, 245,* 526-532.

Reffay, C. & Chanier, T. (2003). How social network analysis can help to measure cohesion in collaborative distance-learning. In B. Wasson, S. Ludvigsen, and U. Hoppe (Eds.), *Designing for change in networked learning environments. Proceedings of the International Conference on Computer Support for Collaborative Learning 2003.* (pp. 343-352). Dordrecht, The Netherlands: Kluwer.

Scott, J., Tallia, A., Crosson, J.C., Orzano, A. J., Stroebel, C., Cicco-Bloom, B., O'Malley, D., Shaw, E., & Crabtree, B. (2005). Social network analysis as an analytic tool for interaction patterns in primary care practices. *Annals of Family Medicine, 3,* 443-448.

Scott, J. (1991). *Social network analysis: A handbook.* Thousand Oaks, CA,US: Sage Publications, Inc.

Wassermann, S. & Faust, K. (1994). *Social network analysis: methods and applications.* New York: Cambridge University Press.

Behavioural adaptation: friend or foe?

Oliver Carsten
Institute for Transport Studies, University of Leeds
Leeds, UK

Abstract

Behavioural adaptation has been a much-debated topic in driver behaviour with the controversy rooted in the Risk Homeostasis process advocated by Wilde. It is rather interesting to note that the debate has very much been a heated and localised one in road traffic and particularly in driver studies, and has not extended much to other modes. This paper reviews the debate, looking particularly at attempts to caste the process of behavioural adaptation in terms of a model. It asks whether the models proposed models are coherent, whether they state anything more than the obvious "users will use systems for their own personal goals", and whether they can act as predictors of behaviour. Has the concept reached the end of its useful shelf life or can it formulated in a way that can aid product designers to foresee misuse and to produce more satisfying and therefore more used systems?

Introduction

Behavioural Adaptation can be treated as a concept or set of concepts, which argues (perhaps rather obviously) that drivers adapt their behaviour to a new situation or new system, often in ways that are not intended by the situation or system implementers.. These adjustments in behaviour have sometimes been termed "indirect effects" (e.g. Draskóczy, Carsten, & Kulmala, 1998). But many proponents of Behavioural Adaptation have tended to have a higher scale of ambition for the concept, and have argued that is a "theory" (see for example Wilde, 1982; Koornstra, 2009). Even those who treat behavioural adaptation as a process or phenomenon, tend to have an implicit theoretical construct in mind as an explanation of the process. Indeed it is hard to see how there could be a concept of the process without an underlying theoretical construct.

A theory can be defined as model that can be can be formalised in terms of a set of predictive relationships, which can then be tested for validity, reformulated, and so on. In other words, a theory has to provide predictions that can be empirically tested. It also has to provide mechanisms that link cause and effect. Otherwise a theory is not proving enlightenment but merely observing the existence of an inexplicable phenomenon.

In D. de Waard, J. Godthelp, F.L. Kooi, and K.A. Brookhuis (Eds.) (2009). *Human Factors, Security and Safety* (pp. 401 - 409). Maastricht, the Netherlands: Shaker Publishing.

Early days

Although he was not the first to observe that driver response could reduce or even eliminate the benefits of safety-related implementations (see for example Peltzman, 1975), nevertheless Wilde can be seen as the father of the claimed theory of Behavioural Adaptation in that it was his formulation — and his argument that in general driver behaviour fully counteracted safety improvements to produce an overall homeostasis — that truly sparked the debate (Wilde, 1982). This debate has been burning in the literature with varied intensity ever since. The "model" of Wilde (1982) is shown in Figure 1. It can be seen that the author certainly viewed the theory as a model, and he subsequently postulated how target risk was determined in one of a set of propositions about the theory of risk homeostasis:

> "The target level of traffic accident risk is determined by four classes of motivating (subjective utility) factors: (1) the expected advantages of comparatively risky behaviour alternatives, (2) the expected costs of these, (3) the expected benefits of comparatively cautious behaviour alternatives, and (4) the expected costs associated with the latter. Some of the motivating factors are economic in nature, others of a cultural, social or psychological kind..." (Wilde, 1988, p. 443)

In other words, he is proposing a kind of economic utility model as determining the level of target risk.

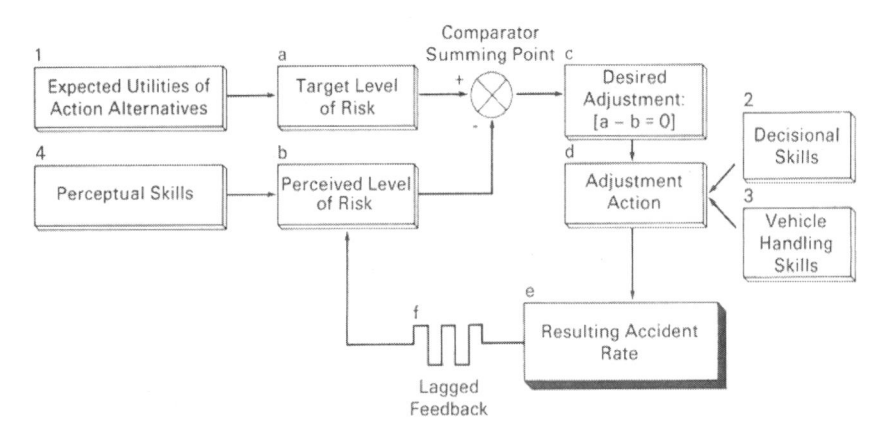

Homeostatic model relating accident rate to driver behavior and vice versa

Figure 1. Risk homeostasis process (Wilde, 1982)

If there is a utility model determining target risk, then it should be feasible to formulate that model in mathematical terms, or alternatively to derive the utilities and trade-offs through the standard game-type approach used in Stated Preference studies. However, Wilde denies that it is possible to derive quantitative utilities:

"[T]he target level of risk is thought of as a construct that people arrive at in an intuitive manner, not as the result of explicit calculation of probabilities of particular outcomes and their positive or negative values." (Wilde, 1988, p. 443)

The OECD review

In 1990 an OECD expert group produced a review entitled "Behavioural Adaptations to Changes in the Road Transport System" (OECD, 1990). A definition of behavioural adaptation was provided:

"Behavioural adaptations are those behaviours which may occur following the introduction of changes to the road-vehicle-user system and which are not intended by the initiators of the change." (p.23) It was further stated: "For behavioural adaptation to occur, it must be assumed that there is feedback to road users, that they can perceive the feedback (but not necessarily consciously), that road users have the ability to change their behaviour, and that they have the motivation to change their behaviour." (p. 23)

However, the report deliberately side-steps the issue of mechanisms: "The mechanisms that have been proposed to explain behavioural adaptation are controversial, and it is not the purpose of the report to select one or other of these explanations. In fact, the report will provide a basis for discussing behavioural adaptation outside the context of any one theory." (p. 20) Thus the report provided much valuable conformation of the phenomenon, but did not even propose, let alone confirm the underlying mechanisms. One of the report's recommendations was: "There is a need for further developments of theories of road user behaviour to assist in the understanding and prediction of behavioural adaptation. At the same time there is a need for those conducting evaluations of road safety programmes to incorporate theory testing in their research to provide a basis for the development of theory." (page 119)

This leaves a problem: without even any proposed mechanisms, there is no possibility of creating a predictive and testable theory. And without a theory, there is no real potential for empirical studies to confirm or repudiate the predictions made by theory. And much of the literature has been posed in such a way that it is not really possible to deduce from the mechanisms proposed what would be the likely outcomes in terms of behavioural adaptation. For example, one might expect a predictive model to indicate how the behaviour of different groups of drivers might be affected by a particular safety measure. But even those theoretical models that can be found in the literature are often at such a high level of generality that it is impossible to make any such deductions. Thus Jiang, Underwood, and Howarth (1992) in their "theoretical model for adaptations to change", tell us that objective risk is not the same as subjective risk and that cognition of risk depends on the possibility of perception of risk. They go on to state that responses to subjective risk will be affected by motivations which can be expressed in the form of a utility maximisation function, but they also state that road users "usually have no clear quantitative description of how their actions can affect the benefit they want to maximise." (p. 259). We are thus little the wiser in terms of any verifiable predictions.

The "hierarchical risk model" of Van der Molen and Bötticher (1988) argues that it can be assumed that responses to risk can be seen as the outcome of a utility maximisation strategy, but also that it is important to distinguish between decisions at the three levels of Michon's (1985) model of the driving task. The authors propose that a separate evaluation process takes place at each of the three levels — strategic, tactical and operational. At the strategic and tactical levels, decisions are affected by judgements and those judgements are in turn informed by both motivations (safety and "other") and expectations which draw on perception. At the operational level a less conscious process operates in which operational behaviour is determined both by the manoeuvring plan (from the tactical level) and, in emergency situations, by the perception of immediate risk.

Here the strategic and tactical levels are to all purposes mirror images of each other, so that in terms of factors affecting decision-making they can be treated as one. There are two parallel processes affecting judgements and decisions:

1. A safety stream in which decisions are based on motivations about safety and accident expectations (i.e. current probabilities of accident outcomes of each alternative);
2. An "other" stream with a similar process affected by non-safety motivations and alternatives.

The model certainly is more complex, than for example Wilde's (1982), but as the authors admit it is certainly not complete in that it omits any errors in decision-making or operation. It also begs the question of how motivations and expectations are determined. Again it is not really possible to use it to construct a set of predictions about how different groups of drivers will adapt their behaviour to specific implementations or situations.

What are the possible mechanisms?

Conceptually, it is possible to think about a number of possible mechanisms for behavioural adaptation. *Utility maximisation* has already been mentioned. One form of utility maximisation is risk compensation: I feel protected from crash outcome in my SUV, so I'll drive faster. But utility maximisation need not necessarily just concern risk. It might for example concern journey time: going though this traffic calming zone has delayed me, so now I'll go faster to catch up on lost journey time.

Motivational factors might also be relevant (Rothengatter, 1988). We might have the effect of emotion: the warnings from this in-car system make me feel irritated, so I behave more aggressively. Or alternatively pleasure may be a motivating factor: I like driving fast, so I'll go as fast as I feel capable of going on this stretch of winding road.

Cognitive factors can have an impact. Thus system design, in terms for example of ease of use, will affect drivers propensity to use a driver assistance system. Other such factors could be ease of learning, workload, system reliability (which would affect trust) and automation induced complacency, which might lead to over-reliance on a system.

It is also important to note the impact of situational context (Cacciabue & Saad, 2008). Drivers will be reluctant to deviate from their "normal" responses to a given situation. They also have to comply more or less to the behaviour of the traffic around them.

But, in spite of the variety of possible factors, the literature has tended to focus on driver manipulation of risk as the major motivation. And the primary mechanism for manipulating risk has been viewed as speed choice. This dominance of risk management through speed as the essential motivation and of speed as the principal mechanism can be illustrated by the application of the task capability interface model of Fuller (2005a) as shown in Figure 2. Although the model itself treats the driver as a participant in a complex traffic environment, the actual discussion and application of the model has concentrated almost exclusively on speed choice as the means to manipulate risk:

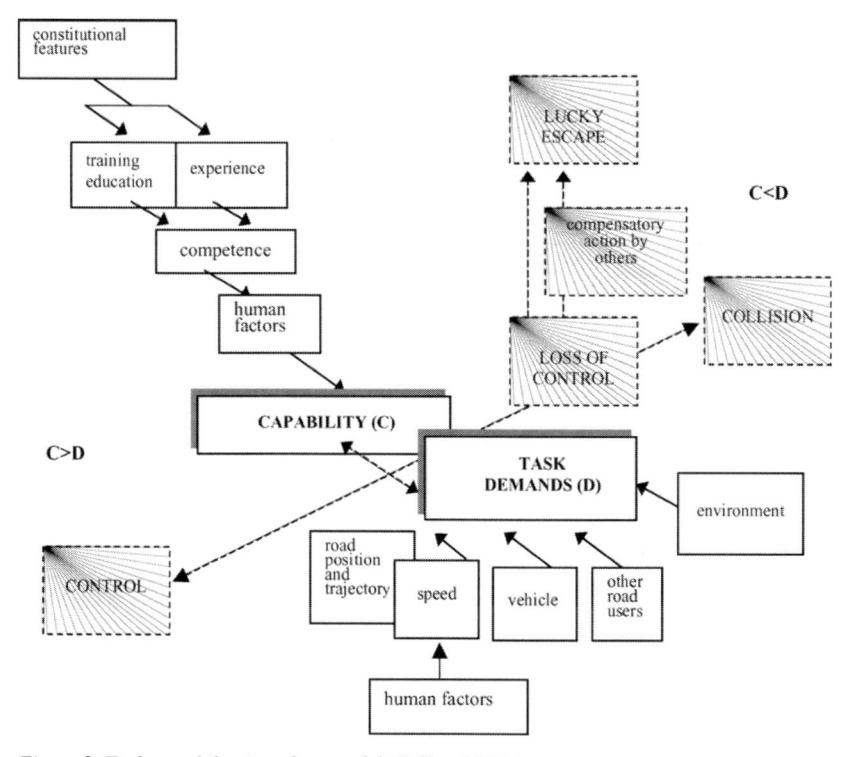

Figure 2. Task-capability interface model (Fuller, 2005a)

Based on the goals of a particular journey, self-appraisal of capability and effort motivation, a driver 'selects' a range of difficulty which s/he is prepared to accept and drives in such a way as to maintain experienced difficulty within that range. Manipulation of speed is the primary mechanism for achieving this, although undertaking or dumping other tasks secondary to the primary driving task may be used on occasion. (Fuller, 2005b, p. 85-86)

Tried and tested

There are the occasional honourable exceptions to the concentration on speed and risk. One such exception is the model of Rudin-Brown and Noy (2002), shown in Figure 3. This model proposes that behavioural adaptation to new in-vehicle systems will be influenced by personality in the form of locus of control and sensation seeking as well as by trust in the system. Trust in turn will be affected by personality as well as by the actual reliability of the system as experienced in driving. Now this is not a complete model of behavioural adaptation — it is really only suited to driver assistance systems rather than to road safety interventions as a whole — but it does offer the prospect of predicting the direction and relative magnitude of adaptation effects. It is thus possible to derive hypotheses from the model and then to test those hypotheses in experimental or real-world situations.

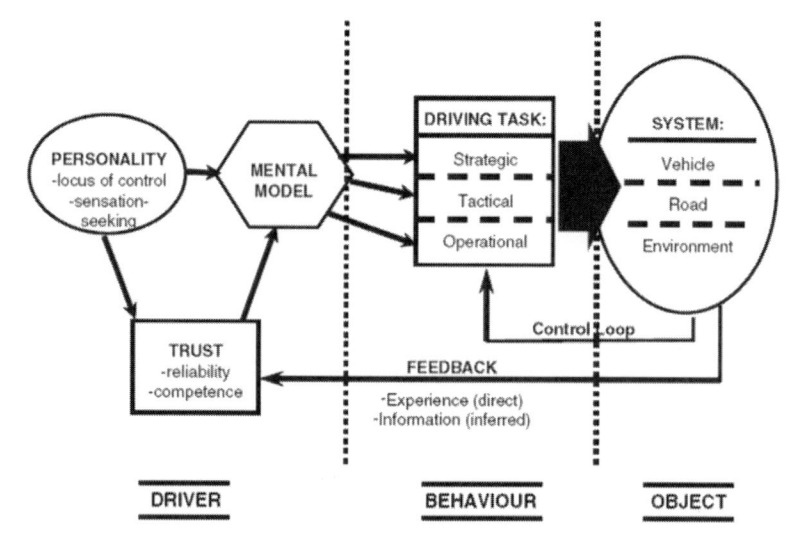

Figure 3. Qualitative mode of behavioural adaptation (Rudin-Brown & Noy, 2002)

That this model could indeed provide reasonable predictions was confirmed in a subsequent test-track study of Adaptive Cruise Control (Rudin-Brown & Parker, 2004). In that study, drivers' willingness to engage in a secondary task while driving with ACC, was significantly influenced by their score on sensation seeking, with high sensation seekers substantially more willing to engage in the secondary task with a consequent negative effect on their ability to detect a hazard. High sensation seekers also had more variation in lane position when driving with ACC. Response to ACC failure was related to locus of control such that externals were slower to respond than internals. The model therefore shows promise.

On the other hand, risk homeostasis continues to haunt us. The latest model proposed by Ray Fuller (see Figure 4) is extremely similar to that of Wilde (1982). It proposes a process of "allostasis" rather than "homeostasis", indicating a process of changing (i.e. variable) response rather than one in which the target level is always constant. But apart from that important modification, the model seems to reduce driving to a balance between demand and capability in which the single important control mechanism is speed choice.

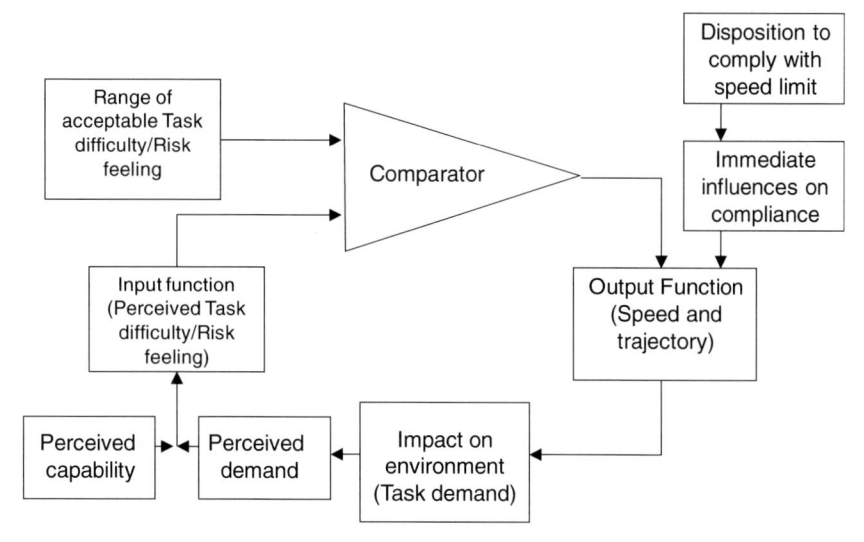

Figure 4. Fuller's risk allostasis model (presented at the International Conference on Traffic and Transport Psychology, 2008)

A challenge from 1996

In 1996, Graham Grayson carried out a review of the literature on behavioural adaptation. He concluded that review by pointing out both the need for and the lack of a theory of behavioural adaptation that could serve as a predictive tool is guiding system designers and those intending to deploy new technologies:

> [W]ith advances in technology it is likely that changes in the transport system in the future will be markedly different from those of the past, because they have the capability to alter the nature of the driving task itself. These differences will be of kind, rather than of degree. The research evidence that is available to advise on the possible effects of such interventions is very limited, and the potential problems are considerable. Innovation in this field is increasingly technology-driven; safety is not usually the main objective, and human factors tend to play a minor role. To some extent this is understandable, but it should also give cause for concern. New technologies offer the prospect

of great benefits, but only if accompanied by a proper understanding of how road users will respond over time to the changes involved. It would not be unreasonable to conclude that the issue of behavioural response to new technology poses research challenges of the highest order. To say that 'more research is needed' is a familiar ending, but one which in this case seems fully justified. (Grayson, 1996, p. 22)

It could be argued that this challenge is just as relevant today as it was in 1996. Remarkably little progress has been made in outlining and subsequently validating a rigorous model of behavioural adaptation.

Conclusions

In the driving domain, we are still a long way from having a reasonably comprehensive and useful predictive model that could guide vehicle and system design, system deployment, road design, infrastructure-based measures, public campaigns, driver training and so on. We are even further from a validated model. But that is not to argue that we should not take up the challenge posed by Graham Grayson in 1996. The new European ITERATE project, which starts in 2009, will be attempting to propose and calibrate a model that can be applied to driving as well as to train and ship operation.

Watch this space!

References

Cacciabue, P.C. & Saad, F. (2008). Behavioural adaptations to driver support systems: a modelling and road safety perspective. *Cognition, Technology & Work, 10,* 31-39.

Draskóczy, M., Carsten, O., & Kulmala, R. (1998). *Road safety guidelines.* Deliverable B5.2 of Project CODE. Atkins Wootton Jeffreys, Birmingham, UK.

Grayson, G.B. (1996). *Behavioural adaptation: a review of the literature.* TRL Report 254. Transport Research Laboratory, Crowthorne, UK.

Fuller, R. (2005a). Towards a general theory of driver behaviour. *Accident Analysis and Prevention, 37,* 461-472.

Fuller, R. (2005b). Driving by the seat of your pants: a new agenda for research. *Behavioural Research in Road Safety 2005: Fifteenth Seminar* (pp. 85-93). London: Department for Transport.

Jiang, C., Underwood, G., & Howarth, C.I. (1992). Towards a theoretical model for behavioural adaptations to changes in the road transport system. *Transport Reviews, 12,* 253-263.

Koornstra, M.J. (2009). Risk-adaptation theory. *Transportation Research Part F, 12,* 77–90.

Michon, J.A. (1985). A critical review of driver behaviour models. In L. Evans and R.G. Schwing (Eds.), *Human behavior and traffic safety* (pp. 485-520). New York: Plenum Press.

OECD (1990). *Behavioural adaptations to changes in the road transport system*. Paris: Organisation for Economic Cooperation and Development, Road Transport Research.

Peltzman (1975). The effects of automobile safety regulations. *Journal of Political Economy, 83*, 677-725.

Rothengatter, T. (1988). Risk and the absence of pleasure: a motivational approach to modelling road user behaviour. *Ergonomics, 31*, 599-607.

Rudin-Brown, C.M., & Noy, Y.I. (2002). Investigation of behavioral adaptation to lane departure warnings. *Transportation Research Record, 1803*, 30–37.

Rudin-Brown, C.M., & Parker, H.A. (2004). Behavioural adaptation to adaptive cruise control (ACC): implications for preventive strategies. *Transportation Research Part F, 7*, 59–76.

Van der Molen, H.H., & Bötticher, A.M.T. (1988). A hierarchical risk model for traffic participants. *Ergonomics, 31*, 537-555.

Wilde, G.J.S. (1982). The theory of risk homeostasis: implications for safety and health. *Risk Analysis, 2*, 209–225.

Spanning attention: what really goes on between perception and action?

Mark S. Young
Brunel University
Uxbridge
UK

Abstract

In human factors and ergonomics, the most widely accepted model of attention in information processing is Wickens' (1980) multiple resources theory. Nevertheless, in its purest form, the multiple resource model does have its flaws, such as how to explain problems with mental underload, and why two tasks that ostensibly draw on separate resource pools can still interfere with each other. Meanwhile, in the psychological literature, Baddeley's (1986) working memory model has many echoes to multiple resource theory, yet there is surprisingly little acknowledged overlap between the two. This paper presents a review and critique of these key models in the cognitive and human factors literature, and draws synergies across them to propose a more parsimonious role for attention in the human information processing chain. Whilst neither multiple resources nor working memory alone offer all the answers, the key conclusion from this discussion is the need for a general reservoir of attentional resources which feeds the modality-specific subsystems. Although taking a theoretical perspective, the issues are relevant to a variety of applied contexts where dynamic human performance depends on the balance between task demands and available resources.

Introduction

For nearly 30 years now, performance modelling and prediction in human factors and ergonomics has been largely dominated by one approach: the multiple resources theory put forward by Wickens (1980, 2002). Wickens (ibid.) deduced the existence of separate pools of resources as a response to, and to account for, observations of dual-task performance which were expected to result in interference. Whilst multiple resources theory (MRT) has rightly become the seminal applied model in this field, for the closely related discipline of cognitive psychology it is working memory (WM) which attracts all of the pure, theoretical research. Baddeley's (1986, 2003) model was developed as an advance on more simplistic categories of memory, but as an active system it seems to account for similar phenomena as MRT, and even has a familiar structure to it. Where MRT has visual/spatial and auditory/verbal pools, WM invokes the visuospatial sketchpad and the phonological loop. Thus whilst the two models hail from different backgrounds and purport to address different objectives, there is significant overlap in their theoretical makeup. Nevertheless, in searching the literature from either field, it is difficult to find one even acknowledging the other, let alone discussing its implications. Moreover, although MRT has clearly enjoyed a great deal of success in predicting performance in applied dual-task scenarios where demands compete for resources, some authors have pointed out its limitations beyond this application.

This paper takes a journey through the evidence for both MRT and WM, looking for a common thread in an effort to resolve the situation, and perhaps suggest a more parsimonious approach which can also account for some of the current limitations. Firstly, the fundamental background is covered with a more detailed review of the MRT and WM models. Next, a critique of each approach is given in turn, explaining some of the limitations alluded to above and highlighting specifically those works where there is an explicit overlap between the two models. Finally, the paper comes full circle in going back to basics, looking at the wider information processing model to determine how, if at all, MRT and WM can coexist.

Background

Multiple resources theory

Resource theories have a longer history than Wickens' (1980) model; the classic and often-cited early work is that of Kahneman (1973), who proposed a unitary capacity model of attention. Essentially, the capacity model proposes a single resource view of attention – that is, attention is viewed as one whole pool of resources. The ability to perform two separate concurrent activities depends upon the effective allocation of attention to each. Interference between tasks depends upon the demands which each separately impose – when task demands drain the pool, performance will suffer. Later research found some flaws with the single resource approach. As Wickens (2002) points out in his historical review of MRT, there were several experiments whereby two tasks were effectively time-shared (i.e., performed concurrently) even when the difficulty of either was manipulated. This was seen as a limitation of single resource theory, which predicted that difficulty manipulations should eventually lead to performance effects on one or both tasks. Furthermore, neurophysiological evidence was emerging to suggest that the two brain hemispheres had their own pools of resources for verbal and spatial processing (e.g., Friedman et al., 1982; Wickens, 1980). Thus emerged multiple resources theory (Wickens, 1980, 2002).

Multiple resources theory posits that there are separate pools of resources along three dichotomous dimensions. The first dimension is processing stages – early vs. late. Perception and central processing are said to demand separate resources from response selection and execution. The second dimension is input modalities – auditory vs. visual. Performance of two simultaneous tasks will be better if one is presented visually and the other presented aurally, rather than using the same modality for both. Finally, the theory states that there are separate resources for whether a task is processed verbally or spatially. This dichotomy also holds for response execution, whereby less dual-task interference occurs if one task is responded to vocally and the other demands a manual response. Thus, there will only be a trade-off between task difficulty and performance to the extent that two concurrent tasks share resources on these dimensions (Wickens, 2002) – interference is a joint function of difficulty (resource demand) and shared processing mechanisms (resource competition).

Whilst the model clearly implies an advantage for multiple tasks to draw upon separate pools, the compatibility principle also implies that codes and modalities should remain consistent *within* a task. That is, if a task has visual input and spatial processing, it should have a manual and spatial response (e.g., driving). Whilst Wickens (2002) notes that it is more common to envisage visual-spatial and auditory-verbal tasks, it is clearly possible to design visual-verbal and even to a certain extent auditory-spatial tasks. Needless to say, it is difficult to imagine a task with a verbal-spatial response, so there are some limitations with this task design principle. Wickens (2002) adds another nuance to the theory, by further subdividing the visual modality into focal and ambient (peripheral) vision, to give the model a fourth dimension. Wickens himself had difficulty in finding evidence for separate pools in central and peripheral vision (Weinstein & Wickens, 1992), concluding that peripheral tasks were simply more demanding than those in central vision. Nonetheless, more recent research suggests that different resources can be allocated to focused and ambient attention (Marsh et al., 2004). Wickens (2002) now feels that the balance of research is accumulating in favour of this distinction.

Wickens (2002) is keen to point out that MRT is not a theory of attention per se, and whilst it is often associated with work overload, it is not just a theory of workload either. Rather, MRT is a theory for predicting and explaining performance in overload situations, which depends on quantitative aspects of workload (resources) as well as qualitative aspects of demand (multiple pools). We will return to this discussion of attention and workload later in the paper.

Working memory

The concept of working memory seems to have been a natural evolution from psychological models of short- and long-term memory (STM and LTM). Where there was a great deal of evidence (including neurophysiological) to support a distinction between STM and LTM, there remained gaps in the models to explain learning and decision-making. Rather than a passive storage system to serve between input and output, what was needed was an active processor – hence WM was born as a link between perception, LTM and action (Baddeley & Hitch, see Baddeley, 2003, for a more detailed review).

The WM model centres on three components: the central executive, and its two slave subsystems, the visuospatial sketchpad and the phonological loop. The phonological loop accounts for short-term verbal memory (such as classic digit span tasks), and is usually associated with auditory tasks as articulatory rehearsal. Similarly, the visuospatial sketchpad performs a parallel role for manipulating and processing spatial information. Again, this is normally assumed to be visual, but Baddeley (2001) points out that such information could equally well come from motor or haptic inputs. Finally, the central executive plays a crucial role in scheduling and allocating attention to the subsystems, depending on the task at hand, and has the capacity to focus, divide or switch attention.

Baddeley (2001, 2003) admits that in the original conception, the central executive was imagined as something of a catch-all for those processes and activities which could not be accounted for in the subsystems. More relevant to the present paper, it was conceived as a limited capacity pool of general processing resources, but later developed into an attentional controller for controlled and automatic processing, borrowing from Norman and Shallice's (1986) supervisory activating system. In other words, there is some initial appraisal of the task, and then a decision is made about whether to activate a schema for automatic processing, or allocate attentional resources to one of the subsystems. Furthermore, the kind of dual-task performance observed by Wickens (1980) as the foundation of MRT also supports the structure of WM (Engle, 1996). Baddeley (2001) reports examples of chess players who can effectively share an articulatory task, whereas a concurrent visuospatial task does interfere with performance. Moreover, a non-visuospatial central executive task also caused deterioration in performance, and to an even greater extent.

In a more recent development, Baddeley (2001) noted that there was no real facility in the WM model to integrate information from the subsystems and provide an interface to LTM. Whilst the central executive was assumed to perform this role, several studies led to the conclusion that the executive did not have storage capacity, causing problems for this explanation. So a new subsystem was introduced – the episodic buffer, which acts as a limited capacity storage system with a multimodal code to interface between the systems. It seems to be the episodic storage parallel for the semantic processing role of the central executive. More importantly, this means that the central executive now moves beyond its memory function to become a purely attentional system.

Critique

It is clear from the summaries presented above that the distinction between multiple resources theory and working memory is somewhat blurred; in particular, the verbal and spatial processing codes seem to correspond with the phonological loop and the visuospatial sketchpad. Wickens' (2002) MRT has more to do with processing and response than traditional filter perspectives on attention (e.g, Broadbent, 1958), thus aligning it with the WM approach. Meanwhile, Baddeley's (1986) central executive was initially conceived as an attention scheduler, giving WM a direct relevance to attentional resources. Unfortunately, there is little literature to distinguish the two models – even the relevant sections of the authors' respective textbooks (Baddeley, 1986; Wickens et al., 2004) do not substantially acknowledge each other. With little else to choose between the two, the following sections of this paper evaluate the shortcomings of MRT before putting it in a head-to-head with WM.

Issues with multiple resources theory

In the present author's opinion, there are two key related areas where MRT has difficulty providing explanations: some specific aspects of dual-task performance, and underload.

Resource models of attention provide a useful framework for defining mental workload, since it essentially becomes a ratio of demands against capacity (Wickens, 2002; Young & Stanton, 2005). Under this assumption, when the demands of one or more tasks exceed the limited capacity of the operator, overload is said to occur and performance deteriorates. However, the fixed and limited capacity model is less able to explain the effects of underload – the phenomenon whereby extremely low levels of workload are equally detrimental to performance. At the theoretical level, there is no reason why excessive levels of capacity should predict negative consequences. Young and Stanton (2002) put forward a modification to the resource model in their Malleable Attentional Resources Theory (MART), in which underload causes resources to shrink in line with the task demands, arguing that a resource explanation is more parsimonious than appealing to wider concepts such as effort mobilisation. Performance problems in underload are usually manifest when demands increase and the operator is required to step up their workload – under MART, then, underload is actually an overload issue, which would be consistent with the accepted understanding of vigilance tasks as presenting high demand (e.g., Warm et al., 1996). Wickens (2002) acknowledges MART but notes there has been little validation of the theory outside of the original authors' research. Nevertheless, Young and Stanton (2002) explain how MART predicts the inverted-U performance curve so often seen in workload studies, and the implication that increasing demand will improve performance to a point has been supported elsewhere (Becic et al., 2007). Whether or not MART holds across resource pools remains to be seen; so far it has been observed with secondary task studies drawing on the same pools of resources as the primary task.

The dual-task paradigm is a way of assessing the relative contributions of each resource pool. Operators are presented with two concurrent tasks and are instructed to maintain performance primarily on one of the tasks. The tasks may be designed to tap either the same resource pool (to assess interference or spare attentional capacity), or different pools (to assess timesharing ability). For instance, Stokes et al. (1990) detailed how parallel processing may be enhanced in a dual task paradigm by exploiting separate resources, particularly across modalities. Indeed, this is the very raison d'être of MRT.

However, there is some debate as to whether single resource models are more appropriate than multiple resource theory. Liao and Moray (1993) argued that a single channel model is of more use in real world situations, which generally have more than two tasks. However, they also stated that the multiple resource approach remains a superior model in purely dual-task scenarios. But a significant body of research calls into question the 'pure' multiple resources account of dual-task performance, whereby tasks that ostensibly draw upon separate resources still interfere with each other. In many cases this involves an auditory/verbal secondary task causing performance decrements on a visual/spatial primary task – usually in the driving domain. Most prominent amongst these is the wealth of literature on the effects of using a mobile phone – whether hand-held or hands-free – or even conversing with passengers while driving (e.g., Haigney & Westerman, 2001; Merat & Jamson, 2005; Ranney et al., 2005; Rizzo et al., 2004), but there are also numerous laboratory studies with similar implications (Lee et al., 2007; Moss & Triggs, 1995; 1997; Young & Stanton, 2007; Zeitlin, 1995). There is also evidence that cross-modal links can be beneficial, whereby an auditory stream is contiguous with, and orients attention towards a visual stream (Grimes, 1991; Spence & Driver, 1999). As with examples of cross-modal interference, these are not consistent with MRT predictions.

The conclusion to be drawn from this set of studies, as has been drawn by many of the authors cited above, is that there exists a general reservoir of attentional resources, which feeds the specific pools. It is not simply a return to a single channel processor, since in most cases the cross-modal effects are not equivalent to intramodal interference – but they are not insignificant either. The idea of general and specific pools is both old and new – Wickens (1980) suggested that STM processes in concurrent tasks draw on a general set of resources, whatever their modality and processing stage; a more recent Resource Competition Framework (Oulasvirta et al., 2005) explicitly proposes a general limit on capacity which can be shared amongst pools. Such notions have clear echoes with the WM model, as we shall see next.

Working memory vs. multiple resources

To suggest a modification of MRT, which incorporates a general capacity associated with specific pools, is wholly reminiscent of the WM structure, with its central executive and modality-specific subsystems. The argument is thus already lending itself towards WM as the favoured model, but before dismissing MRT, it is worth reviewing whether WM can assume the full role of (modified) MRT in all its detail.

Firstly, there is overwhelming agreement in the WM literature that the capacity of WM – which is a prime determinant of performance – is really a reflection of attentional resources (Carnas & Barrouillet, 2004; Conway & Engle, 1994; 1996; Engle, 1996; 2002; Qi & Xia, 2007). Furthermore, the consensus is that this capacity consists of a general and limited pool (cf. Engle, 1996), which resides in the central executive (Baddeley, 2001). While the central executive is seen as the general store, the slave subsystems are thought to be modality specific (Morrison, 2005), with no interference between the visuospatial sketchpad and the phonological loop (Bruyer & Scailquin, 1998). Thus to a certain extent, WM can supplant the dimensions of input modality and processing code in MRT – with the crucial difference that its attentional resources are, in the first instance, modality independent. It does indeed seem to reflect a general reservoir with exclusive pools

All is not perfect, however. In a separate debate, there remains some confusion over whether the subsystems of WM act as a processor, as storage, or as both. Baddeley (2003) clearly sees WM fulfilling both roles, but May (2001) is more conservative, discussing the relative merits of the subsystems as passive storage facilities or active processors. This has implications for the role of attentional resources, since the MRT perspective is clearly as a processor, with little emphasis on storage. Conway & Engle (1994) argued that working memory (storage) capacity is indirectly related to attentional resources, as retrieval from working memory depends upon the ability to inhibit irrelevant information, which is in itself resource demanding. This implies a distinction between the capacities of working memory and attentional resources, even though one is reflected in the other. Duff (2000; Duff & Logie, 2001) seeks to clarify this situation by suggesting another multiple resource view, where storage is based on independent resources for the slave subsystems, while processing draws on a central shared set of resources.

Presumably this central set – and indeed the main processor itself – is part of the central executive. However, it is not clear from the literature whether the executive 'holds' resources, or merely allocates them. Certainly there is plenty of evidence for executive-level tasks, which are performed directly without delegation to the subsystems, and which do affect subsystem activities. The chess players noted by Baddeley (2001) are one such example, with performance being affected by a random digit generation task. There are similar studies in the human factors literature. Harms (1991) found that driving demands were correlated with calculation times on mental arithmetic tasks. Liu & Wickens (1994) successfully used a time estimation task as a measure of spare capacity in a dynamic decision-making task. In an attempt to develop an instantaneous measure of subjective MWL, Tatersall & Foord (1996) found that such metacognitive demands interfered with the primary task. Each of these secondary tasks is characteristic of the role of the central executive, and taken together these studies suggest that the central executive itself draws on resources just as much (if not more so) as the subsystems. Whilst this does not rule out the executive as a store of resources, it does suggest that the pool lies elsewhere.

In a similar vein, there is one paper (Brown, 1997) which directly pits multiple resources theory against working memory as alternative explanations of interference effects in timing – a function normally ascribed to the central executive. Usually in dual-task studies, in the absence of any instructions to prioritise one task over the other, the interference effect will be symmetrical; that is, both tasks will be equally affected. However, the interference effect when trying to maintain timing (i.e., make a response every 3s) with a concurrent nontemporal task (such as visual search, or tracking) seems to be asymmetrical. Brown (1997) found that timing performance was adversely affected by a concurrent task, but performance on search or tracking tasks was not affected by the simultaneous timing task.

MRT does not have a specific allocation for time estimation tasks (and indeed this has been used as another argument for a general purpose resource pool; Shinohara, 2002), and certainly cannot explain the asymmetry. The central executive of WM, on the other hand, can offer an explanation, since it controls attentional and coordinational functions, such as allocating attention between dual tasks. Simple visual search or tracking tasks only use the visuospatial sketchpad. Therefore, the nontemporal tasks do not suffer with a concurrent timing task, as the visuospatial sketchpad is essentially still dedicated to a single task. However, the mere introduction of a dual-task scenario draws on the coordination skills of the central executive, which also looks after temporal activities, hence the interference effect on timing. Brown (1997) concludes that working memory and multiple resources both attempt to explain similar phenomena and rely on similar concepts, but working memory is distinct in its provision for general purpose resources. Brown (1997) also hints at the speculation for a general pool of resources in MRT.

So the discussion does seem to be favouring a WM model as the more parsimonious explanation for the myriad laboratory and applied findings, as opposed to a modified version of MRT with a general resource reservoir. Nevertheless, WM alone cannot account for all of the dimensions of MRT – it primarily covers processing codes and (to a lesser extent) input modality, but there is certainly no presumption of processing stages being included. WM focuses on central processing, not response. Hence some authors have advocated a more hybrid solution, with multiple resource versions of working memory. Klapp and Netick (1988) started the trend, proposing more independent subsystems for WM along the lines of auditory/visual input and manual/vocal response. More recent papers have tried to incorporate WM principles into MRT (e.g., Uhl, 2003; Samman et al., 2005), in particular by employing a central executive-type mechanism as a processing modality or general pool of resources (Oulasvirta et al., 2005; Uhl, 2003).

In moving forward, then, the challenge for a hybrid model would be to see if it can account for those phenomena not adequately explained by either MRT or WM alone. Returning to the earlier challenges for MRT, the hybrid model would certainly allow for a general reservoir of resources, whether through a central executive mechanism or otherwise, which would explain instances of cross-modal interference. The underload problem does remain a challenge, though, as this has not been addressed by WM research either. Nonetheless, an explanation along the lines of MART holds promise, particularly as it is consistent with the original formulation of resource theory (Kahneman, 1973) as well as recent research drawing parallels between WM capacity and such declines in cognitive resources (Caggiano & Parasuraman, 2004; Morrow, 2007). In the closing section of this paper, the arguments thus far will be summarised in an effort to draw together an approach to future research.

Conclusions

At the top of this paper, it was stated that despite being from separate backgrounds and with different theoretical objectives, the MRT and WM models are structurally similar. It was also posited that the respective literatures for MRT and WM are almost akin to like poles of a magnet – they have similar properties, but repel efforts to bring them together. Having journeyed through the literature now, it is clear that this is not entirely fair. Indeed, the key protagonists of each do in fact recognise the existence of the other, even if at a somewhat superficial level. Wickens (1980) suggested that the different resources in the brain hemispheres could reflect verbal/spatial WM respectively, and later he even reflected that the dimension of processing code echoed the subsystems of WM (Wickens, 1991). Meanwhile

Baddeley (1993) readily admits the significance of attention in WM, going so far as to dub the model 'working attention'.

Moreover, several studies reviewed here point towards a theoretical merging of the two models, whether that means incorporating elements of WM into MRT (Brown, 1997; Oulasvirta et al., 2005; Samman, et al., 2005; Uhl, 2003) or vice-versa (Engle, 1996; Klapp & Netick, 1988). Which model prevails as the 'home' will largely depend on the ability to account for the shortcomings in each individual approach. To summarise, WM is a more theoretical than applied model, with some debate over its role as processor or storage facility. In order to compete with the applied relevance of MRT, it would need to emphasise the processing function (although Engle, 1996, does argue that WM performance on dual-tasks does predict real-world cognitive ability). Furthermore, whilst it clearly accounts for processing codes and to a lesser extent input modalities, it lacks any real treatment of response selection and execution as a processing stage. On the other hand, the main shortcoming with MRT is the absence of a general pool of resources, which WM ostensibly offers in its central executive.

So what would a hybrid model look like? At one extreme, a reductionist approach would just look at the conceptual overlaps between WM and MRT, collapsing across processing codes and essentially reverting to something more like a single-resource model. At the other extreme, incorporating all of the unique aspects from each model would result in an expanded set of multiple resources, complete with a central executive or supervisory attentional system.

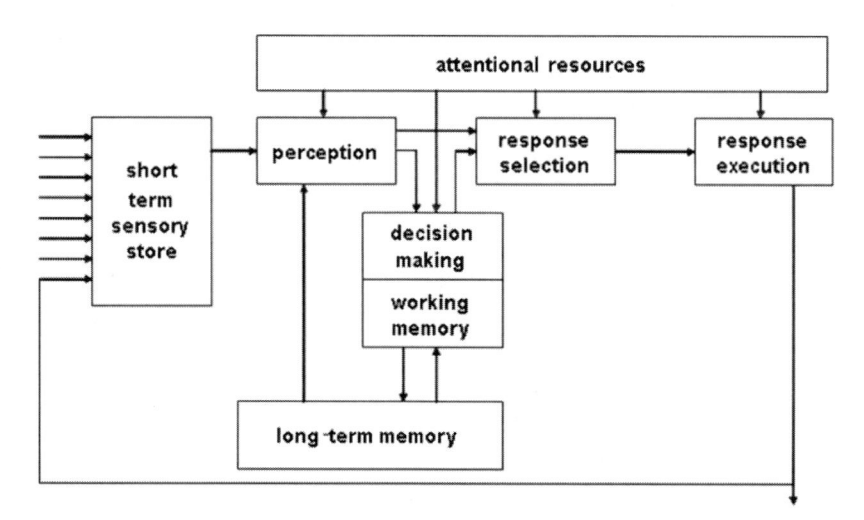

Figure 1. Generalised information processing model (adapted from Wickens et al., 2004)

It is the contention of this paper that a refined version of MRT should be put forward for testing in future applied work in human factors and ergonomics – the key point being that it needs to incorporate a general reservoir of resources, including some capacity for executive function. None of this is revolutionary – as reviewed earlier, similar notions have been advanced before, and can even be seen in the generalised information processing model – such as exemplified in Wickens' own textbook (Wickens et al., 2004; see Figure 1). But thus far there seems to have been few explicit attempts to really impress such a model on the discipline.

Even so, there remain some effects that are difficult to explain by either model. Perhaps most significant amongst these is the underload problem, which eludes both MRT and WM (but see Caggiano & Parasuraman, 2004, for a view on the related problem of vigilance). Baddeley (2003) admits that WM has no link to physiological arousal, which has similar effects on performance as workload (Young & Stanton, 2002). Possibly MART would offer a way out of this particular conundrum for the resources approach, but more research on this hypothesis is necessary. Finally, there are yet more aspects of information processing which have barely been acknowledged in any of these theories. Not least amongst these is the realm of tactile input, which has been suggested should have its own specific pool of resources (Brill, 2007; Samman et al., 2005). Similarly, there is evidence for other atypical sources of demand which can drain attentional resources, such as physical workload (Perry et al., 2008), pain (Eccleston, 1995), anxiety (Hayes et al., 2008) and depression (Christopher & MacDonald, 2005).

With several questions remaining to be answered, future research should be directed at elucidating the role of a general pool of resources, which feeds an expanded set of modality-specific inputs, processing, and responses. Lab-based research using asymmetrical interference, such as the WM experiments of Brown (1997), or crossover studies examining the effects of physical workload on mental demand, could help to distinguish such a model from purely single- or purely multiple-resource frameworks. Similarly, technological advances in tactile and haptic displays can be used to present cognitive tasks in this modality, helping to determine the existence of a separate pool for tactile processing. But given the long history of empirical research comparing single- and multiple-pool explanations, which is still far from being resolved, introducing a 'third way' into the same paradigm may not help to clarify the waters. Instead, current trends in neuroergonomics (e.g., Parasuraman, 2003) offer a promising way forward – in what may be seen as a return to the neurophysiological roots of the WM and MRT models. It has taken a roundabout journey to get there, then, but this paper suggests that both psychologists and ergonomists already have good accounts of what goes on between perception and action – but for the detail, they must meet in the middle.

References

Baddeley, A. (1986). *Working Memory*. Oxford: Clarendon Press.
Baddeley, A.D. (1993). Working memory or working attention? In A.D. Baddeley (Ed.), *Attention: Selection, awareness, and control: A tribute to Donald Broadbent* (pp. 152-170). New York: Clarendon Press.
Baddeley, A.D. (2001). Is Working Memory still working? *American Psychologist, 56*, 849-864.

Baddeley, A. (2003). Working memory: looking back and looking forward. *Nature Reviews: Neuroscience, 4*, 829-839.

Baddeley, A.D. & Hitch, G. J. (1974). Working memory. In G.A. Bower (Ed.), *Recent advances in learning and motivation, Vol. 8* (pp. 47-90). New York: Academic Press.

Becic, E., Kubose, T., Kramer, A.F., Dell, G.S., Garnsey, S.M., & Bock, K. (2007). Aging and the effects of conversation with a passenger or a caller on simulated driving performance. *Driving Assessment 2007. Proceedings of the 4th International Driving Symposium on Human Factors in Driver Assessment, Training and Vehicle Design*, Stevenson, Washington, July 9th-12th. Iowa: The University of Iowa.

Brill, J.C. (2007). A comparison of attentional reserve capacity across three sensory modalities. *Dissertation Abstracts International: Section B: The Sciences and Engineering, 68*(3-B), 1965.

Broadbent, D. (1958). *Perception and Communications*. New York: Pergamon.

Brown, S.W. (1997). Attentional resources in timing: Interference effects in concurrent temporal and nontemporal working memory tasks. *Perception & Psychophysics, 59*, 1118-1140.

Bruyer, R. & Scailquin, J.C. (1998). The visuospatial sketchpad for mental images: Testing the multicomponent model of Working Memory. *Acta Psychologica, 98*, 17-36.

Caggiano, D.M. & Parasuraman, R. (2004). The role of memory representation in the vigilance decrement. *Psychonomic Bulletin & Review, 11*, 932-937.

Carnas, V. & Barrouillet, P. (2004). Adult counting is resource demanding. *British Journal of Psychology, 95*, 19-30.

Christopher, G. & MacDonald, J. (2005). The impact of clinical depression on working memory. *Cognitive Neuropsychiatry, 10*, 379-399.

Conway, A.R.A. & Engle, R.W. (1994). Working memory and retrieval: A resource-dependent inhibition model. *Journal of Experimental Psychology: General, 123*, 354-373.

Conway, A.R.A. & Engle, R.W. (1996). Individual differences in working memory capacity: More evidence for a general capacity theory. *Memory, 4*, 577-590.

Eccleston, C. (1995). Chronic pain and distraction: An experimental investigation into the role of sustained and shifting attention in the processing of chronic persistent pain. *Behaviour Research and Therapy, 33*, 391-405.

Engle, R.W. (1996). Working memory and retrieval: An inhibition-resource approach. In J.T.E. Richardson, R.W. Engle, L. Hasher, R.H. Logie, E.R. Stoltzfus, and R.T. Zacks (Eds.), *Working Memory and Human Cognition* (pp. 89-119). New York: Oxford University Press.

Engle, R.W. (2002). Working memory capacity as executive attention. *Current Directions in Psychological Science, 11*, 19-23.

Grimes, T. (1991). Mild auditory-visual dissonance in television news may exceed viewer attentional capacity. *Human Communication Research, 18*, 286-298.

Haigney, D. & Westerman, S.J. (2001). Mobile (cellular) phone use and driving: a critical review of research methodology. *Ergonomics, 44*, 132-143.

Hayes, S., Hirsch, C., & Mathews, A. (2008). Restriction of working memory capacity during worry. *Journal of Abnormal Psychology, 117*, 712-717.

Kahneman, D. (1973). *Attention and Effort*. Englewood Cliffs, NJ: Prentice-Hall.

Klapp, S.T. & Netick, A. (1988). Multiple resources for processing and storage in short-term working memory. *Human Factors, 30*, 617-632.

Lee, Y.-C., Lee, J.D., & Boyle, L.N. (2007). Visual attention in driving: The effects of cognitive load and visual disruption. *Human Factors, 49*, 721-733.

Liao, J., & Moray, N. (1993). A simulation study of human performance deterioration and mental workload. *Le Travail Humain, 56*, 321-344.

Marsh, R.L., Hicks, J.L., & Cook, G.I. (2004). Focused attention on one contextual attribute does not reduce source memory for a different attribute. *Memory, 12*, 183-192.

Merat, N. & Jamson, A.H. (2005). Shut up I'm driving! Is talking to an inconsiderate passenger the same as talking on a mobile telephone? In L. Boyle, J.D. Lee, D.V. McGehee, M. Raby, and M. Rizzo (Eds.), *Proceedings of the 3rd International Driving Symposium on Human Factors in Driver Assessment, Training and Vehicle Design*, Rockport, Maine, June 27-30 2005 (pp.426-432). Iowa: University of Iowa.

Morrison, R.G. (2005). Thinking in working memory. In K.J. Holyoak and R.G. Morrison (Eds.), *The Cambridge handbook of thinking and reasoning* (pp. 457-473). New York: Cambridge University Press.

Morrow, D.G. (2007). External support for pilot communication: Implications for age-related design. *International Journal of Cognitive Technology, 12*, 21-30.

Moss, S.A. & Triggs, T.J. (1995). Assessing limitations in cognition and attention by manipulating task emphasis and difficulty. In A.C. Bittner and P.C. Champney (Eds.), *Advances in Industrial Ergonomics and Safety VII* (pp. 187-193). London: Taylor & Francis.

Moss, S.A. & Triggs, T.J. (1997). Attention switching time: A comparison between young and experienced drivers. In Y.I. Noy (Ed.), *Ergonomics and safety of intelligent driver interfaces.* (pp. 381-392). Mahwah, NJ: Lawrence Erlbaum Associates.

Norman, D.A. & Shallice, T. (1986). Attention to action: Willed and automatic control of behaviour. In R.J. Davidson, G.E. Schwarts, and D. Shapiro (Eds.), *Consciousness and self-regulation: Advances in research and theory, Vol. 4* (pp. 1-18). New York: Plenum Press.

Oulasvirta, A., Tamminen, S., Roto, V., & Kuorelahti, J. (2005). Interaction in 4-second bursts: The fragmented nature of attentional resources in mobile HCI. *CHI 2005, April 2-7, Portland, Oregon, USA*.

Parasuraman, R. (2003). Neuroergonomics: research and practice. *Theoreitcal Issues in Ergonomics Science, 4*, 5-20.

Perry, C.M., Sheik-Nainar, M.A., Segall, N., Ma, R., & Kaber, D.B. (2008). Effects of physical workload on cognitive task performance and situation awareness. *Theoretical Issues in Ergonomics Science, 9*, 95-113.

Qi, Z. & Xia, W. (2007). Working memory span: Resource constraint, memory decay or switching mechanism? *Acta Psychologica Sinica, 39*, 777-784.

Ranney, T.A., Harbluk, J.L., & Noy, Y.I. (2005). Effects of voice technology on test track driving performance: Implications for driver distraction. *Human Factors, 47*, 439-454.

Rizzo, M., Stierman, L., Skaar, N., Dawson, J.D., Anderson, S.W., & Vecera, S.P. (2004). Effects of a controlled auditory-verbal distraction task on older driver vehicle control. *Transportation Research Record, 1865*, 1-6.

Samman, S.N., Sims, V., & Stanney, K.M. (2005). Multimodal Working Memory: The unfolding story. *Proceedings of the Human Factors and Ergonomics Society 49th Annual Meeting*, Orlando, Florida, September 26-30, 2005 (pp. 337-341) (Available in CD-ROM format). Santa Monica: Human Factors and Ergonomics Society.

Shinohara, K. (2002). The role of attentional resources and working memory in time estimation. *Japanese Psychological Review, 45*, 195-209.

Spence, C. & Driver, J. (1999). Multiple resources and multimodal interface design. In D. Harris (Ed.), *Engineering Psychology and Cognitive Ergonomics Volume Three: Transportation Systems, Medical Ergonomics and Training* (pp. 305-312). Aldershot: Ashgate.

Stokes, A.F., Wickens, C.D., & Kite, K. (1990). *Display technology: human factors concepts.* Warrendale, PA: Society of Automotive Engineers Inc.

Uhl, M. L. (2003). *The effects of driver distraction on simulated driving: Are multiple resource theory predictions mediated by central executive involvement?* MA Thesis, California State University, Northridge.

Warm, J.S., Dember, W.N., & Hancock, P.A. (1996). Vigilance and workload in automated systems. In R. Parasuraman and M. Mouloua (Eds.), *Automation and human performance: Theory and applications* (pp. 183-200). Mahwah, NJ: Lawrence Erlbaum Associates.

Weinstein, L.F. & Wickens, C.D. (1992). Use of nontraditional flight displays for the reduction of central visual overload in the cockpit. *International Journal of Aviation Psychology, 2*, 121-142.

Wickens, C.D. (1980). The structure of attentional resources. In R. Nickerson (Ed.), *Attention and Performance VIII* (pp. 239-257). Hillsdale, NJ: Lawrence Erlbaum.

Wickens, C.D. (1991). Processing resources and attention. In D.L. Damos (Ed.), *Multiple-task performance* (pp. 3-34). London: Taylor & Francis.

Wickens, C.D. (2002). Multiple resources and performance prediction. *Theoretical Issues in Ergonomics Science, 3*, 159-177.

Wickens, C.D., Lee, J.D., Liu, Y, & Gordon-Becker, S.E. (2004). *An Introduction to Human Factors Engineering* (2nd edition). Upper Saddle River, NJ: Pearson.

Young, M.S. & Stanton, N.A. (2002). Attention and automation: new perspectives on mental underload and performance. *Theoretical Issues in Ergonomics Science, 3*, 178-194.

Young, M.S. & Stanton, N.A. (2005). Mental workload. In N.A. Stanton, A. Hedge, K. Brookhuis, E. Salas, & H.W. Hendrick (Eds.), *Handbook of Human Factors and Ergonomics Methods* (Ch. 39). London: Taylor & Francis.

Young, M.S. & Stanton, N.A. (2007). Miles away. Determining the extent of secondary task interference on simulated driving. *Theoretical Issues in Ergonomics Science, 8*, 233-253.

Zeitlin, L.R. (1995). Estimates of driver mental workload: A long-term field trial of two subsidiary tasks. *Human Factors, 37*, 611-621.

Performance in manual process control – mediation of time pressure and practice effects by the structure of control behaviour

Stefan Röttger & Dietrich Manzey
Berlin Institute of Technology
Berlin, Germany

Abstract

An explanation of how practice and time pressure may affect performance in dynamic control tasks is offered in the Contextual Control Model (COCOM, Hollnagel, 1993): Performance depends on the orderliness of operator behaviour, i.e. how well control actions are planned in advance. Whether an operator can develop proper plans to guide his/her control actions depends on a number of contextual factors, e.g. the subjectively available time (which is decreased by time pressure) and the familiarity with the task (which increases with practice). COCOM further predicts that due to recurring patterns of actions, the structure of a sequence of systematically chosen control actions will be more regular than actions taken in a trial-and-error fashion without a guiding plan, which will show more random patterns.

The predictions of COCOM were tested in an experiment with 40 participants who manually controlled five parameters in a process control simulation. While earlier studies of COCOM relied on subjective ratings to assess orderliness, we examined the regularity of the structure of control actions with emc_{dfr}, an objective measure of regularity which is based on information theory (Röttger, Klostermann & Manzey, 2007). Results show that both performance and the amount of regularity in the operator behaviour increase with practice and decrease under time pressure, and that regularity of control actions mediates practice and time pressure effects on performance. These are the first objective data to yield empirical support to COCOM's predictions regarding time pressure and practice effects on operator behaviour.

Introduction

In order to optimize the design of human-machine systems, it is necessary to predict, and thus to understand how operators act in various situations that may arise while they are performing their tasks. A possible way to gain insight into the processes of operator behaviour is to develop models of operators. Such models are often developed from a theoretical perspective and have a rather small empirical basis. A prominent example is the Contextual Control Model (COCOM) from Hollnagel (1993, 2000).

Structure of Control Behaviour in the Contextual Control Model

Within COCOM it is assumed that different ways or "modes" of controlling a dynamic system can be distinguished. These modes differ in the degree of orderliness of the human control behaviour. Orderliness can be described as the degree to which an operator acts in a systematic fashion because his or her behaviour is based on plans, appropriate procedures, and anticipations of future system states (Hollnagel, 2000). Orderliness is a central determinant of control performance: the higher the orderliness of control behaviour, the higher the level at which tasks can be accomplished. Another consequence of a high orderliness is that control actions contain frequently recurring patterns (Hollnagel, 2000). Thus, changes of orderliness are accompanied by changes in the structure of operator behaviour, i.e. in the features of an operator's action sequences. In the following paragraph, the four different control modes suggested by Hollnagel and their consequences for operator performance and for the structure of operator behaviour are briefly described.

The low end of the continuum of orderliness is marked by the so-called "scrambled" control mode, in which control actions are taken without any planning in an unpredictable trial-and-error fashion. No recurring patterns can be observed in the scrambled mode – the behaviour is basically random and thus highly irregular. A higher (but still rather low) degree of orderliness characterizes the "opportunistic" control. In this mode, only little anticipation and planning occur and the course of action is mainly guided by the most salient signals of the environment. Behavioural patterns in this mode of control mainly reflect regularities of the task environment, e.g. a sequence of alarms that reiterates over and over again. A moderate level of orderliness is present in the "tactical" control mode: control actions more or less follow rules or known procedures, goals beyond the present needs are pursued, and the choice of the next action is based on anticipations of the system state in the near future. In a sequence of tactical actions, rules and procedures will be reflected as recurring patterns. A high degree of orderliness can be found in "strategic" control actions. Plans underlying strategic control sequences have a wider time horizon than in the tactical mode, and take into account a higher number of goals and side-effects. Therefore, the structure of the control behaviour in the strategic mode is more complex than the structure of behaviour in the tactical mode, and observable patterns are longer.

According to COCOM, orderliness of control actions and thus control modes will change depending on the subjective representation of contextual factors such as the subjectively available time (which is closely related to, but not necessarily identical with the objectively available time) or the familiarity of the situation. Transitions between control modes are assumed to occur in a linear fashion from a given mode to the next higher or next lower mode of control (i.e. from scrambled to opportunistic to tactical to strategic, and conversely). In addition, transitions between scrambled and tactical mode are considered to be possible.

Studies testing COCOM's predictions

So far, there are only two articles reporting empirical tests of assumptions of the Contextual Control Model. Stanton, Ashley, Roberts and Xu (2001) investigated teams operating a simulated gas distribution system and categorized the teams's control behaviour according to the control modes proposed by Hollnagel (2000). They report a high inter-rater reliability of categorizations (Spearman's Rho = 0.79), which shows that the control modes described in COCOM can indeed be observed and reliably distinguished from each other. In addition, a high proportion of transitions between control modes matched the predictions of COCOM (275 out of 280).

However, Stanton et al. (2001) do not report whether the raters were blind towards the hypothesis of the study. Therefore, the exceptionally high congruence between theory and data regarding transitions between control modes may be due to the fact that the raters, being aware of the hypotheses, were biased towards identifying control mode transitions as predicted by COCOM.

The second article reporting empirical tests of some of COCOM's predictions was published by Feigh, Pritchett, Denq and Jacko (2007). In this study, the influence of time pressure on performance and self-assessed control mode in a simulated airline operations management task was investigated. While there was no effect of the objectively available time for the solution of the airline rescheduling task, there was a small but significant correlation between subjectively perceived time pressure and self-assessed control mode ($r = .23$, $p = .037$). As the rating procedure required participants to assign higher values to less orderly modes of control, this correlation indicates that a higher level of perceived time pressure is accompanied by lower levels of orderliness. With regard to the performance that can be achieved in the different modes of control, significantly better solutions of the airline rescheduling problems were found in trials in which participants acted in the opportunistic or strategic mode as compared to the scrambled mode of control. These results are clearly in line with the predictions of COCOM.

Some of the findings reported by Feigh et al. (2007), however, can not be explained by the model. For example, participants reported to act in the tactical and even in the strategic mode of control in the condition with the least objectively available time, and performance did not differ between scrambled and tactical control mode. Unfortunately, Feigh et al. (2007) do not report any psychometric analysis of their control mode rating scale, so it remains unclear whether the rather weak effects and equivocal results are due to a limited reliability of the control mode self-assessments, or whether they point to a limited validity of the Contextual Control Model.

The work of both Feigh et al. (2007) and Stanton et al. (2001) illustrates the main difficulty associated with attempts to empirically test COCOM: the measurement of orderliness. Although expert ratings and self-ratings have the advantage that a number of relevant contextual factors can be considered, they are very tedious to make and are easily biased towards the expectations of the raters, whatever these expectations may be.

Regularity emc$_{dfr}$ as an indicator of orderliness

Röttger, Klostermann and Manzey (2007) suggested that a more objective approach to investigate orderliness of control behaviour may be to use methods from information theory. Information theory was originally developed to investigate the transmission of messages comprising discrete symbols such as letters or Morse code. Most measures from information theory are essentially based on the probabilities / the relative frequencies of the symbols contained in a message. If behaviour can be described as a sequence of discrete actions (e.g. opening a valve, switching off a pump), measures from information theory can be used to assess the features of such action sequences, e.g. their randomness, complexity or regularity. Röttger et al. (2007) argue that the complexity of action sequences, as measured by the the effective measure complexity (*emc*, Grassberger, 1986; Schlick, Winkelholz, Motz & Luczak, 2006), is closely related to the orderliness of control behaviour and would therefore be a particularly well suited, objective indicator of orderliness. However, emc is based on estimates of entropies of the observed action sequences, which are rather unreliable and above all biased unless enormous amounts of data are available for the estimation procedure (Schürmann, 2004). This bias in the entropy estimates can lead to paradoxical results such as very high *emc* values indicating highly orderly behaviour for purely random data (this effect will be explained in slightly more detail in the section "Measures" below).

A possible way of dealing with this problem is to avoid absolute *emc* values, and to compare instead how much *emc* of a real control sequence differs from *emc* of a synthetic random sequence with the same length and the same number of different actions (Röttger et al., 2007). In this case, the bias in entropy estimation is roughly comparable for both *emc* values, and the difference between them mainly reflects the observed behaviour's deviation from random, and thus scrambled control actions. Accordingly, this difference is referred to as the effective measure complexity's deviation from random, or *emc$_{dfr}$* in short. This difference does not reflect the complexity (as defined by Grassberger, 1986) of an action sequence anymore, but rather it's regularity: The higher the proportion of frequently recurring patterns in a sequence of control actions, and the longer these patterns, the higher *emc$_{dfr}$*. As the occurrence of frequently recurring patterns in operator behaviour is one of the features of orderliness (Hollnagel, 2000), more orderly control behaviour should lead to higher values of *emc$_{dfr}$*.

An important limitation of this approach is that the maximum difference will be obtained between *emc* values of random sequences and *emc* values of simple repetitive sequences, e.g. pressing the same button over and over again. As such perseverations as well as random behaviour can both occur in the scrambled mode of control, scrambled behaviour could theoretically lead to both maximum as well as minimum *emc$_{dfr}$* values. Thus, a prerequisite for obtaining valid results from the application of *emc$_{dfr}$* to the analysis of sequences of control actions is that study participants do not show perseverating behaviour.

In a study with 79 participants working on a simulated manual process control task, Röttger et al. (2007) tested the applicability of *emc$_{dfr}$* to operator behaviour. They found a strong correlation (r = .54, p < .001) between process control performance and regularity of control behaviour. As one of COCOM's assumptions is that a close relationship exists between orderliness of control behaviour and control performance, this correlation suggests that emc_{dfr} may be a valid indicator of orderliness and that perseverating behaviour does not occur in an extent that would compromise its validity.

The purpose of the study presented in this paper was to use emc_{dfr} for a closer examination of three predictions that can be derived from the Contextual Control Model: 1) That more familiar situations are accompanied by more regular behaviour and better performance, 2) That time pressure leads to a decrease of behavioural regularity and performance, and 3) That the effects of familiarity and time pressure on performance are mediated by the orderliness of the control actions as indicated by the amount of regularity in the action sequences.

Method

Participants

20 male and 20 female students of the Berlin Institute of Technology with a mean age of 26.9 years (SD = 3) participated in the experiment. Their academic backgrounds were in the areas of Engineering, Psychology, or Human Factors. They were blind towards the hypothesis of the experiment and had not previously taken part in experiments with the same task as used in the present study.

Task

The experiment was conducted using ManuCAMS (Manzey, Bleil, Bahner-Heyne, Klostermann, Onnasch et al., 2008), a PC-based simulation of a manual process control task. ManuCAMS is a modified version of the Cabin Air Management System CAMS (Hockey, Wastell, & Sauer, 1998), which simulates a life support system of a spacecraft consisting of five subsystems that are critical to maintain atmospheric conditions in the space cabin (i.e. sufficient levels of oxygen, pressure, carbon dioxide, temperature, and humidity). The hypothetical scenario throughout the experiment was that four automatic controllers of the life support system had failed and that the participants were in charge of keeping the levels of oxygen, pressure, carbon dioxide, and temperature within their normal operation ranges. To this end, participants were required to manually control the different components of the life support system (e.g. switch on and off the CO_2 scrubber or open and close the pressure relief valve). Feedback as to the state of the atmospheric parameters was given in charts showing the development of the parameters as a function of time.

Design

The design of the experiment is illustrated in figure 1. Both familiarity of the situation and time pressure were manipulated within participants. On day 1 of the experiment, familiarity was manipulated by practice. The first block of the experiment was considered to constitute a completely unfamiliar situation to the participants, while there was a considerably higher amount of familiarity with the experimental task during the second block of the experiment. On day 2, time pressure was induced by accelerating the ManuCAMS simulation. Compared to the blocks with normal simulation speed, the parameters' rate of change was twice as high in the time pressure condition, so participants had only half of the accustomed time to detect deviations from the normal operation ranges and to respond to these events. As effects of increasing familiarity with the task could have occurred on the second day of the experiment as well, the order of the accelerated and normally speeded block was balanced across participants.

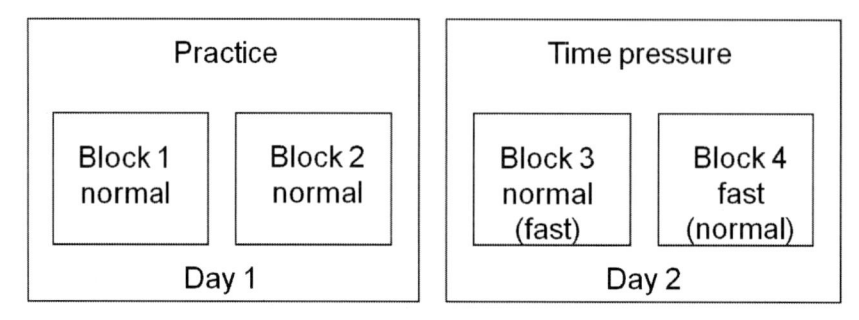

Practice		Time pressure	
Block 1 normal	Block 2 normal	Block 3 normal (fast)	Block 4 fast (normal)
Day 1		Day 2	

Figure 1. Experimental design. Practice was manipulated on day 1 of the experiment, time pressure on day 2, both within participants. Order of time pressure conditions (normal / high) on day 2 was balanced across participants.

Procedure

On day 1 of the experiment, participants received a standardized instruction regarding design, functionality, and handling of the life support system. No information was provided on the dynamics and interactions of parameters and on strategies for an effective manual control. Having received these instructions and having signed an informed consent, participants performed two blocks of manual control with a duration of 30 minutes each. To obtain baseline values for physilogical workload measurements (not reported here), participants were asked to sit back, relax and passively watch the autonomously running ManuCAMS for a period of 5 minutes prior to each block. On the beginning of the experimental session of the following day, participants did not receive further instructions but were informed that the dynamics of the parameters would be altered in one of the blocks, and whether this would apply to the first or second block of day 2. Apart from this, block duration and procedure were identical to those of day one. Having completed the experimental blocks, participants were debriefed and received course credit or €20, respectively, for their participation.

Measures

Dependent variables of the experiment were the performance of the participants in the manual process control task and the amount of regularity in their action sequences. Performance was defined as the mean number of parameters that were kept within their normal operation range. The number of parameters being within range was determined for each second of a block and the average of these values was used as overall performance indicator for the block as a whole.

To assess the regularity of the participants' behaviour, emc_{dfr} of the action sequences was calculated. Calculations were carried out as given in equations (1) through (4). First, the Nth order entropies of the action sequences were determined from the probabilities p_i of the different actions (1st order entropy), pairs of actions (2nd order entropy), subsequences of three actions (3rd order entropy) and so forth up to the highest order N_{max}. The formula for entropy estimation is given in equation (1).

$$H_N = -\sum_{i=1}^{k^N} p_i \log p_i \quad (1)$$

The number of different actions in a sequence is denoted with k, so there can be a maximum number k^N different combinations of N actions, which represents the upper bound of the index i. Relative frequencies were used as estimates of the probabilities p_i. The parameter N_{max} was set to 5 for the analysis of the present experiment.

An example may be appropriate here. When controlling ManuCAMS, participants could carry out one action at a time out of $k=12$ possible actions (e.g. switching on CO_2 scrubber, opening pressure relief valve, etc.). Participants' actions were recorded with a numerical code in a log file, so a part of an action sequence as recorded in this experiment could look like this: 1 – 8 – 2 – 5 – 1 – 8 – 4 – 2 – 5 – 12. To calculate the first order entropy of this sequence, the relative frequency of each action must be determined and multiplied by its logarithm. The sum of these products is the first order entropy (see table 1). The same procedure can be carried out for pairs of actions to obtain the second order entropy (also demonstrated in table 1), triplets for 3rd order entropy, etc.

Table 1. Calculation of first and second order entropy of example action sequence.

Action(s)	Frequency	Relative frequency p_i	$log_2 p_i$	$p_i log_2 p_i$	$-\Sigma p_i log_2 p_i$
1	2	0.2	-2.322	-0.4644	
2	2	0.2	-2.322	-0.4644	
4	1	0.1	-3.322	-0.3322	
5	2	0.2	-2.322	-0.4644	
8	2	0.2	-2.322	-0.4644	
12	1	0.1	-3.322	-0.3322	$H_1 = 2.522$
1 – 8	2	0.2	-2.322	-0.4644	
8 – 2	1	0.1	-3.322	-0.3322	
2 – 5	2	0.2	-2.322	-0.4644	
5 – 1	1	0.1	-3.322	-0.3322	
8 – 4	1	0.1	-3.322	-0.3322	
4 – 2	1	0.1	-3.322	-0.3322	
5 – 12	1	0.1	-3.322	-0.3322	$H_2 = 2.59$

In a second step, conditional entropies h_N were derived from the entropy values as shown in equation (2).

$$h_N = H_{N+1} - H_N \quad (2)$$

Conditional entropies of order N are a measure of the remaining uncertainty regarding the next element in a sequence if N preceding actions are known. emc was calculated from these conditional entropies according to equation (3).

$$emc = \sum_{i=1}^{N_{max}-1} h_i - h_{min} \quad (3)$$

A high effective measure complexity will be obtained from sequences with high lower-order conditional entropies and markedly reduced higher-order conditional entropies, which is the case if it is difficult to predict the next action in a sequence unless a high number of preceding actions is known. However, a sharp drop of the conditional entropy with increasing order will be observed even in random sequences due to biased entropy estimates if the length of the analysed sequence is smaller than k^{Nmax}. In the present experiment, k^{Nmax} amounts to 12^5, but only 200 to 10^3 actions could be observed in a single block. Therefore, the effective measure complexity's deviation from random was used to assess the regularity of the participans' action sequences. emc_{dfr} is calculated as given in equation (4).

$$emc_{dfr} = emc(random\ sequence) - emc(observed\ sequence) \quad (4)$$

The random sequence has the same length and the same number of k different elements as the observed sequence.

Analysis

The predictions of COCOM were tested by means of regression analyses. Independent variables were coded with 0 and 1 for absence and presence of time pressure, or no experience and one block of experience with the task for practice, respectively. In order to show that the effects of time pressure and practice on performance are mediated by the orderliness of behaviour as indexed by emc_{dfr}, it has to be tested whether a significant reduction of the regression weights of time pressure and practice as predictors of performance occurs when emc_{dfr} is added to the regression as a second predictor. Mediation analysis was conducted with the SPSS® implementation of the modified Sobel-test published by Preacher & Hayes (2008). Because the prerequisite of stationarity of the behaviour was violated in the time pressure condition, separate mediation analysis had to be carried out for the first and second half of the blocks with doubled simulation speed. As the results of these periods are very similar, they are reported for the first half of the time pressure condition only.

Results

Effects of practice

The unstandardized regression weights of practice and emc_{dfr} on performance are shown in figure 2. Practice was positively related to both emc_{dfr} ($b = .35$, $p = .002$) and performance ($b = .21$, $p = .018$), and a better performance could be observed in participants with a higher emc_{dfr} ($b = .29$, $p = .001$). When the variance in the performance data was explained by the regularity of the participants' behaviour, the regression weight of practice predicting performance significantly decreased ($p < .05$) and became insignificant ($b = .11$, $p = .208$). Thus, the practice effect on performance is completely mediated by the regularity of the control actions.

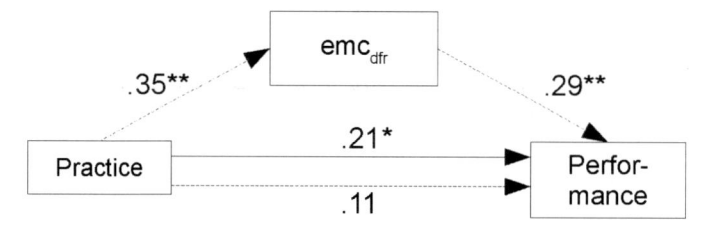

Figure 2. Unstandardized regression coefficients in the mediation of practice effects on performance by the regularity of control behaviour. The difference between regression coefficients of practice predicting performance is statistically significant ($p < .05$). *$p < .05$, **$p < .01$.

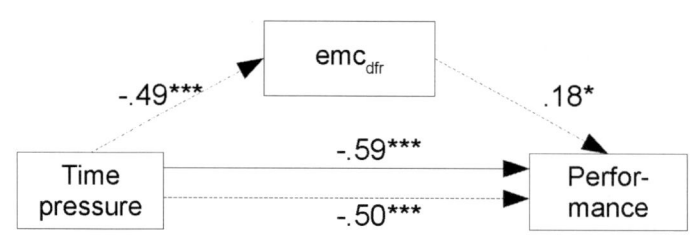

Figure 3. Unstandardized regression coefficients in the mediation of time pressure effects on performance by the regularity of control behaviour. The difference between regression coefficients of time pressure predicting performance is statistically significant ($p < .05$). *$p < .05$, ***$p < .001$.

Effects of time pressure

A similar pattern of results was observed for the effect of time pressure (see figure 3). Time pressure led to a significant ($p < .001$) decrease in both emc_{dfr} ($b = -.49$) and performance ($b = -.59$), and again a positive relation between regularity and performance could be observed ($b = .18$, $p = .014$). After adding emc_{dfr} to the regression of performance on time pressure, the regression weight of time pressure was significantly smaller ($p < .05$), but still significant ($b = -.50$, $p < .001$). This pattern of results indicates a partial mediation of the time pressure effect on performance by the regularity of the action sequences.

Discussion

The findings of the study are in line with COCOM's predictions. If a situation becomes more familiar in the course of practice, behavioural patterns become more regular and performance increases, which was reflected by a positive correlation of practice with both emc_{dfr} and performance. The negative correlation of emc_{dfr} and performance with time pressure further represents supporting evidence for one of COCOM's assumptions: A decrease in subjectively available time leads to a deterioration of orderliness and performance. The central role of orderliness for the explanation of performance variability was confirmed by a mediation analysis, which revealed that practice does not directly affect performance, but rather the orderliness of behaviour which in turn influences performance. Parts of the effect of time pressure on performance can be explained by orderliness as well, but there are other features of control behaviour influencing performance under time pressure that are not reflected by the regularity of the actions. One of these features probably is a correct timing of the actions, which is important to effectively control the dynamic parameters of the life support system, and which was more difficult to achieve in the block with accelerated simulation speed.

Directions for further research can be derived from theoretical as well as more practical considerations. From a theoretical perspective, it would be interesting to find out the mechanisms of the observed changes in orderliness. Practice, for example, could improve the orderliness of control behaviour in at least two different ways: One way could be to make participants more efficient monitors of the life support system because they learn to predict when a parameter will reach the boundary of its normal operation range. Thus, the participants' actions follow more closely the "rhythm" of the process, and the source of the regularity would be rather exogenous as in the opportunistic mode. Another way of increasing the amount of regularity in the action sequence would be to learn how the parameters of ManuCAMS interact and to develop strategies for a more efficient control of the life support system which exploit these interactions (e.g. running the CO_2 scrubber decreases cabin pressure). The repeated use of these strategies would result in more regular activity patterns, and the source of the regularity would be rather endogenous, as in the tactical or strategic mode.

From a practical perspective, emc_{dfr} appears to be an interesting candidate for the assessment of the operator functional state (OFS; Hockey, Gaillard & Burov, 2003): It has been shown to be related to performance and to be influenced by factors that are known to induce or alleviate workload, and although it is concerned with a person's behaviour, it is independent from the specific content or meaning of the behaviour and can thus be applied to a variety of tasks. However, despite the markedly reduced amounts of data required to compute emc_{dfr} as compared to emc, there are still several hundred actions needed for a reliable calculation of emc_{dfr}. The registration of such a high number of actions would take too much time to be feasible in the context of online monitoring of OFS to predict potential performance decrements.

However, there are other indicators of operator behaviour which might be used for this purpose. A possible alternative could be an assessment of the regularity of eye-movements reflecting the allocation of attention and the structure of information sampling behaviour. Eye-movements produce a much higher number of discrete events (e.g. fixations on certain displays or regions of interest) per unit time and, thus, provide a suitable basis for a more frequent calculation of emc_{dfr}. Further research could be conducted to investigate to what extend emc_{dfr} of fixation patterns is correlated with operator performance and sensitive to stressors such as time pressure or task load.

To sum up, the present study reports the first objective evidence for the assumptions of the Contextual Control Model regarding time pressure and practice effects on operator behaviour. Moreover, these results show that emc_{dfr} may be a suitable approach to reveal and explain effects of workload-related factors on operator performance in general. In situations which may cause operators to change their way of solving a task in order to maintain their level of performance and physiological arousal (Hockey, 1997), emc_{dfr} could complement traditional task performance metrics and physiological measures of operator functional state, and may give an objective indication of changes in OFS while performance and physiological parameters remain constant.

References

Feigh, K.M., Pritchett, A.R., Denq, T.W., & Jacko, J.A. (2007). Contextual Control Modes During an Airline Rescheduling Task. *Journal of Cognitive Engineering and Decision Making, 1*, 169-185.

Grassberger, P. (1986). Toward a quantitative theory of self-generated complexity. *International Journal of Theoretical Physics, 25*, 907-938.

Hockey, G.R.J. (1997). Compensatory control in the regulation of human performance under stress and high workload: A cognitive-energetical framework. Biological Psychology, 45, 73-93.

Hockey, R., Gaillard, A.W.K., & Burov, O. (2003). Operator Functional State: The Assessment and Prediction of Human Performance Degradation in Complex Tasks. Amsterdam: IOS Press.

Hockey, G.R.J., Wastell, D.G., & Sauer, J. (1998). Effects of sleep deprivation and user interface on complex performance: A multilevel analysis of compensatory control. *Human Factors, 40*, 233-253.

Hollnagel, E. (2000). Modeling the orderliness of human action. In N. B. Sarter and R. Amalberti (Eds.), *Cognitive engineering in the aviation domain* (pp. 65-98). Mahwah, NJ: Lawrence Erlbaum Associates.

Manzey, D., Bleil, M., Bahner-Heyne, J.E., Klostermann, A., Onnasch, L., Reichenbach, J., & Röttger, S. (2008). *AutoCAMS 2.0 Manual* (Technical Report No. 2008-01). Berlin: Berlin University of Technology, Chair for Work, Engineering and Organizational Psychology. Retrieved October 1st, 2008, from http://www.aio.tu-berlin.de/fileadmin/a3532/Berichte/Manual_AutoCAMS_2.0_190908.pdf

Preacher, K.J. & Hayes, A.F. (2008). Asymptotic and resampling strategies for assessing and comparing indirect effects in multiple mediator models. Behavior Research Methods, 40, 879-891.

Röttger, S., Klostermann, A., & Manzey, D. (2007). An information theory-based approach to measure orderliness of control behaviour. In M. Rötting, G. Wozny, A. Klostermann, and J. Huss (Eds.), *Prospektive Gestaltung von Mensch-Technik-Interaktion* (pp. 131-136). Düsseldorf: VDI-Verlag.

Schlick, C.M., Winkelholz, C., Motz, F., & Luczak, H. (2006). Self-Generated Complexity and Human–Machine Interaction. *IEEE Transactions on Systems, Man and Cybernetics, Part A: Systems and Humans, 36*, 220-232.

Schürmann, T. (2004). Bias analysis in entropy estimation. *Journal of Physics A: Mathematical and General, 37*, L295-L301

Stanton, N.A., Ashleigh, M.J., Roberts, A.D., & Xu, F. (2001). Testing Hollnagel's Contextual Control Model: Assessing Team Behavior in a Human Supervisory Control Task. *International Journal of Cognitive Ergonomics, 5*, 111-123.

Acknowledgement to reviewers

The editors owe debt to the following colleagues who helped to review the manuscripts for this book:

Arne Axelsson, HFN, The Swedish Network for Human Factors, Linköping, Sweden
Wolfram Boucsein, University of Wuppertal, Wuppertal, Germany
Anne Marie Brouwer, TNO Defense and Security, and Safety, Soesterberg, The Netherlands
Rino Brouwer, TNO Defense and Security, and Safety, Soesterberg, The Netherlands
Peter Essens, TNO Defense and Security, and Safety, Soesterberg, The Netherlands
Stephen Fairclough, Liverpool John Moores University, Liverpool, UK
Michael Falkenstein, IfADo Leibniz Institut für Arbeitsforschung an der TU Dortmund, Dortmund, Germany
Joanne Harbluk, Transport Canada, Ottawa, Canada
Maarten Hogervorst, TNO Defense and Security, and Safety, Soesterberg, The Netherlands
Jettie Hoonhout, Philips Research, Eindhoven, the Netherlands
Samantha Jamson, University of Leeds, Leeds, UK
Dan Jenkins, Brunel University, Uxbridge, UK
Hans Korteling, TNO Defense and Security, and Safety, Soesterberg, The Netherlands
Ben Lewis Evans, University of Groningen, Groningen, the Netherlands
Bernd Lorenz, Eurocontrol, Budapest, Hungary
Natasha Merat, University of Leeds, Leeds, UK
Ulla Metzger, Deutsche Bahn AG, München, Germany
Björn Peters, VTI, Linköping, Sweden
Nick Reed, TRL, Wokingham, Berks, UK
Jürgen Sauer, University of Fribourg, Fribourg, Switzerland
Jan Maarten Schraagen, TNO Defense and Security, and Safety, Soesterberg, The Netherlands
Frank Steyvers, University of Groningen, Groningen, the Netherlands
Sonja Straussberger, Airbus, Toulouse, France
Lex Toet, TNO Defense and Security, and Safety, Soesterberg, The Netherlands
Nic Ward, Montana State University, Bozeman, MT, USA
Alistair Weare, TRL, Wokingham, Berks, UK
Clemens Weikert, Lund University, Lund, Sweden
Mark Young, Brunel University, Uxbridge, UK
Rolf Zon, NLR National Aerospace Laboratory, Amsterdam, The Netherlands

In D. de Waard, J. Godthelp, F.L. Kooi, and K.A. Brookhuis (Eds.) (2009). *Human Factors, Security and Safety* (p. 437). Maastricht, the Netherlands: Shaker Publishing.